P9-AFP-644

The Handbook of Personal Area Networking Technologies and Protocols

This handbook offers an unparalleled view of wireless personal area networking technologies and their associated protocols. It lifts the lid on their growing adoption within the consumer electronics, home automation, sports, and health and well-being markets.

Bluetooth low energy, ZigBee, EnOcean, and ANT+ are comprehensively covered, along with other WPAN technologies including NFC, Wi-Fi, Bluetooth classic and high speed, and WHDI. It also features 802.11ac, the Internet of Things, Wireless USB, WiGig, and WirelessHD.

The handbook shows how white space radio, cellular, and femtocells have inadvertently blurred the boundaries between personal and wide area communications, creating disruptive topologies through technology convergence. It explores how pervasive WAN technologies have spawned a new generation of consumers through the Lawnmower Man Effect and explains how our personal space has become integral to social media streams, including Twitter, Facebook, and Pinterest.

An essential read for students, software engineers and developers, product planners, technical marketers, and analysts.

Dr. Dean Anthony Gratton is a bestselling author and columnist. Dean has worked extensively within the wireless communications R&D industry and has an accomplished career in software engineering. He was an Editor of the *Specification of the Bluetooth System: Profiles, v1.1*, and participated in defining the initial Bluetooth Personal Area Networking profiles. He was also active in the NFC technology and marketing committees. Dean is a contributor to several industry periodicals, where he has written many influential articles, sharing his thoughts and challenges on wireless industry news and opinions.

"Once again, Dean demonstrates his knowledge of the wireless standards landscape by providing a comparison of the many different technologies available. This is a valuable starting point for designers trying to unravel the complexities of the wireless world, helping them to determine the best option for their products."

Nick Hunn, independent consultant

The Handbook of Personal Area Networking Technologies and Protocols

DEAN ANTHONY GRATTON

CAMBRIDGE
UNIVERSITY PRESS

University Printing House, Cambridge CB2 8BS, United Kingdom

Published in the United States of America by Cambridge University Press, New York

Cambridge University Press is part of the University of Cambridge.

It furthers the University's mission by disseminating knowledge in the pursuit of education, learning and research at the highest international levels of excellence.

www.cambridge.org
Information on this title: www.cambridge.org/9780521197267

First published 2013

Printing in the United Kingdom by TJ International Ltd. Padstow Cornwall

A catalog record for this publication is available from the British Library

Library of Congress Cataloging in Publication data
Gratton, Dean A.
The handbook of personal area networking technologies and protocols / Dean Anthony Gratton.
pages cm
Includes bibliographical references and index.
ISBN 978-0-521-19726-7 (hardback)
1. Personal area networks (Computer networks) – Handbooks, manuals, etc. 2. Computer network protocols – Handbooks, manuals, etc. 3. Ubiquitous computing – Handbooks, manuals, etc. I. Title.
TK5105.76.G73 2013
004.6′2–dc23 2013012162

ISBN 978-0-521-19726-7 Hardback

For Sarah, my darling wife and best friend,

"*you embody my hope and my immortality*"
– D.

"You see, wire telegraph is a kind of a very, very long cat. You pull his tail in New York and his head is meowing in Los Angeles. Do you understand this? And radio operates exactly the same way: you send signals here, they receive them there. The only difference is that there is no cat."

– Albert Einstein

"The future success of wireless technology rests upon it becoming as overlooked as electricity."
– Dean Anthony Gratton

Contents

About the Author

Dr. Dean Anthony Gratton is a bestselling author and columnist.

Dean has worked extensively within the wireless communications R&D industry and has an accomplished career in software engineering. He has enjoyed a variety of roles and responsibilities in addition to being an Editor of the Specification of the Bluetooth System: Profiles, v1.1 (the original specification). He has participated in defining the initial Bluetooth Personal Area Networking profiles, and was active in the Near Field Communication (NFC) technology and marketing committees. His wireless research work has been patented.

Dean has developed, architected, and led teams across several new product developments for mobile phones, DigitalTV, Broadband (triple-play), femtocells, Bluetooth, Wi-Fi, ZigBee, NFC, and Private Mobile Radio.

Dean is a Community Editor-in-Chief for Eden (engageSimply.com) as well as being a popular columnist and contributor to a number of industry periodicals, where he has written many contentious articles sharing his thoughts and challenges on wireless industry news, opinions, and gossip. He continues to provide an authoritative published and vocal presence within the wireless communications industry.

Dean has become an influential social media persona (as @grattonboy), with an increasing Twitter following, and has been listed in the 50 "Top Dogs" of Twitter (bullsandbeavers.com). He is also listed as one of the "Top Marketing Book Authors on Twitter" in *Social Media Marketing Magazine* (smmmagazine.com), and is listed in the "Top 5" Twitter Elite in the United Kingdom, as rated by TweetGrader.com.

Dean holds a BSc (Hons) in Psychology and a Doctorate in Telecommunications.

You can contact Dean at thepanhandbook@deangratton.com and follow him on Twitter (@grattonboy) to enjoy his risqué humor, witty shenanigans, social media, and technology-related tweets. You can also read more about his work at deangratton.com and teamgratton.co.uk.

Making *My* Book Social

I have personally embraced social media for several years now and have been very fortunate in garnering some influential clout along the way. As such, I have become an influential social media persona myself as @grattonboy. I've been listed in the 50 "Top Dogs" of Twitter and have been regarded as one of the "Top Marketing Book Authors on Twitter" in *Social Media Marketing Magazine*.

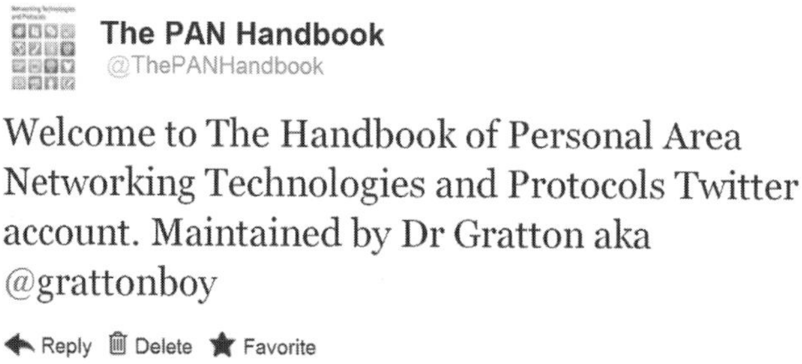

Welcome to The Handbook of Personal Area Networking Technologies and Protocols Twitter account. Maintained by Dr Gratton aka @grattonboy

Reply Delete Favorite

Figure 1. The @ThePANHandbook Twitter account, where my book has its own brand, along with its own identity, to encourage engagement and to share thoughts, industry news, and gossip surrounding the technologies that are featured within.

So, I want to make *my* book social; I mean social in terms of *social media*. If you like, I want to provide my book with its very own voice and I would like my readers to engage and correspond with my book. Those who are fans of social media will understand the associated benefits and holistically believe in the encouragement of *brand awareness* and *brand identity* – after all, it's consumers like *you* that define our brands today. You see, social media has inadvertently opened a door. In fact, I dare say it has knocked down walls and barriers that have normally closeted and protected big name brands. It has enabled everyday consumers to engage in two-way communication with them, preventing them from hiding behind the doors of traditional monolog-based promotion and filtering their consumer feedback into fan mail to bolster their corporate egos. What's more, these same brands can no longer afford a monolog engagement, but instead fundamentally need to embrace engagement with their consumer-base through social media. As consumers, we have become vocal about what we want, and social media has enabled us to engage with the brands that touch and shape our lives – now we are empowered through social media to touch and shape *them*! I'm introducing my book as a brand with its own identity to encourage engagement and to share thoughts, industry news, and gossip surrounding

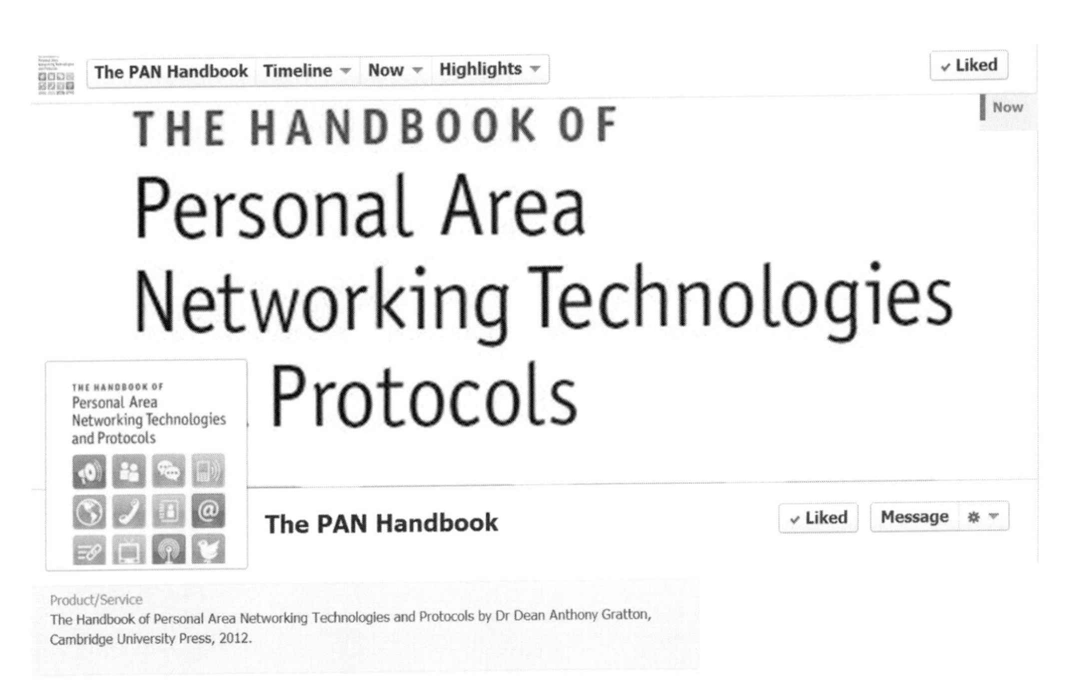

Figure 2. The PAN Handbook's Facebook Page (facebook.com/ThePANHandbook).

the technologies that are featured within. Incidentally, every technology that features in this book also has its very own brand identity on Twitter and Facebook – it's a perfect platform to engage your audience and build an army of *brand ambassadors*.

As such, *The Handbook of Personal Area Networking Technologies and Protocols* has its very own social media presence; after all, my book has an identity, a brand that you can engage with to learn more about its content and where you can discuss in general the technologies that uniquely form the wireless PAN. I've created these platforms to

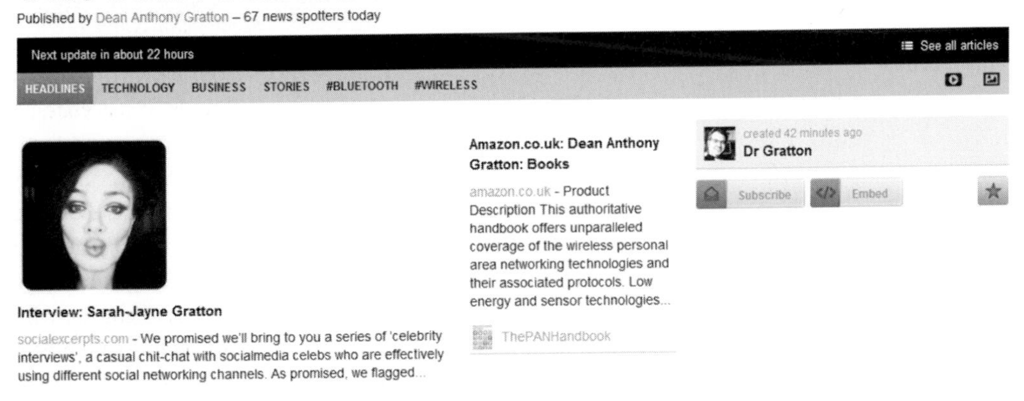

Figure 3. *The PAN Handbook News*. (Source: paper.li.)

offer you every opportunity to engage with me and to learn more about the book. I hope to dispel the rumors and hype that often beset the technology industry.

You can follow @ThePANHandbook via Twitter (see Figure 1) or become a fan of the book via its Facebook Page, facebook.com/ThePANHandbook (see Figure 2). You can even get your daily fix of all things wireless through the book's very own online newspaper, *The PAN Handbook News* (courtesy of paper.li), as illustrated in Figure 3.

Finally, you may be a little surprised to see this type of narrative surrounding social media in what is, indeed, a book that is dedicated to the personal area networking space. Well, you'll be pleasantly surprised to discover more about how our personal area networking space has been drawn into the social media domain later in Chapter 4, "Introducing the Lawnmower Man Effect (LME)."

Acknowledgements

Crikey, what was I thinking? Another book!

But who am I kidding? I love the challenge that the technology industry provides; I love the volatility, diversity, and inherent freedom of wireless technology. I romance with the notion of what wireless technology can potentially achieve and has achieved; and I just love the crafting process and seeing these pages fall into place – it's simply wonderful. You will soon discover more about my motivation and psyche (verging on madness) when reading the upcoming section "Before We Begin."

But for now, I like to think that I have written a book that encompasses a healthy review of the wireless personal area networking domain; one which I believe is long overdue. I've scanned Amazon and Barnes & Noble and I'm unable to discover similar books. After all, there's no other book which offers the density of all these technologies in one place – this is by far the most exhaustive coverage yet of the wireless personal area networking space!

Admittedly, I have had some highs and lows during this adventure, and, whilst at the keyboard, I have endured a breadth of emotions, which have been compounded by, on occasions, a need to do something else – damn Twitter! Nevertheless, I'm so pleased that these words have finally reached these pages. My journey hasn't been lonely – far from it: a large healthy glass of red wine (on occasions too much) has been in proximity whilst I've tapped my fingers skillfully across the keyboard listening to Elvis Presley, Neil Diamond and Dusty Springfield, who all made their respective guest appearances, albeit on my Windows Media Player playlist.

I should start by thanking Phil Meyler, Sabine Koch, Sarah Finlay, and Mia Balashova, the Cambridge University Press team, for their incredible patience – yes, it's taken some time and I have missed several (okay, many) deadlines, and I did manage to see my way through two editorial assistants during my seemingly endless overrun, with Mia being my third! So, so sorry, Phil!

I should also like to thank Nick Hunn and Bruno de Latour for their sanity check mechanisms – a prerequisite when tackling such a technical volume.

As usual, the content that follows has required the support of a number of amazing people and companies without whom this book would have never achieved what it needed to achieve for the wireless personal area networking space. So here goes: the people and companies that have made this short-range RF phenomenon become more than just a dream and, indeed, more than just a seemingly endless number of PowerPoint slides and empty promises.

In true Hollywood style, my guest stars all appear in alphabetical order . . .

AlertMe (alertme.com)
Jody Haskayne, Corporate Communications Director.
Amimon Incorporated (amimon.com)
Uri Kanonich, Senior Director Product Management at Amimon.
ANT Wireless (thisisant.com)
Catherine Gardiner Aylesworth, Marketing Manager at Dynastream Inc.;
Rod Morris, Director of ANT Wireless at Dynastream Inc.;
Sebastian Barnowski, Team Lead, Applications Engineering.
Barclays (barclays.co.uk)
Thomas Gregory, Head of Digital Payments at Barclaycard;
Louise Chan, Head of Commercial Management, Digital Payments at Barclaycard.
Bluetooth SIG (bluetooth.com)
Lindsay Peattie, Marketing Manager – Global Brand and Events;
Mike Foley, former Executive Director, Bluetooth SIG.
D-Link (dlink.com)
Andrew Mulholland, Marketing Manager (for Europe), D-Link UK.
Ember (ember.com)
Ravi Sharma, Director of Marketing.
EnOcean (enocean.com)
Graham Martin, Chairman & CEO, EnOcean Alliance;
Dr. Wolfgang Heller, Product Line Manager, EnOcean GmbH;
Marián Hönsch, Software Applications Engineer, EnOcean GmbH;
Zeljko Angelkoski, Marketing Manager Global Communications, EnOcean GmbH;
Angelika Dester, PR Manager, EnOcean GmbH.
Garmin (garmin.com)
Justin McCarthy, Media Relations, Garmin International;
Jake Jacobson, Social Media Manager, Garmin International.
Hewlett Packard (hp.com)
Dean L. Sanderson, Product Manager at Hewlett Packard.
IPSO Alliance (ipso-alliance.org)
Geoff Mulligan, Chairman, IPSO Alliance;
Pete St. Pierre, President, IPSO Alliance;
Jessica Barnes, Director of Marketing and Management, IPSO Alliance;
Kate Easton, Marketing Director, IPSO Alliance.
Jabra (jabra.com)
Anja Brøgger Winther, Channel Marketing Manager at GN Netcom;
Suzaan Sauerman, Marketing Director at GN Netcom.
Netgear (netgear.co.uk)
Carly Hill, eCom Channel Manager, United Kingdom;
Sylvain Clemenz, Senior Retail, eCom Channel Manager, France.

NFC Forum (nfc-forum.org)
Ruth Cassidy, Vice President of Communications, Virtual Inc.
Nokia (nokia.com/press)
The Nokia Brand Clinic.
Nordic Semiconductor (nordicsemi.com)
Anne Strand, Marketing Communications Manager.
Orange UK (orange.co.uk)
The Orange Press Office at GolinHarris;
Chris Heeley, Project Delivery Manager, France Télécom Group.
RosettaStone (personalrosettastone.com)
John Bottorff, General Manager at Objecs LLC.
Sony (sony.net)
Masayuki Takezawa, Deputy General Manager, FeliCa Business Division at Sony Corporation;
Yuka Matsudo, FeliCa Business Division at Sony Corporation.
Texas Instruments (ti.com)
Katrine Leander Brophy, Marketing Communications Manager at TI, Norway.
Weightless SIG (weightless.org)
Professor William Webb, CEO, Weightless SIG;
Alan Woolhouse, Director at Cambridge Startup Ltd.
WirelessHD (wirelesshd.org)
John Marshall, Chairman, WirelessHD.
Wireless Home Digital Interface (whdi.org)
Leslie Chard, President WHDI LLC;
Alexandra Crabb, Director of Media Relations at Ink Communications, Inc.;
Marta Twardowska, Relations Manager at Ink Communications, Inc.
Wi-Fi Alliance (wi-fi.org)
The Wi-Fi Alliance team.
WiGig Alliance (wirelessgigabitalliance.org)
Carlos Cordeiro, Chief Standards Architect at Intel Corporation;
Martyn Gettings, Account Manager, Proactive PR.
WirelessUSB (usb.org)
The USB-IF Administration Team.
Zeebox (zeebox.com)
Anthony Rose, Co-founder and CTO.
ZigBee Alliance (zigbee.org)
Joseph Reddy, Systems Architect, Texas Instruments;
Kevin Schader, Director of Communications, ZigBee Alliance.

Before We Begin

I've been writing books now for over ten years and I entered into this profession completely by accident following a heated discussion with the wife (Sarah). You know how people often say things happen at "the right place at the right time"? Well, for me that's exactly what happened and how I secured my first book contract with *Bluetooth Profiles: The Definitive Guide* back in September 2001. The amazement and sheer disbelief that I experienced – "Crikey, I did this!" – and seeing my first book pieced together in a bumper volume inspired me to write and want more. As for the heated discussion with the wife, I've been duly advised to save that story for another occasion, perhaps over several glasses of red!

I have historically titled this section of the book the "Preface" – maybe a hangover from my early days of writing. I don't personally think there's anything inappropriate with such a title, but I want to start anew by personalizing this section and sharing with you my motivation for writing this book. So, this is where I will now offer you an insight into my psyche and, on occasion, madness. Perhaps I will start by sharing more about me; you know, my specific experience and know-how, and ultimately what qualifies me to write such a book.

Okay. So, my experience spans over 25 years and I'll keep my account of it here brief! I have worked with networking technologies for around 22 years and have exclusively worked with wireless technology for close to 18. I have worked as a software engineer, developing products first-hand; as a software architect devising unique, creative, and effective methods for enabling networking and wireless technologies, which has led to some designs becoming patented; and I have managed and led teams for a host of consumer electronic products, including mobile phones, DigitalTV, broadband (triple-play), femtocells, Bluetooth, Wi-Fi, ZigBee, NFC, and Private Mobile Radio.

And it's still an ongoing adventure!

Incidentally, I initially didn't embark upon a career in software engineering, as my first passion was cooking, although I did dabble with computers and electronics at an early age – at 14, my first computer was an Oric-1 48k! I felt, at the time, that my destiny lay elsewhere, and I started work as a cook at my local Butlins' Barry Island holiday resort in the Vale of Glamorgan, South Wales, in the hope of following in my grandfather's footsteps as a culinary wizard. I know what you're thinking, "Butlins?", but everyone has to start somewhere, right? Anyway, it all came to an abrupt end and resulted in my first software engineering role, which I still have the fondest memories of. The position was with a small company based in Cardiff, South Wales – a world away from the wireless communications saga I find myself relishing in today. In the late 1980s and early 1990s I was writing software for the gaming industry; in particular for *Amusement With Payout* (AWP) and *Skill With Payout* (SWP) machines for the United Kingdom, Europe, Africa, and Russia. I was primed to undertake a degree in computer science, but was offered the job on the condition that I withdraw my studies. Okay, so I took the cash! As such, my early career was nothing short of a steep learning curve, as I came to terms with my first software engineering role. But it was a time that was equally so rewarding, as I witnessed some of my crazy notions mature into real products – it was breathtaking. This was also a time sensationalized by Hollywood-like stories, for example when one of my subtle bugs resulted in the Chair of the company being held at gunpoint in Russia – I spent several days sifting through my source code realizing that the fraction of a fraction of a fraction was causing a miscalculation in the jackpot prize that, by the way, occurred over a year! Let's just say that the Chair escaped unscathed, made a safe return back to base, and was too exhausted and relieved to be angry, much to my relief. Oh, and the excitement over "fake bugs," known as *Manufactured Malfunctioning Features* (MMFs). It was a craze in the early 1990s to write gaming machines that appeared to be faulty. I would write software that gave the impression to the player that the game was defective, but it was all part of the gambling experience. It seemed like a marvelous idea at the time, but it all backfired, as bars, clubs, and other venues were returning the machines to the operators, as they too saw the games as flawed.

It was such fun.

So, my first engineering role provided me with a sufficient foundation from which I could take on other challenges – I was ready, or so I thought. . . . I was eventually, after five years, wrenched from my role in Cardiff, and from what I considered to be my comfort blanket, kicking and screaming prior to moving to Cambridge, England, with my now wife in the mid1990s. I managed to secure a number of software engineering roles in the Cambridgeshire area, which ultimately led me to some of my most diverse and challenging experiences in technology. But it wasn't until I secured my role at a North London company in 1999, where I was involved with some of the first Bluetooth products on the market, that my passion and innate belief in wireless technology was seeded. The potential I saw was huge and an enormous leap from the remote control, but fundamental questions needed to be addressed, such as "How should the consumer experience this freedom?" These were questions predominately surrounding the user experience, such as "What should the consumer see at the user interface level?" This had an impact across all consumer electronic products that undertook the wireless theme, and I wanted to encourage a simple, seamless, and transparent experience, especially when there wasn't an obvious user experience to share, such as for Bluetooth headsets, where you had to rely on a sequence of colored and blinking LEDs to pair and connect.

It proved to be an arduous task then and, unbelievably, it still is today.

Crikey, we took the cable away and introduced a new level of complexity to a product which, for all intents and purposes, worked just as well with a cable. A number of problems emerged as I tried to piece together a simpler and more transparent way of making sure wireless technology delivered on the promises the sales and marketing team were spontaneously piecing together – it was a very early stage for wireless technology, and the technology was running even before it could walk.

A time plagued with so much hype.

It was not until 2004, when I received an opportunity to work for a Dutch company based in Leuven, Belgium, where I would become responsible for a breadth of wireless technologies and their introduction to a range of consumer electronic products, that I began truly to comprehend and appreciate the impact of taking away the cable. I was responsible for scoping the software and hardware impact, its feasibility, prototyping, and ultimately the cost of introducing wireless into a number of consumer electronic products. I found myself as a lone ambassador, singing the praises of the potential and opportunity that wireless technology presented. It was primarily my first and uncensored exposure to the wireless PAN space. I was so eager to be one of the first to architect an MP3 player with wireless technology, banishing the routinely cabled headset that traditionally accompanied the products. But my task was fraught with difficulties by an immature wireless technology space and overall poor user experience, which was further compounded by an up-cost in product retail price that was difficult to justify to marketing let alone the end-consumer.

Nowadays, costs have decreased due to the prevalence and acceptance of wireless technology, whereas confidence in the ability of manufacturers to enable consumer electronic products wirelessly has grown. But let's remind ourselves that the wireless technology journey is ongoing, and that as this technology propagates a wavering path across an unknown troposphere of bureaucracy and mass adoption, all governed by that dreaded price-tag, it has yet to reach its final destination.

I'm guessing at this point that I have unashamedly revealed my maturity and my associated experience with the technology and the industry overall. Admittedly, I now feel old. Anyhow, I write for a number of technology periodicals, which some of you might be familiar with, in which I share my thoughts and opinions about the wireless technology industry. As such, I regularly challenge and tackle issues that need to be shared, but more often faced head-on, so that industry pundits can't blinker the general consumer into believing features, facts, and figures that may have been tweaked to offer a gloss that isn't necessarily there (that hype thing again).

In essence, *The Handbook of Personal Area Networking Technologies and Protocols* is a compilation of my experience of working directly within the wireless communications industry for a wealth of some well-known and steadfast companies from both an engineering and a leadership perspective. By the way, I'm not talking about the fluffy tassels either; I'm talking about the bits, nibbles, and bytes that form the essential software building blocks of some of the most significant wireless products. So yes, I have opened the wireless equivalent of Pandora's Box; but don't worry, I have been heavily medicated for your safety.

I chose to write this book to condense the most popular and upcoming technologies and place them all into one volume – a book that would have proved beneficial to me on my technological discovery with wireless technology and one which I'm confident will prove valuable to your own personal journey too, as we all grapple with new wireless experiences. What's more, I hope to see future editions of this book providing expanded information in the existing chapters as well as adding new chapters for the new and upcoming technologies that (even as I write) are teetering on the edge of being launched into wireless stardom.

And let's not forget that whilst I have worked in the industry for a long time, the personal area network domain is still relatively new, as you will soon discover within these pages. The wireless technologies that now empower our everyday products are still evolving and adapting as they, of course, propagate their wavering paths across an unknown troposphere.

To be continued . . .

– Dean Anthony Gratton

About This Book

Any sufficiently advanced technology is indistinguishable from magic.
– Arthur C. Clarke

I have provided some blurb about my background and experience, and ultimately my motivation, as to why I have written this book in the section entitled "Before We Begin." So now I want to discuss, in more detail, what I have been up to and what I have actually packed into this gorgeous hardback. I would often write this section at the start of a project with a clear view of what I wanted to achieve, then I would revisit this section at the end of the project to ensure that ultimately I'd achieved what I set out to accomplish.

I'll be honest with you (intoned with a Welsh twang – just think Rob Brydon), what I often start with doesn't necessarily end up in the completed manuscript. Until the manuscript is handed over, the book remains a living and breathing document, and so does the table of contents for that matter. I think, as a consequence, it's a more solidified and stronger end result.

As I have already mentioned elsewhere, the wireless personal area networking domain is relatively new and it's still an exciting technology sector, as I have personally experienced and am still witnessing. It remains an exciting field of innovation, with manufacturers, engineers, and developers continuing to develop new products in this arena. It's had a modest lifespan of over a decade or so and has only now started realistically to enable consumers to simplify connectivity between everyday consumer electronic products. Our journey with the short-range RF domain has only just begun (sorry, I didn't necessarily mean that to sound like a Carpenters' classic). Anyhow, what I want to say is: wireless technology is still far from perfect, as there are still some idiosyncrasies, wrinkles, if you like, that need ironing out. *The Handbook of Personal Area Networking Technologies and Protocols* provides an exhaustive reference for the wireless PAN enthusiast and innovator. A reference that will undoubtedly define the foundation from which new wireless product development can be derived and should evolve and adapt.

In essence, the target audience for this book typically comprises individuals from those fields who are continually seeking new perspectives on the subject matter. Moreover, students and professionals who wish to embark upon a career in this enriched domain will seek references that inform them how best to understand, what to expect of, and how to develop applications, products, and services. As such, *The Handbook of Personal Area Networking Technologies and Protocols* offers the reader a consolidated wealth – a wireless bible of personal area technologies all documented in a single resource.

The book also addresses disruptive topologies through technology convergence, which is a slow but inevitable fate for the personal area domain. Existing/evolving and newer technologies have inadvertently created borderless topologies that engulf the personal area network, such as cellular (3G, 4G – LTE Advanced), white space radio, femtocells, and so on, all providing consumers with every opportunity to connect to an IP-enabled network. You will learn more about what I have penned as the *Lawnmower Man Effect* (LME) later in Chapter 4, "Introducing the Lawnmower Man Effect," a term that classifies the ability afforded to a new generation of consumers who seek to have a permanent connection to anything, anytime, anywhere. Essentially, we are witnessing a rapid development of new products and services which have diversified the market sector. The telecommunications industry continually offers a breadth of services and products for consumers who wish to remain connected, irrespective of location.

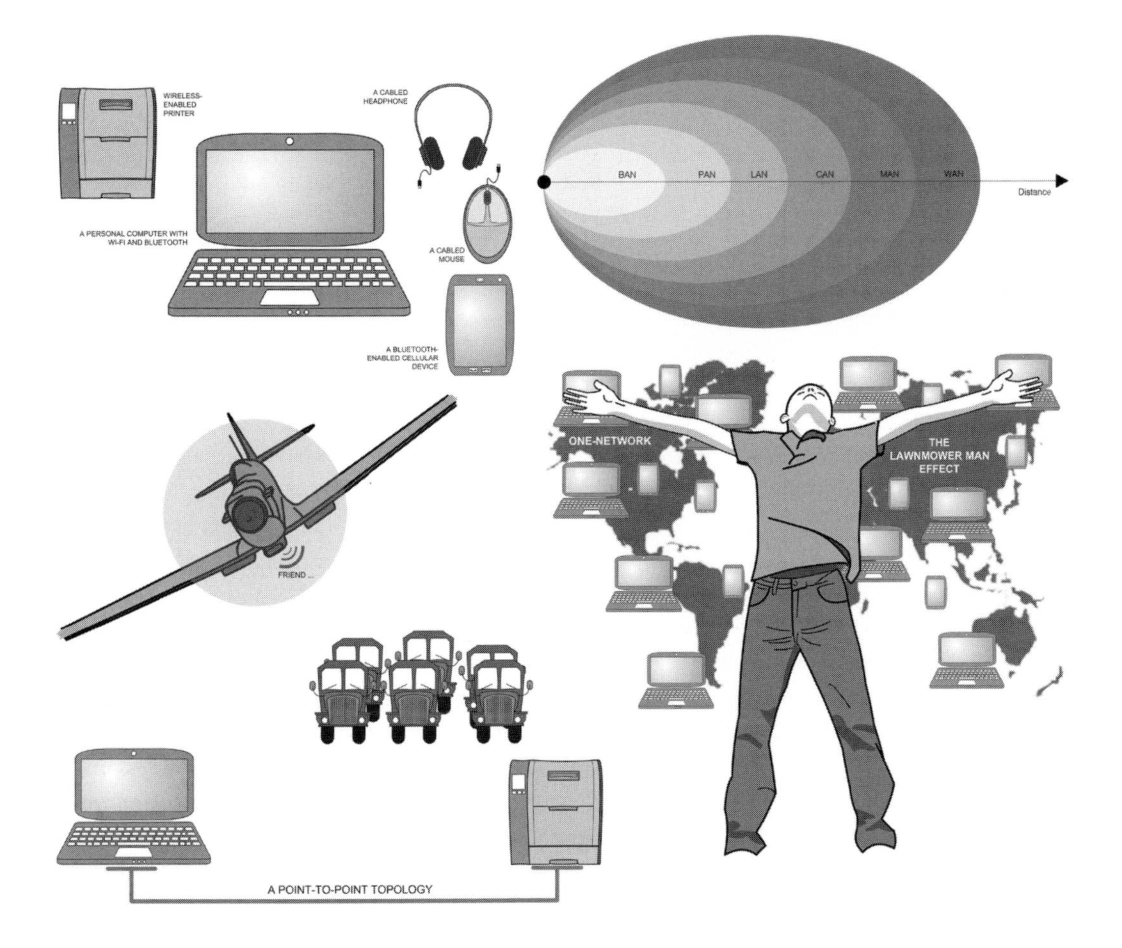

So, *The Handbook of Personal Area Networking Technologies and Protocols* has been structured into four *parts* or sections, and in Part I, "What's In Your Area Network?," I provide a close-up view of the personal area space and explore what a personal area network is. I look at how disruptive topologies through technology convergence are diversifying our everyday ability to remain connected despite our location. I also introduce the Lawnmower Man Effect, which explains how, as consumers, we can nowadays traverse digital systems all captained from our personal networking space.

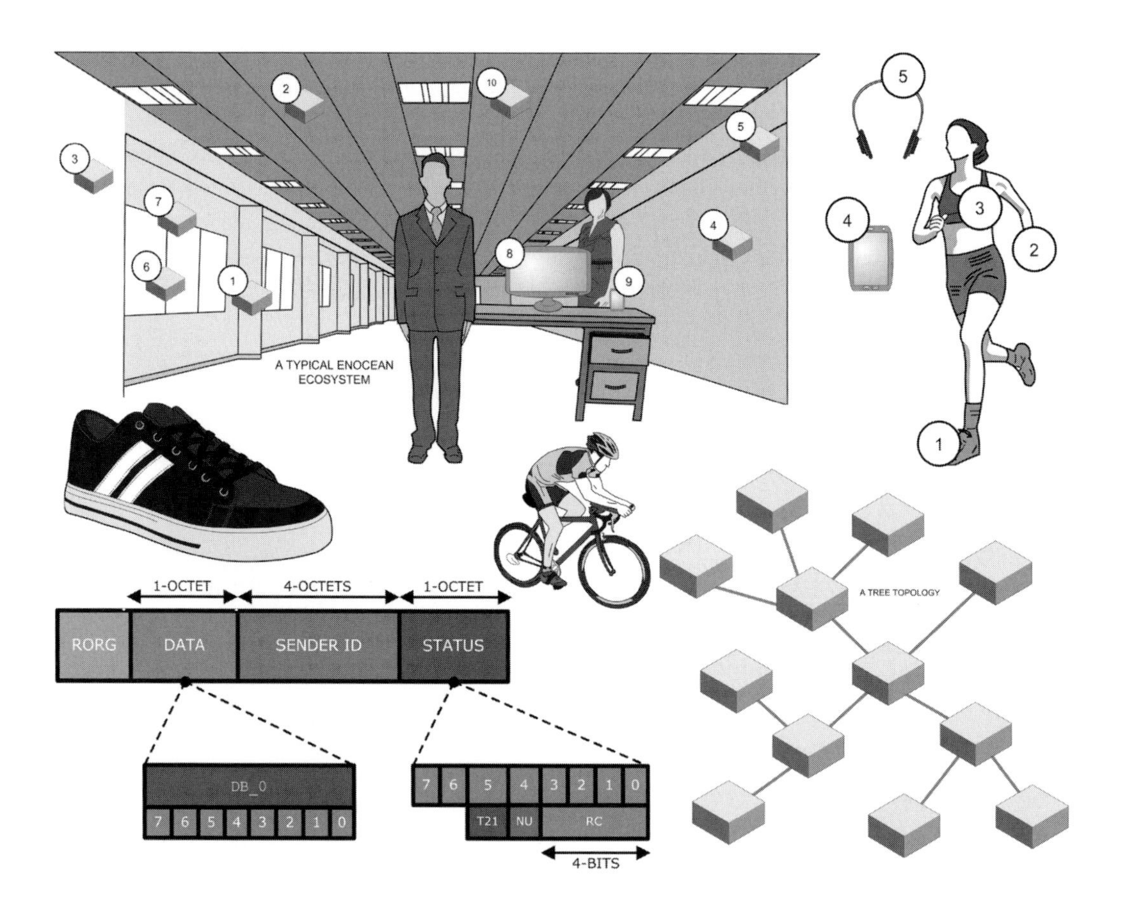

In Part II, "The Wireless Sensor Network," I showcase the wealth of low power and sensor technologies that will ultimately shape and define how technology is used within homes, industry, and commercial buildings. The notion of a "smart home," now referenced today as *home/building automation*, or *domotics*, has been used for more than thirty years and only today are we witnessing mass adoption. Nowadays, in an increasingly eco-conscious society, we need to understand better the consequences of the long-term impact of our relentless consumption of limited energy resources. I review the overall energy impact of consumer electronic products and look at several techniques and initiatives that help and encourage manufacturers to provide greener products.

As we move to Part III, "The Classic Personal Area Network," I review the most common and well-known technologies that we have taken for granted and commonly associate with wireless connectivity. Firstly, I look to unravel the basic ingredients of the PAN space whilst delving deeper into the wireless technologies that have shaped what I have coined the "classic technologies." In this section we review in hardcore detail a collection of familiar technologies that have enabled all of us to synchronize our data and extend our PAN, LAN, and so on to that all-important WAN.

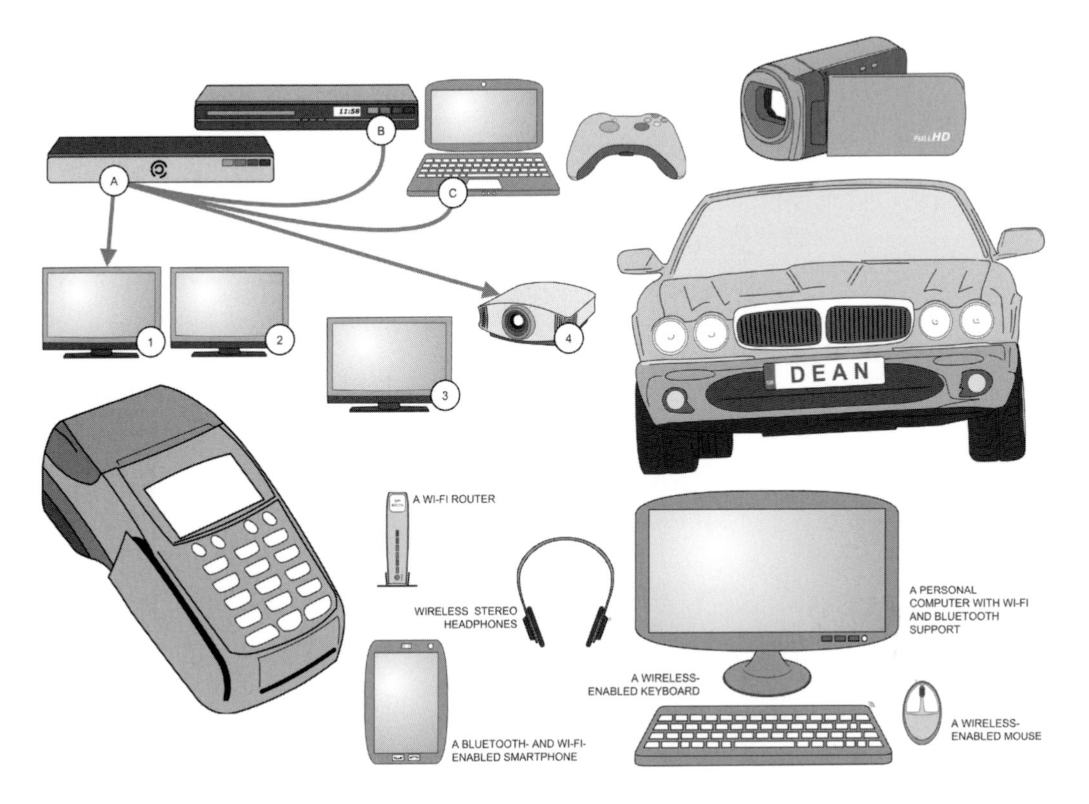

In Part IV, "Future Technologies and Conclusions," I conclude with an evaluation of new and upcoming technologies. The wireless technology industry is renowned for moving slowly with the introduction of new technologies, but manufacturers are always eager to deliver interim products invariably labeled as "*x*-Ready" or "Draft-*x*" in an attempt to accelerate consumer uptake and awareness. So, I take a peek at some of the new technologies that are awaiting their turn in the wireless hall-of-fame. And finally, I summarize my thoughts and conclusions – I paraphrase our journey so far and use an anecdotal narrative to reinforce what we've learned and perhaps discover what we should expect to happen next!

Part I

What's In Your Area Network?

1 It's a Small Wireless World

A number of books have often dedicated their first chapter to a narrative covering the historical aspects of wireless technology. It's a process that seemingly brings the reader up to date; a snapshot, if you like, of where we are and how far wireless technology has evolved. Nonetheless, whilst we acknowledge that wireless technology has reached a level of maturity with venture capitalists, product developers, engineers, and consumers alike, this chapter will instead focus on the changing perception of wireless technology and consider how it is fundamentally perceived by consumers. A comprehensive historical introduction to wireless technology would have certainly been relevant 20 years or so, maybe even as little as a decade, ago, but many consumers have now abandoned their cables for the transparency, simplicity, and ease of use that is often purported by advocates of the technology, making such lengthy historical introductions unnecessary.

Anyhow, wireless technology is still a relative newcomer in enabling a wealth of consumer electronic products, although its penetration isn't as pervasive as some would like. In short, we no longer need to dedicate a lengthy introduction and discussion surrounding wireless technology to explain the historical perspective of its conceptualization and ultimately its formation; nor do we need to discuss the ingenuity of the pioneers who were instrumental in paving the way for today's voice and data communication capabilities. Undoubtedly, these pioneers will never be forgotten, and will most certainly be remembered in other various guises and texts. Instead, in this chapter, we may use historical references merely to reinforce models of perception and primarily to illustrate how consumers have adopted wireless technology so comfortably. Likewise, we analyze consumers' perceptions of the small wireless world, offering perspectives from both the consumers themselves and the consumer electronics industry. This will provide you with an insight that will enable you to form a better understanding of consumers and their ability to adapt to the new technologies that are becoming increasingly inherent in today's innovative products. With this in mind, let's jump straight into one of the primary issues that arguably plagued wireless technology from its onset; that is, the attempt to overcome several shortcomings and hopefully become increasingly confident in delivering and architecting better products that continue to support an ethos of transparency, simplicity, and ease of use.

1.1 Let's Keep It Simple (Stupid!)

One ingredient that has often escaped the recipe for so many consumer electronic products is *simplicity*. This attribute alone should be instilled, force-fed, and, to be honest, beaten into innovators, developers, manufacturers, or whomever decides to embark upon developing products that are ultimately wireless-enabled. It may be blatantly obvious (to some) that, when you remove the ability to connect using a cable, you inevitably introduce a degree of complexity. If the product is too complex to use, the consumer *will* discard it. Therefore, in a team comprising developers, product engineers, marketers, business managers, and so on, there must be a number of questions that should be debated and addressed prior to embarking upon any product conceptualization, its design and subsequent development, and ultimately its marketing. More specifically, these questions and their associated answers should become a prerequisite of any prospective development life-cycle and should be explored exhaustively. Likewise, during the development life-cycle, progress should be systematically cross-referenced against these answers to ensure that no deviations from the initial objectives have occurred. In the following section, we offer a basic model or, if you like, a "rule of thumb" when embarking upon new wireless product development. The model offered here should be used primarily as a guide when assessing new development for wireless-enabled products.

1.1.1 Applying ICE to Your Wireless Product Development Life-cycle

In an attempt to establish a foundation upon which an approach to new wireless product development can be solidified, some basic guidelines are provided here. As part of your conceptualization process, issues surrounding *Interoperability*, *Coexistence*, and *Experience* (ICE) all need to be addressed and clarified. These facets are not new, but are often overlooked, and their purpose here is to enable you to form a better understanding and to encourage you to solidify your product conceptualization.

1.1.1.1 Interoperability

Interoperability is a term regularly used to characterize how a product from one manufacturer interoperates with a product from another manufacturer. If a consumer purchases a product that is Wi-Fi- or Bluetooth-enabled, for example, then the consumer will expect that product to interoperate with other like-enabled equipment, irrespective of manufacturer. It's probably an overemphasis, as it is a notion that was established in the very early days of wireless development and can easily be forgotten in a market where manufacturers wish to dominate. Similarly, you should ask whether you require your product to interoperate with other manufacturers or whether you are simply looking to create a standalone or unique ecosystem.

1.1.1.2 Coexistence

Interoperability and coexistence were indisputably the "troublesome two" during the early onset of wireless development, and they remain significant factors today. In the beginning, Wi-Fi and Bluetooth were typically pitted against each other as competing technologies. In fact, the press and many feature articles presented situations where Wi-Fi and Bluetooth endured problems when sharing the same radio spectrum, let alone the same room! Both technologies have advanced suitably and moved amicably forward, and Wi-Fi and Bluetooth now provide effective interference techniques to overcome any coexistence and interference issues. More importantly, they are now seen as two separate, independent technologies offering unique applications. Your choice of short-range RF technology will therefore become a significant consideration when embarking upon new product development.

Nowadays, issues of coexistence and interoperability are well understood, as there are several schemes available that overcome any shortcomings. Nevertheless, whilst this facet of the ICE model reflects the ability to cooperate with multiple manufacturers and their radio spectrums effectively, it's also about ensuring you have selected the right wireless technology for your product. It may be a cliché, but it's a chicken and egg scenario. Undoubtedly, in your technical feasibility you have defined what your intended product will do and specified its future-proofing. Likewise, you have also selected the appropriate technology that will effectively meet your product and marketing requirements; its durability in a potentially harsh environment and other factors, such as battery life and so on, are all key contributors. The wireless personal area networking technologies mentioned in this book all deliver expectations that may suit your intended application and audience, but due diligence should be afforded to the frequency, range, and durability in an environment where other technologies are likely to be present.

1.1.1.3 Experience

The user experience should overwhelmingly form the most important part of your product design. It can't be made any clearer – if questions such as "How will the consumer interact with my product?" and "How should the consumer experience my product?" are not addressed at the product definition or prior to the development onset, then the premise of user interaction and/or its interface merely becomes an afterthought, where potentially you may introduce a degree of complexity that can possibly render the product unusable or too complex to operate. Typically, standards bodies or groups who have defined the technology offer terminology that should remain consistent across manufacturers as well as fulfill expectations at the user interface. That said, a manufacturer should be abundantly aware that a product will be operated more often by a complete novice who has no expertise. The term "out-of-the-box" describes an ideal experience in which the product is unpacked, switched on, and simply operates as intended.

1.1.2 Interference and the Packet-hungry

A cable, as a minimum, offers the consumer a secure, fixed, reliable, and, on occasions, a powered connection. Alas, bestowing a product with wireless technology inevitably results in it falling upon the mercy of an air-interface, which unfortunately becomes subject to a myriad of worldly influential elements. The global nuances that invisibly occupy our space, which, in turn, arguably impose operating conditions that are disruptive to the wireless communication pathway – more commonly known as interference, are heightened by an ever increasing number of packet-hungry eavesdroppers who, with their hidden agendas, are always too pleased to know more about our wireless-established conversations. An additional factor to consider in making things wireless is, of course, removing the cable; this has resulted in products having to locate a new power source, which is typically derived from batteries.

Naturally, the majority of manufacturers are striving toward the belief that wireless-enabled products should support a mantra of transparency, simplicity, and ease of use, but inescapably our products have become complicated; there's now a degree of unreliability in our air-interface communication: we have to change the batteries regularly to sustain power within the device, and we have introduced a security compromise by removing the cable.

The motivation for innovators differs across industries; however, some tout a tenuous argument that cables are cumbersome, clutter our environment, and are too complicated to use. Installing cables in a large fixed environment can prove to be an arduous task, and we'll come back to this point later on in this chapter. Others may have become privy to the knowledge that wireless technology *is* founded upon a sincere belief – that of affording the consumer the gift of simplicity. Numerous manufacturers continue to develop new silicon, implement enhanced security schemes, devise new low power techniques, and participate in numerous standards committees to evolve and support their wireless technology flavor. On the one hand, we may assume these companies merely have an economic death wish; more realistically, on the other hand, there's something about wireless that offers a sense of freedom. We are not suggesting that cables literally bind you, but with Wi-Fi, for example, and the explosion of hotspots populating streets, restaurants, offices, metros, airports, and so on, we have quite literally become unconstrained by our location. Likewise, simple applications such as synchronizing your cellular phone with your desktop or notebook has become a small matter of bringing your cellular phone within the proximity of your desktop computer, for example. Both devices wirelessly detect each other and automatically synchronize – (normally) no user intervention is required.

Let's return our attention to the cable argument. Removing the cable from your environment is a valid contention for wireless, as it has been the founding belief for so many wireless technologies. In particular, the notion has been pivotal in the marketing of Bluetooth wireless, which has been touted as a cable replacement technology from its inception. Likewise, Wi-Fi technology was hyped as an alternative to a fixed cabled infrastructure, and, as such, Wi-Fi-enabled equipment is often installed where a cabled alternative is difficult to deploy, especially within a large building or perhaps across a large university campus, for example. Nowadays, connecting to a Wi-Fi *Access Point* (AP) has become akin to plugging in an Ethernet cable, something which has become very evident in today's technology fashion; for example, Apple's iPad, iPhone, and its MacBook Air, along with numerous Netbooks and smartphones that are so readily available, simply require a Wi-Fi connection. All these products rely on the idea of wirelessly connecting, which is tantamount to an increasing trend to sustain that all-important wireless connection. Let's not forget that Apple suggested that, with a ubiquitous wireless presence, we no longer need to rely upon other physical mediums (CD-ROM drives, for example) to retrieve data, software, or any other applications. This is further compounded by the many "application" stores offered by so many software providers, such as Apple, Microsoft, and Google.

1.1.3 Cutting the Cable

So, let's move beyond our *Personal Area Networking* (PAN) space for a moment, and shift our attention to the *Wider Area Network* (or WAN) to strengthen the "cutting the cable" argument. Naturally, deploying cables across continents and oceans is undoubtedly a challenging task. This is something that has already been done, but often these cables are subject to the elements surrounding them and can be prone to accidental severing. Nevertheless, the British firm, Global Marine Systems, is laying a new transatlantic fiber-optic link called the Hibernian Express.[1] The new link aims allegedly to "shave six milliseconds" off the existing leader's (Global Crossing) AC-1 cable, which currently supports a transatlantic connection of 65 milliseconds. The saving of just one millisecond can be worth up to $100m "a year to the bottom line of a large hedge fund." Increasingly, worldwide communications rely on some kind of fixed infrastructure where, for example, cables have already been deployed; however, utilizing fixed satellite communications to support the wider network has become somewhat easier to manage and deploy (not wanting to over-simplify unnecessarily the infrastructure or the technology). Fixed satellites comfortably support voice and data communications across the world, enabling the remotest of areas to sustain that all-important digital communication channel.

What seemed to be a tenuous argument for wireless technology has now become a significant factor in its future success. Likewise, the promise of simplicity also seems to be seeping into numerous products and applications. But applying security to these products has become a cumbersome process since the initial cutting of the cord (see Figure 1.1). In the chapters that follow, we tackle issues of security and encryption and examine how industry bodies and manufacturers overcome these unfortunate shortcomings. In the meantime, the packet-hungry eavesdroppers will continue to challenge innovators, and numerous standards bodies will continue to strive to ensure that connecting and pairing products becomes as simple as connecting a cable.

In taking a forward step and reviewing the innovation landscape, it seems, to a greater extent, that many innovators and manufacturers haven't forgotten the simplicity needed to ensure the future success of wireless technology. In fairness, it's an arduous and challenging task to achieve seamlessness between many different products, and let's not forget that plugging in one end of a cable to device **A** and the other end to device **B** has, over many decades, become second nature to a generation of consumers – it's become an unconscious process. Fundamentally, this is something we need to mimic in the new generation of wireless-enabled products. In short, we need to strive for the

[1] Williams, C. "The £300m Cable that will Save Traders Milliseconds," 2011.

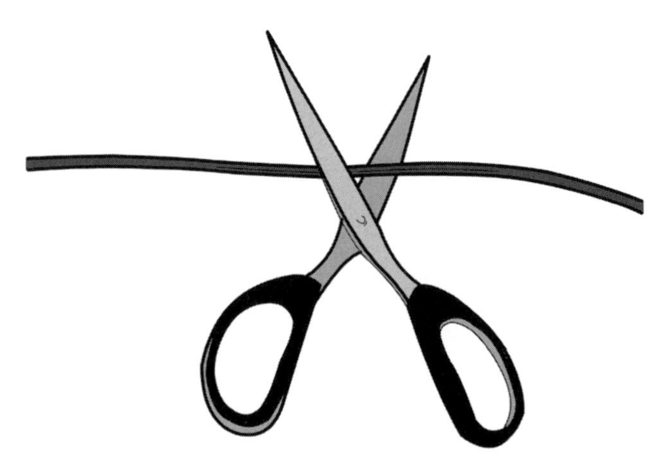

Figure 1.1. Cutting the cable is not necessarily a straightforward task.

same simplicity in all wireless-enabled products and deliver an uncompromising sense of unconscious usability.

The fundamental premise of wireless communications is to simplify connectivity between personal area devices. What's more, this premise should extend to simplifying communication within buildings or over distances where normally a fixed infrastructure would be difficult to deploy. Nonetheless, cutting the cable introduces an unmeasured sense of complexity.

Looking back thus far, we have discussed some of the factors that have, in part, been the motivation behind wireless technology. These have included simplicity, where deploying a fixed cabled infrastructure may be troublesome; an ability to use and operate a device easily; sustaining communication in both a personal and wide area perspective, in turn enabling communities around the world to remain connected no matter how far apart they are; and the need to drive toward an ecosystem supporting transparency and ease of use across all consumer electronic products. We now have a better understanding of some of the motivation behind "making" wireless, but let's turn our attention to understanding the audience – the community of consumers who will ultimately invest and provide longevity for a wealth of products and their associated technologies. Ultimately, we need to convince the purchasing population that wireless technology is here to stay, that it will become more pervasive, and that its use will increasingly become commonplace.

1.2 Understanding the Audience

More and more, manufacturers are wirelessly enabling their products and, as a result, cables are gradually being thought of as a second choice for connecting devices. Yes, a new generation of consumers has begun to look at their products very differently – we all yearn for the day when we can simply bring our television *Set-top-box* (STB) into proximity with our DVD player and our 7.1 stereo speakers (and other entertainment related products) and witness all our devices behave by seamlessly interoperating with each other. A day that shall be devoid of user manuals, configuration, and set-up – a truly holistic out-of-the-box experience, which is currently lacking in many consumer electronic products (again, it's another reminder of the simplicity factor and addressing those primary questions!). In successfully marketing wireless-enabled products to the consumer, marketers and manufacturers alike need to establish an added-value perspective – both a qualitative and quantitative factor, if you will, that convinces the consumer to "buy-in." We are already armed with numerous one-liners to hook the consumer, but the *Low Hanging Fruit* (LHF) needs to be devilishly tempting for the masses to bite into it. More often than not, adding new technology into a product inflates the purchase price, which, in turn, may deter rapid consumer adoption. With this in mind, we provide a cursory review of the marketing techniques used to increase product sales, which encourage the consumer to buy-in, at which point, of course, marketers offer that important added-value! In moving forward, the following sections impart a succinct perspective of the typical psychological makeup behind the ordinary consumer and offer an insight that will allow a better understanding of the intended audience. But first, we look at a technique that marketers have employed to convince ordinary consumers that this wireless stuff is all brand new and, perhaps, cleverly mislead them into a belief that wireless technology has only largely been conceived within the last 20 or so years.

1.3 Making Wireless Technology *New*

In fact, wireless communications has been operating now for over 100 years, and whilst we delicately tiptoe around a history lesson, it is merely curiosity, along with the need to demonstrate the aforementioned marketing technique, that momentarily draws us into an historical anecdote. Nevertheless, the purpose of highlighting the technology's longevity is to demonstrate that wireless communication and its associated techniques, algorithms, modulation schemes, conceptualizations, use case scenarios, and so on, were, to a greater extent, founded in the early twentieth century and witnessed extensive growth during World Wars I and II.

1.3.1 The Nineteenth-century Wireless Secret

Again, whilst eluding a lengthy historical narrative, let's reinforce, through example, some fundamental technologies and products that have been in existence for several decades, if not at least a century. Let's take the basic radio; that is, radiotelegraphy or wireless telegraphy, as coined in the mid nineteenth century. It's amazing to witness a notion of *wireless*, or at least a definition used to characterize an application using a transport medium, which wasn't physically connected or wired. Most notably, Albert Einstein, when asked to explain wireless technology, used an analogy of a long cat:

> You see, wire telegraph is a kind of a very, very long cat. You pull his tail in New York and his head is meowing in Los Angeles. Do you understand this? And radio operates exactly the same way: you send signals here, they receive them there. The only difference is that there is no cat. (Albert Einstein)

Figure 1.2. Ground forces would await the aircraft's response via RFID prior to taking action.

In later years, the term "radio" was generally adopted globally, where it characterized a device that provided basic communication from one point to another, that is, without the long cat. Some early use cases of wireless telegraphy were used to offer the first communication applications in the form of Morse code. In the early twentieth century (circa 1910), we eventually witnessed early broadcasting, the delivery of real-time audio content. In a modest number of scenarios, these early pioneers transmitted voice over relatively short distances but, fundamentally, the technology evolved to undertake an instrumental role in the First World War. Incidentally, the terms "radio" and "wireless" were used synonymously; however, in the early 1920s "radio" became widely adopted across the world. In Britain, for example, the word "wireless" was still used to characterize the broadcasting medium, and nowadays we can still witness the use of the word in publications such as *Practical Wireless* (pwpublishing.ltd.uk), and perhaps your grandparents may often refer to the modern day radio as a "wireless."

Infrared (IR) is another example of our nineteenth-century wireless secret. Of course, hitherto use case scenarios of IR have extended to some generic wireless applications, and you will surely correlate IR to such products as your television and sound system remote controls. Infrared radiation was discovered in 1800, and its applications are not limited to telecommunications. Infrared light lends itself well to other application areas, such as defense/military, astronomy, meteorology, medicine, and so on. Nonetheless, IR in data communications is performed over a short distance and has been integrated into a number of personal computers and other mobile devices.

As our final example, we cite *Radio Frequency Identification* (RFID) – a technology currently employed in many industries including healthcare, packaging, manufacturing, aerospace, and defense – which has its origins firmly dated as far back as World War II, as we illustrate in Figure 1.2. We say more about NFC topologies in Chapter 12,

Figure 1.3. The RFID technology has identified the aircraft as friendly, much to the relief of the ground forces!

"Just Touch with NFC." RFID was used in the *Identify Friend or Foe* (IFF) system, a top-secret project developed by the British[2] and installed in British aircraft. When an aircraft passed over radar units on the ground, a bi-directional communication channel would be established to let the ground forces know whether the planes were friend or foe, as we illustrate in Figures 1.2 and 1.3. Nowadays, RFID is still, of course, used in a variety of contexts, which we have already touched upon, but the technology has now become synonymous with *Near Field Communications* (NFC), but more about this later in Chapter 12.

1.3.2 Sourcing New Applications

We have established that some wireless technologies have their roots firmly seated in a number of early applications, but with technologies such as Bluetooth wireless, Wi-Fi, NFC, and so on, consumers are perhaps led to believe that the techniques and ingenuity of the technologies have been conceived within the last couple of decades. In fact, these technologies have derived their ingenuity with techniques and algorithms that have been around for 50 to 100 years. Admittedly, innovators may have tweaked them here and there, but ultimately there's nothing entirely new. What marketers have cleverly done is to reinvent numerous user scenarios (or applications) that have re-engaged the imagination

[2] Roberti, M., "The History of RFID Technology," 2005.

of the consumer. In other words, marketers have derived new applications, that is, new ways to use products and their associated technology in a way that is relevant to today's consumers' expectations and needs.

In summary, thus far, we have demonstrated through various examples that wireless technology has evolved substantially over the last decade or so, and has transformed itself into a stable medium, offering a diverse number of robust applications. Likewise, we have illustrated through other examples that numerous techniques used within today's technologies were, to a greater extent, established in the early twentieth century. The standard radio, IR, RFID, and, of course, a host of other personal area networking technologies, techniques, and modulation schemes have been bubbling around for over 100 years or so. Indeed, these techniques and algorithms have vastly improved over this period and will continue to do so.

So, why exactly is wireless technology perceived to be new?

Consumers are led to believe through effective marketing that wireless technology is fundamentally new; moreover, effective marketing perpetuates consumers' perceptions. With the overwhelming popularity of wireless technology, and telecommunications for that matter, both average and experienced consumers to date have increasingly become wireless-savvy. It is with some irony that wireless has become an infectious buzzword, not necessarily the definition *wireless*, but rather definitions such as Bluetooth, Wi-Fi, NFC, 2G, 3G, and 4G, to name a few, have all aroused some sense of curiosity within the consumer. Naturally, the marketers touting these new products have used these definitions and acronyms to create a broad appeal combined with a sense of innovation and modernity. Likewise, we are all steered into a belief that our personal and working lives will benefit enormously when we purchase the most up-to-date products with the latest technology – surely they are a "must have," as we are often informed! It's a phenomenon that has marketers saturating product flyers and packaging with a plethora of, what may seem to be, unfathomable keywords and three letter acronyms that render the consumer unconstrained by their content. Perhaps we could argue that, to some extent, a minority of consumers may choose to shy away from reading the user manual – perhaps they feel an unmeasured sense of confidence derived from the early adopters and advocates of technologies. Primarily, the ironic perspective of wireless technology is founded upon consumers' perceptions and good old fashioned marketing. The technology press, the manufacturers' marketing teams, and hearsay, perhaps through social media, increasingly play a large role in making wireless technology *new* or at least to *appear* new. But then the most fundamental consideration is that our use of and interaction with technology have also evolved, and we touch upon this in the following section.

1.3.3 Reinventing the Wheel

We should explore this notion a little further; yes, good old fashioned marketing coupled with consumers' perceptions does factor significantly in making wireless technology appear new. However, other variables that should be included to complete our equation are the user scenarios or intended applications in which the product is used. In other words, some bright spark has invented an application or a new way of working which has a hitherto real-world purpose that may serve or improve an everyday function. If you like, when we take ourselves back to the First World War, then the application of radio communications, that is, communication with ground forces, aircraft, and warships, was one that served a very real and immediate purpose. Nowadays, technology marketing is all about creating a solution to a problem (sometimes the problem doesn't even exist and has to be created by the marketers), but, more importantly, the consumer can rationalize and use reason to deduce that this is something that will instantly improve his or her personal or working life; a need that is defined by the marketers becomes instantly satisfied. In the twenty-first century, our priorities in everyday life differ dramatically to those of previous generations, and marketers find themselves creating new scenarios (consumer problems, if you like) in which the use of a wireless product is the only answer. Moreover, "Creating problems for consumers, however unethical it may seem, will undoubtedly increase the sales of a product that does not already fulfil an obvious active problem."[3]

1.3.4 Selling the Sizzle, not the Sausage

For marketers, it has always been traditional to sell consumers the *sizzle* – in other words, "What will this product do for me?" – rather than "How does it do it?", the *sausage*. This is akin to saying, "When I turn the tap on, I want water – I don't want to know where it comes from." You may have already seen product packaging and labeling with elusive descriptions such as "With the new smart Feature Set v2.1, users can now experience greater simplicity and ease of use in any situation." Surely nowadays, sheer humankind curiosity would lead us to ask "How does it do that?" Wireless technology has suffered some notoriety from poor consumer experiences; moreover, this has resulted in consumers choosing to be no longer ignorant about their purchases, especially when they are parting with their cash!

The distinction afforded by the sizzle versus the sausage marketing method has become increasingly blurred, as it is more often difficult to sell a product (the sizzle) without informing the consumer as to how it does whatever it does (the sausage). There are other factors that need to be considered here (other than sheer humankind curiosity and hard earned cash). We are not suggesting that marketers and sales teams provide lengthy technical discussions on how our water is delivered, but instead that the evolution and education of the consumer may point future marketers toward not tarring their entire audience with the same brush. What's more, consumers have (perhaps inadvertently) become wireless-savvy, as a consequence of the technology press printing sensationalistic stories that have rendered wireless technology inferior or unusable, or because some laboratory or hacker has compromised the security of a particular technology, such as Bluesnarfing (Bluetooth) or WarChalking (Wi-Fi). As such, it has arguably become the responsibility of marketing and sales (of a particular manufacturer) to educate consumers accordingly: "Please ensure you have enabled Bluetooth pairing"

[3] Gratton, S. and Gratton, D. A., *Marketing Wireless Products*, 2004.

for your cellular phone, and "Enable *Wi-Fi Protected Setup* (WPA)" for your Wi-Fi access point – this is terminology that needs to be explained so that the consumer can understand why, if they don't carry out these instructions, other people can steal personal information from their device or network. Conversely, consumers have acquired the ability to associate industry symbols or logos to a specific function or application; for example, a Bluetooth logo may offer a particular function to synchronize or use a stereo headset, for example, whereas, a Wi-Fi logo (in a hotspot) would enable the consumer to gain access to the Internet.

1.3.5 A Wireless Utopia

It seems we have again resurrected the discussion surrounding simplicity, as removing the cable has introduced that unmeasured sense of complexity. Furthermore, this may be unique within the wireless communications sector, but more and more consumers are continually scrutinizing their purchases and understanding what they are, in fact, receiving for their money! Many standards bodies insist, as part of their certification program, upon the use of logos and other product information to inform potential consumers about their purchases. We will use the following anecdote concerning the classic example of the Apple iPhone furor to cement the sense of this curiosity by a growing tech-savvy consumer population. Apple offered us the iPhone 2G, which was soon followed by the 3G, and then the iPhone 4. In its latter launch, Apple omitted the "G" from its product labeling, which led to consumers wanting to understand why. Naturally, Apple couldn't label the product "iPhone 4G," as this would imply that the phone would support the next generation of cellular technology, which at the time of the phone's launch didn't exist!

So, product labeling and associated logos help to inform consumers what the product is capable of and what other like-products you might need to purchase to achieve successful interoperation; for example, a cellular phone will support the Bluetooth headset profile, but won't support the audio streaming profile; as such, you need either to purchase a phone that is capable of supporting both profiles or simply purchase a Bluetooth earpiece that supports the headset profile. It's an unfortunate factor of the sales pitch, but consumers must be informed and educated about their wireless products. Likewise, when we make a purchase, we would like to ensure that we don't necessarily have to purchase another product after only six months, for example – our purchase is important to us, and we need to be assured of its use for at least two years or so.

Alas, the wireless utopia we so desire to work seamlessly with a range of consumer electronic products is a little difficult to reach at this time. You may recall our earlier discussion in Section 1.2, "Understanding the Audience," in which we stated that a new generation of consumers will have to await the day that is devoid of user manuals, configuration, and set-up – a true holistic out-of-the-box experience!

2 What is a Personal Area Network?

The *personal area networking* space may nowadays encompass a personal computer and may include a diverse collection of other peripherals or gadgets, where each serves a unique purpose within the personal *topology*. Let's not forget that, first and foremost, a *Personal Area Network* (PAN) is a term used to describe a particular networking topology (we include in Table 2.1 further examples of networking topologies). The *Wireless Personal Area Network* (WPAN) has offered us a diverse collection of short-range radio technologies, which have had an influence on the dynamics of the personal area network, and we will explore this later. The PAN is a relatively new term to the computing industry, probably appearing about two decades ago or so. This chapter aims to provide a conclusive and modern definition of the PAN, by drawing on several attributes and technologies that have already influenced the network space, as well as potentially dissolving its boundaries. In fact, in Chapter 3, "Disruptive Topologies through Technology Convergence," and Chapter 4, "Introducing the Lawnmower Man Effect," we look at the diversification of the PAN domain, where social (media) networking has extended what was once a private space into several virtual communities that have supported technology-enabled socializing. Furthermore, Section 2.3, "Ever-decreasing Circles," looks at the increasingly blurred PAN topology and how the boundaries that separate one topology from another are rapidly dissolving. But first we look at the origin of the *area network* and other networking topologies that have characterized the way we perceive network interconnection and uniquely how we have come to understand them.

2.1 The Origin of the Area Network

The computing industry uses a variety of naming conventions to characterize different forms of networking topologies – we have summarized some of these area networks in Table 2.1. The area networks identified here are typically qualified by distance or proximity, and allude to a particular type of infrastructure – these definitions have been widely used over several decades. The table also identifies the more common network topologies that are used today, but other colloquial references have been generically formed to signify a specialized, ad hoc, or unique form of area network, such as *Vehicle Area Network* (VAN) and *Wireless Video Area Network* (WVAN). Nevertheless, as the popularity of an area network increases, it may become adopted into mainstream computing.

Table 2.1. A list of the more popular area network topologies that have been coined within the computer industry to allude to a particular infrastructure

Network type	Description
Body Area Network (BAN)	A very recent classification, which characterizes a collection of wired or wireless-enabled devices that may be integral to the human body for measuring heart rate, blood pressure, and so on; but may also be integral to clothing (for example, *wearable* technology).
Personal Area Network (PAN)	A collection of fixed or wireless devices, which may comprise a notebook or desktop personal computer; a cellular or smartphone; a printer; a keyboard; and a mouse. Typically, these devices are located in proximity to, but no more than distances of 30 meters or so from, one another, and form a personal or private network – see Section 2.2, "The Personal Area Network."
Local Area Network (LAN)	The first definition used within the computing industry to characterize a number of computers interconnected to form a network within relative proximity, such as an office, for example.
Campus Area Network (CAN)	As the name suggests, a university campus may have several local area networks, spreading across several sites on a campus, which are in moderate proximity. The sites may all be interconnected with a wired or wireless connection. Likewise, a company or corporation may use a similar topology to interconnect its business sites.
Metropolitan Area Network (MAN)	A MAN topology is similar to a CAN, but geographically the site in which the area is interconnected is much larger than a CAN and a MAN may indeed extend to an entire city.
Wide Area Network (WAN)	The wide area network represents an interconnected global community. Typically, the WAN may interconnect numerous PANs, LANs, MANs, and so on across the world, ensuring seamless and transparent connectivity to one other. The WAN is often referred to as the Internet.

The origin and first formation of an *area network* date from approximately September 1977,[1] when one of the first communication protocols, namely the *Attached Resource Computer NETwork* (or ARCNET), was used to interconnect or network a number of computers. Primarily, this typified the computing industry's introduction to the definition *Local Area Network* (LAN), which characterized a network of computers located within a room or over a relatively short distance, within a building, for example. ARCNET gained popularity in the 1980s – however, it was soon replaced by Ethernet. In fact, during the 1980s, Ethernet competed with ARCNET and other communication protocol technologies, such as *token ring*, to establish itself as the dominant connectivity protocol.

Ethernet proved to be extremely popular and is still widely used today; the *Institute of Electrical and Electronics Engineers* (IEEE) continues to maintain the 802.3 standard, which includes the *Physical* (PHY) and *Media Access Control* (MAC) layers. We discuss this in Chapter 13, "The 802.11 Generation and Wi-Fi," where we take a closer look at

[1] "Digital History: ARCNET, the First Local Area Network," (n.d.).

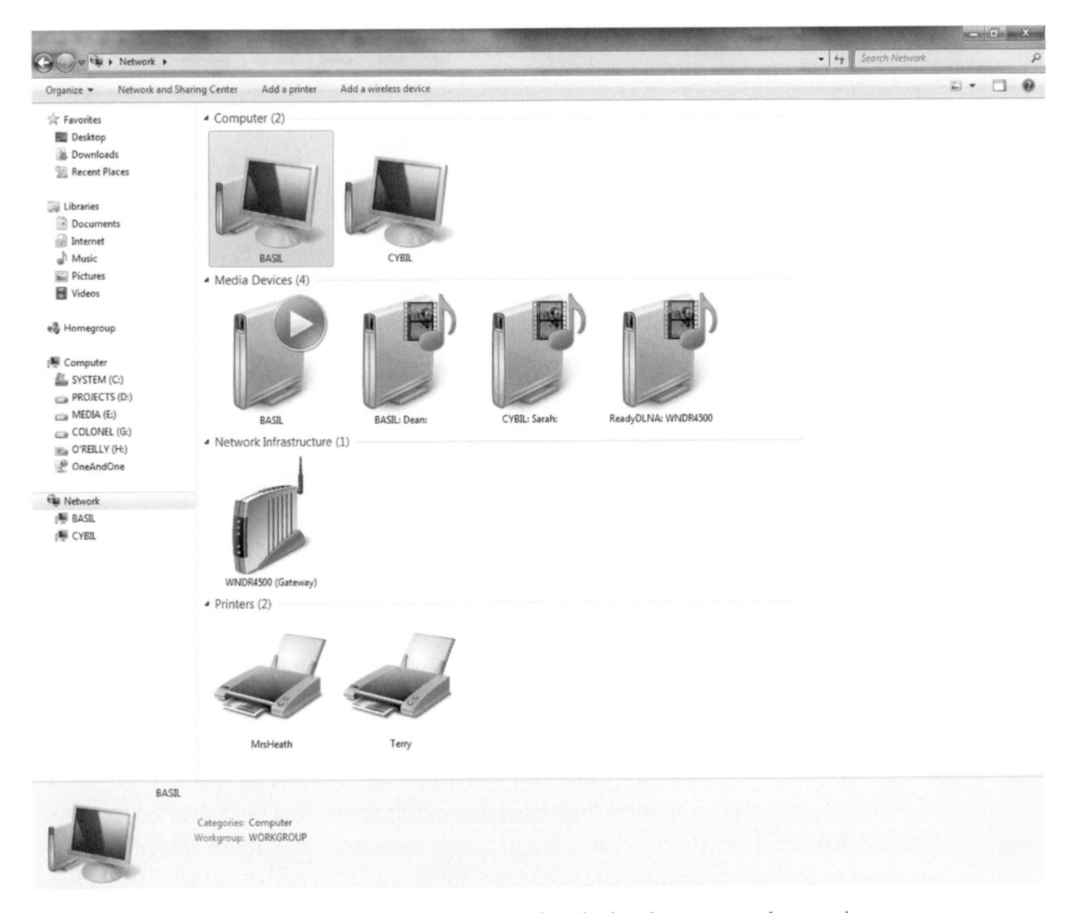

Figure 2.1. In this illustration, we can see that the local area network comprises two computers named Basil and Cybil along with two printers, called MrsHeath and Terry.

Ethernet. In Figure 2.1, we depict the typical infrastructure of a small business local area network, where the computers that are interconnected are discovered through Windows Explorer using the Microsoft Windows 7 (Professional) operating system.

2.2 The Personal Area Network

The personal area network is a collection of fixed or wireless devices, which may comprise a notebook or desktop personal computer; a cellular or smartphone; a headset; a printer; a keyboard; and a mouse. Typically, these devices are located in proximity, for example on a desk at home or in a small office, but on average the distance between them does not exceed more than 30 meters or so – collectively, these devices form a personal or private network, as we illustrate in Figure 2.2. The components that form such a network are personal to you – for someone to gain access to your personal area network, they usually require your permission through a dedicated username and password or passkey.

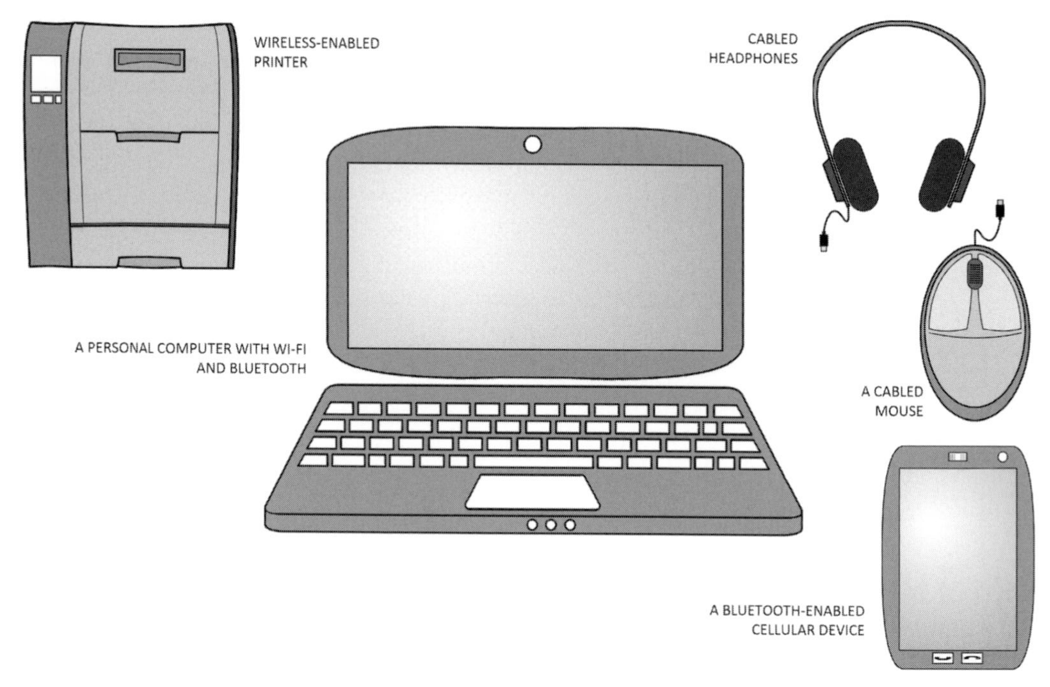

Figure 2.2. A typical personal area network, where a combination of wired and wireless devices may occupy your PAN space.

Likewise, some users who form their PAN may configure and set up dedicated areas that can be shared between users – for example, some users may configure a shared directory on their hard drive to expose safe and non-confidential files (music, video, documents, and so on), or others may configure a specific location on their smartphone to share pictures and video.

2.2.1 Conception

The term "personal area network" was initially coined as far back as 1996 by researcher Thomas Zimmerman in his paper "Personal Area Networks (PAN): Near-field Intra-body Communication," in *IBM Systems Journal*. Arguably, the definition didn't enter mainstream computing until circa 2000 with the introduction of Bluetooth wireless technology. Moreover, Section 2.2, "The Personal Area Network," and Section 2.2.2, "The *Wireless* Personal Area Network," provide more accurate and up-to-date definitions of the personal area network in use today and how ultimately it is perceived by industry contemporaries.

2.2.2 The *Wireless* Personal Area Network

The devices that form your personal area network may be wireless-enabled. Nonetheless, a *wireless* personal area network, as we touched upon earlier, is specifically a PAN in which devices permit connection through a wireless technology. Fixed PAN devices, such as a keyboard or mouse, for example, may be connected to your personal computer through a *Universal Serial Bus* (USB) cable. The definitions PAN and WPAN are occasionally used synonymously, but typically a WPAN as a noun is often used to characterize the number of short-range radio technologies that may indeed enable your PAN to interconnect – this forms the subject matter for most of this book. For instance, Bluetooth, *Wireless Home Digital Interface* (WHDI), and ZigBee are all WPAN technologies that may wirelessly enable your PAN. In short, a PAN represents a topology, whereas WPAN refers to the ability of the topology to interconnect wirelessly. The definition WPAN may also refer to the number of short-range technologies that enable your PAN to interconnect wirelessly.

2.2.3 The Wi-Fi Anomaly

We need to challenge Wi-Fi as a WPAN technology, or at least understand it a little better in context. It is arguably an anomaly since Wi-Fi is very much seen to be integral to the PAN, yet it isn't a WPAN technology (in the strictest sense). In a similar vein to PAN, LAN is a topology representing a number of computers interconnected through a cabled manner over a relatively short distance. *Wireless Local Area Network* (WLAN) is the ability to interconnect these same computers wirelessly. So, in essence, Wi-Fi is WLAN, but more often (to the contention of the industry purists) it can also be perceived to be WPAN. In the majority of personal area networks, a consumer would typically utilize a Wi-Fi connection to extend connectivity to the wider area network or to the Internet. In Chapter 4, we discuss other technologies that extend the PAN into other area networks.

The number of commonplace peripheral devices that are nowadays supporting Wi-Fi technology is increasing rapidly. In one such example, a Wi-Fi-enabled printer might support the ability to connect[2] to a Wi-Fi access point which, in turn, offers its resources to the rest of the local area network. In other words, if your printer is connected to the same access point from which you can access the Internet, for example, then your operating system will typically locate and configure it as a *shared* networked printer. The device will become integral to your LAN, and yet it still forms a component of your PAN, as your printer is typically in your proximity. What's more, the diversification of computers, peripherals, and ancillary devices has increasingly confused the area networking space. This is also evident with *Wide Area Networking* (WAN) technologies, which are nowadays indiscriminately penetrating and supporting the PAN domain. The availability of femtocells and cellular devices, for example are, in turn, creating disruptive topologies (we discuss disruptive topologies in Chapter 3). As a consequence, consumers are able to reach the wide area network irrespective of traditional hierarchical topologies – and some are beginning to question where the PAN begins and ends. In fact, the cellular or smartphone (as illustrated in Figure 2.2) is perceived to be an integral device within your PAN; yet, it has the ability to extend your PAN to the WAN. Similarly, femtocells,

[2] In Chapter 13, "The 802.11 Generation and Wi-Fi," we will learn how *Wi-Fi Direct*-enabled devices remove the need for an access point to interconnect Wi-Fi products.

Table 2.2. A summary of the 802.15's seven working groups. In particular, 802.15.4 provides ZigBee with its air-interface characteristics

Group	Standard	Scope	Status
1	802.15.1	The IEEE 802.15.1 standard was ratified in 2002 and 2005 based on the Specification of the Bluetooth System: Cores, v1.1 and v1.2, respectively.	Withdrawn.
2	802.15.2	A standard that looked at coexistence.	Hibernation.
3	802.15.3	The PHY and MAC high-rate WPAN standard.	Complete.
4	802.15.4 / 802.15.4-2003	The initial low-rate PHY and MAC WPAN standard. The task group was assigned to understand the complexity of a low data rate solution for use on monthly and annual bases. This standard has now been superseded. The first standard was released in 2003, hence 802.15.4-2003.	Superseded.
4a	802.15.4a	The low-rate alternative WPAN.	Approved.
4b	802.15.4b / 802.15.4-2006	Revisions and enhancements to the 802.15.4-2003 standard and was republished in 2006, hence the 802.15.4-2006 standard.	Approved.
4f	802.15.4f	PHY and MAC amendments for active RFID.	
4g	802.15.4g	PHY amendment for the *Smart Utility Network* (SUN).	
5	802.15.5	The mesh networking task group is responsible for identifying the necessary attributes needed within the PHY and MAC layers that would enable a mesh network.	Approved.
6	802.15.6	The BAN task group has been assigned to develop a communications standard specifically targeted for low power usage for the human body in addition to applications to include medical and consumer electronics.	Ongoing.
7	802.15.7	The task group has been assigned to define the PHY and MAC standard for *Visible Light Communication* (VLC).	Ongoing.

white space radio, and cellular technologies remain disruptive, as the boundaries between the topologies identified in Table 2.1 are increasingly becoming blurred. In Chapter 3, "Disruptive Topologies through Technology Convergence," we review in more detail the technologies responsible for shifting our norm and present a better understanding of the potential benefits of disruptive topologies.

2.2.4 802.15 WPAN

The IEEE 802.15 working group comprises seven task groups, as can be seen in Table 2.2. The working group's objective[3] is to establish industry standards and recommend industry practices, whilst addressing issues surrounding interoperability and coexistence for the wireless personal area networking space. The IEEE typically develops standards for the PHY and MAC layers, but these are not always adopted by the industry. For example, the 802.15.1 standard was ratified using the Specification of the Bluetooth System: Cores, v1.1 and v1.2 (as two separate standards, 2002 and 2005, respectively,

[3] IEEE 802.15 Working Group for Wireless Personal Area Networks.

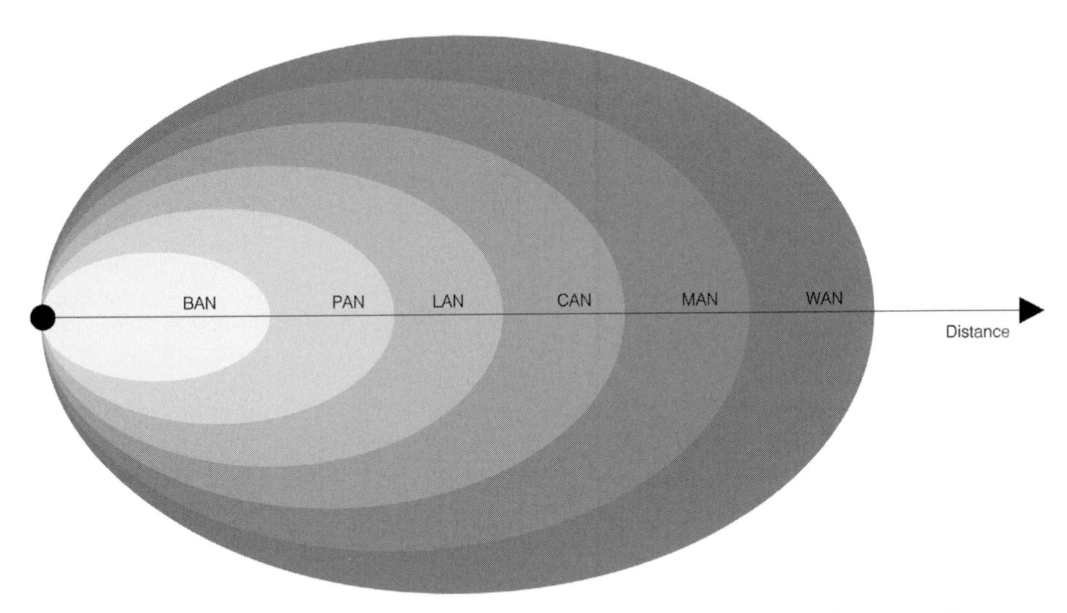

Figure 2.3. Geographically, area networks are qualified by distance or proximity, and allude to a particular type of infrastructure.

as shown in Table 2.2), but the Bluetooth *Special Interest Group* (SIG) withdrew its participation, focusing instead on defining their own protocols and standards for the industry. Conversely, the ZigBee Alliance adopted the 802.15.4 standard, again shown in Table 2.2, and the PHY and MAC layers now underpin ZigBee wireless technology.

2.3 Ever-decreasing Circles

Returning to our topic of disruptive topologies, we should perhaps explore a little further what we have penned as the *blurred topology*, not only within the personal area network, but also across the entire scope, as illustrated in Figure 2.3. It seems that with the introduction of pervasive WAN communication products and technologies, such as white space radio, LTE (3G/4G), femtocells, and so on, our perception of the personal area topology has perhaps become a little confused. Using well-established definitions, such as BAN, PAN, LAN, CAN, MAN, and WAN (as illustrated earlier in Table 2.1), would normally assist many industries to target products appropriately and more effectively, in addition to providing guidance to newcomers to the computing industry. Nonetheless, with wide area networking technologies infiltrating your personal area space, it becomes difficult to know where one topology starts and the other ends.

In Figure 2.3, we provide a gradient perspective of the numerous area networks over distance; the naming conventions used for an area network allude to a particular type of infrastructure. This is a typical illustration used to marry the area networks with purpose and associated distance, but, of course, with disruptive technologies, coupled with the consumer's incessant need to connect to anything, anytime, anywhere, a new holistic area network may be perceived. Inevitably, a *one-network* area or space may evolve (if it hasn't already done so) to accommodate a virtual community connecting individuals in their bedroom/study to another's office/study/bedroom relying on a persistent WAN connection, and this is something we touch upon in Chapter 4, "Introducing the Lawnmower Man Effect."

3 Disruptive Topologies through Technology Convergence

In Chapter 2, "What is a Personal Area Network?," we started to reveal how, with the diversification of pervasive *Wide Area Networking* (WAN) communications, coupled with a breadth and diverse collection of wireless-enabled products with their associated applications, and with the consumer's ability to remain permanently connected to the *one-network*, a new generation of technologies and products has inadvertently spawned an era of *disruptive topologies* through *technology convergence*. In short, a plethora of connectivity opportunities have become openly available and widely encouraged – consumers have every reason and every opportunity to sustain that all-important IP-fix, irrespective of location. We discuss in Chapter 4, "Introducing the Lawnmower Man Effect," how the *Lawnmower Man Effect* (LME) characterizes our ubiquitous ability to traverse digital systems across the globe, the *one-network*, as we illustrate in Figure 3.1.

There are several contributing factors which lead to consumers craving an IP-fix. One such obvious factor may simply be a trend to have immediate access to anything, anytime, anywhere. Another significant factor, which we discuss in Chapter 4, is *social media*. Social media platforms, such as Twitter, Facebook, Pinterest, LinkedIn, Foursquare, and so on, indirectly encourage consumers to maintain their online status and presence in the hope that they never miss a tweet or Facebook message from their virtual community, or perhaps to triumph through the adornment of a new badge awarded to them by Foursquare, as they oust their colleague to become *Mayor* of a popular business venue or restaurant. Traditionally, the PC within your *Personal Area Networking* (PAN) space sustains your online status, but increasingly the prevalence of wireless technology, whether you are connecting via a Wi-Fi connection at a shopping center or whilst sitting on the Brecon Beacons in Wales via a 3G network, extends our virtual attendance when away from the PAN. In essence, this technology allows us to fuel an eternal presence, irrespective of location.

This chapter looks at disruptive technologies and technology convergence, and provides a better understanding of our traditional networking topologies. In essence, 3G, *Long Term Evolution* (LTE), white space radio, and femtocells are some examples of the technologies that are responsible for creating disruptive topologies through technology convergence. Later in the chapter, we consider their effect, and we review the diversity of digital content in Section 3.2, "Technology Convergence."

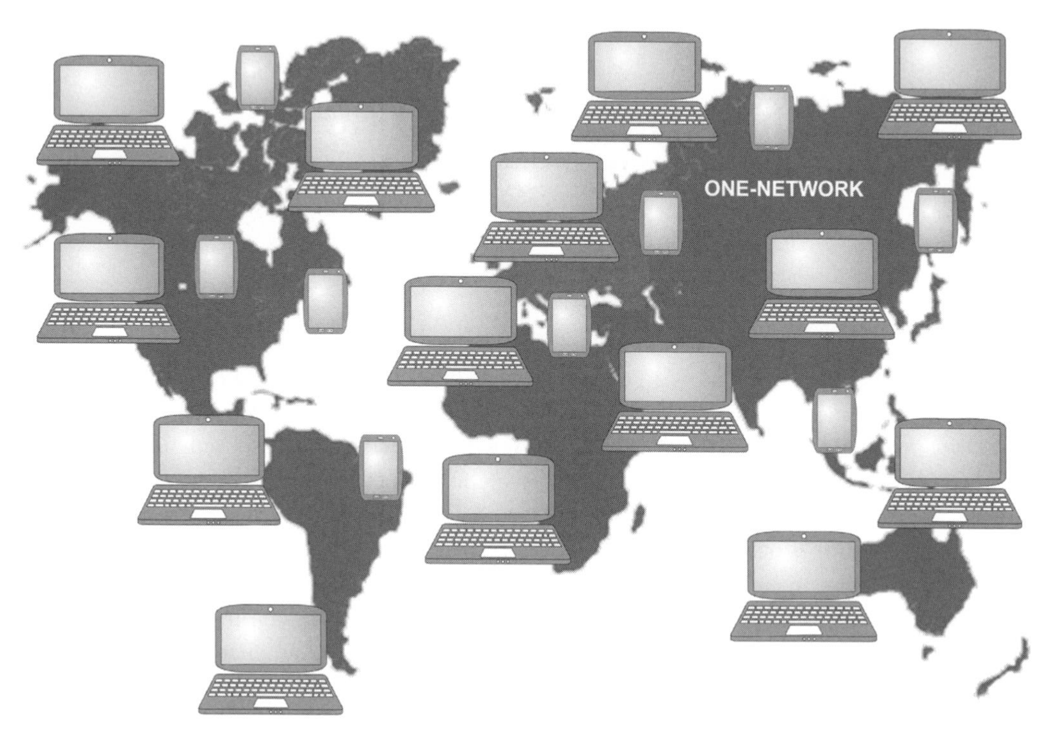

Figure 3.1. The one-network is devoid of boundaries and, as such, consumers are capable of traversing the digital globe with ease, all captained from their personal area network.

3.1 Disruptive Topologies

A disruptive topology is typified by the interference of the boundaries that typically separate an area network, which may allude to distance, scope, and purpose. In Figure 3.2, we offer a conceptual segregation of several area networks over distance. We refer to a "conceptual" boundary here because within the computing industry over the last few decades, an area network has consistently been used to characterize a set of features, or attributes if you like, that are unique to a particular networking topology. The suggestion of a conceptual separation is to suggest that, in reality, these area networks are, more often than not, interconnected, as discussed in Chapter 2. Nonetheless, let's attempt to understand our traditional networking topologies a little better, as they have been widely adopted by the computing industry to demonstrate wireless range over distance and capability, along with a host of other defining attributes.

3.1.1 Understanding Networking Topologies

To gain a better understanding of the argument for disruptive topologies, let's turn our attention to the dismantling of a typical networking topology. A network topology characterizes how a number of devices, such as computers and printers, are joined

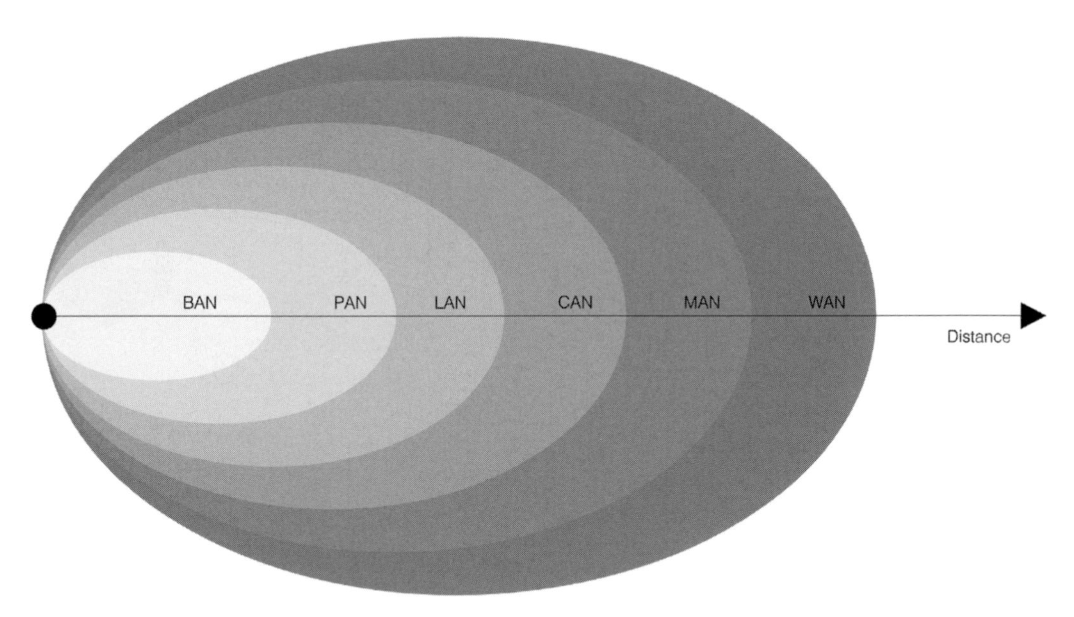

Figure 3.2. The traditional hierarchical topologies that have been used to classify range, capabilities, and features are all increasingly becoming threatened by disruptive topologies and technology convergence.

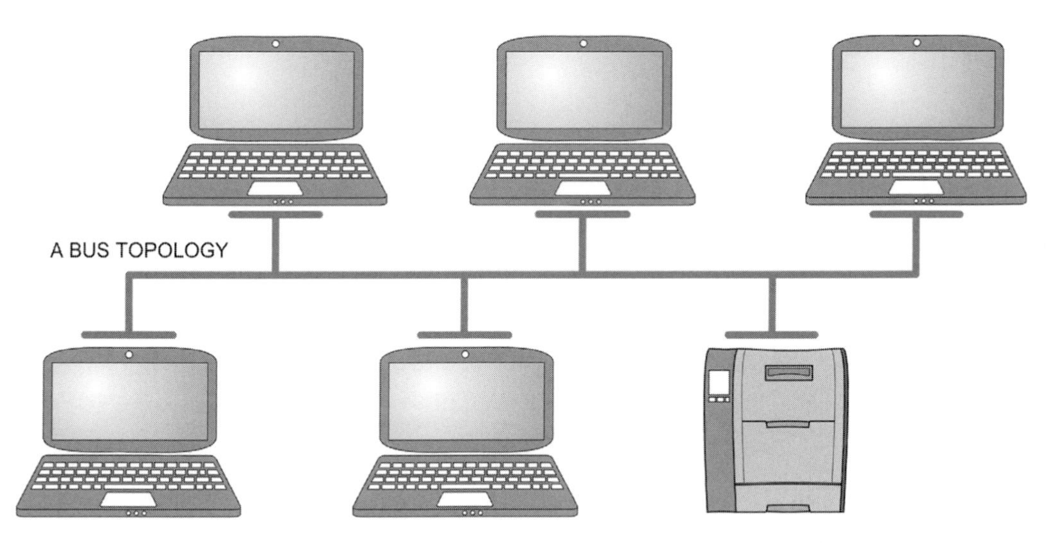

Figure 3.3. A bus topology (also typically used to form a local area network).

or interconnected – either with cables or through a wireless connection to form a network. In particular, a topology prescribes the structure of interconnected devices and/or peripheral products in terms of how they will successfully communicate with each other. A networking topology may typically take the following structure: a *bus* (see Figure 3.3), *ring*, *star* (see Figure 3.4), or *mesh* (see Figure 3.5) topology. Incidentally, a *point-to-point* topology uniquely characterizes the one-to-one endpoint connection of

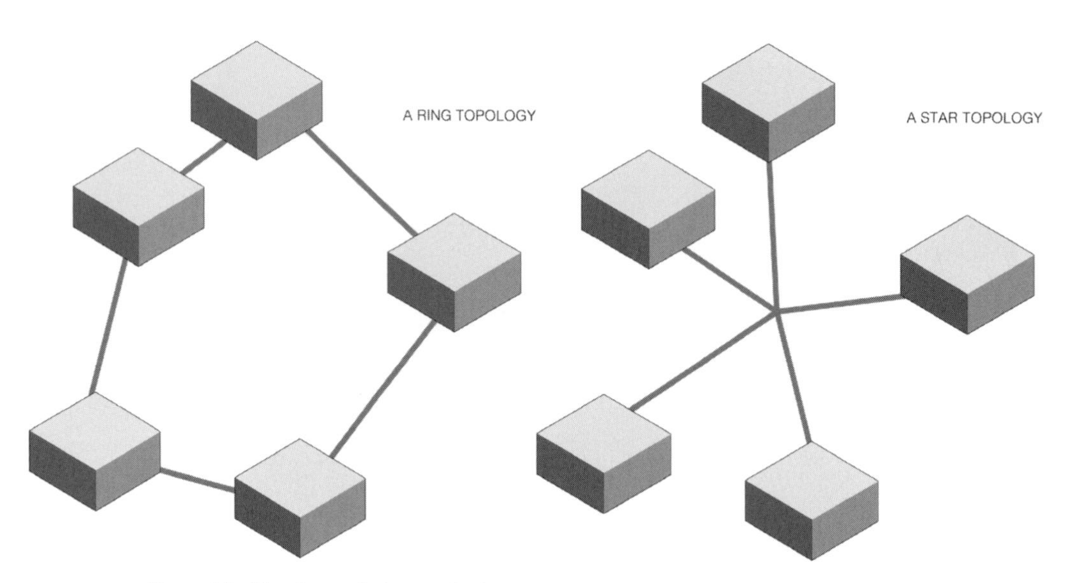

Figure 3.4. The ring and star topologies.

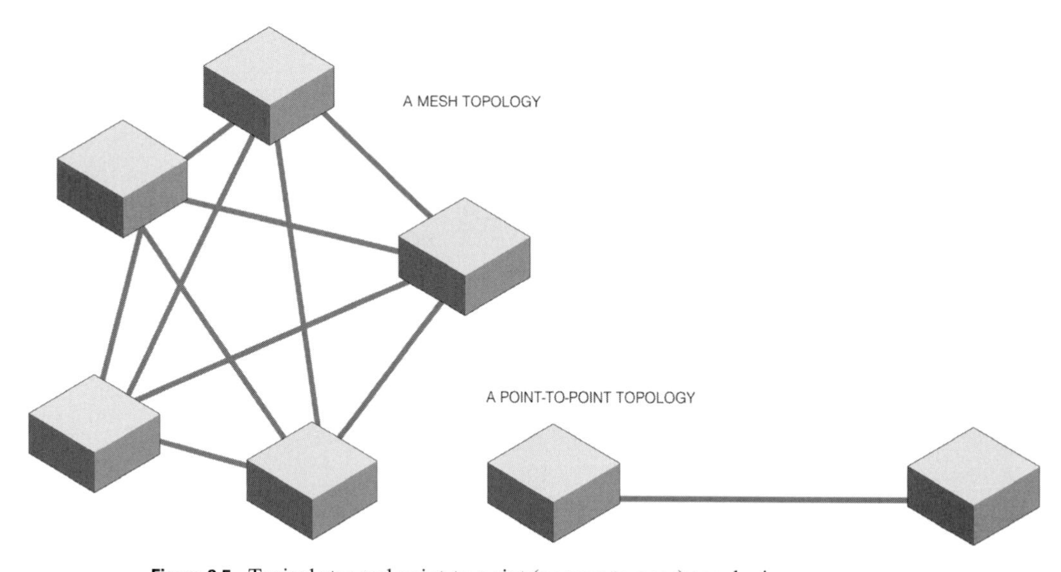

Figure 3.5. Typical star and point-to-point (or peer-to-peer) topologies.

two devices, such as a desktop or notebook computer connected to a printer, as illustrated in Figure 3.6. Similarly, there are several other topologies that rely on these basic structures. The various combinations permit the amalgamation of these key structures, which, in turn, generates *hybrid* topologies that possess unique attributes, allowing them to fulfill a specific scope and purpose. In such hybrid topologies, a combination of basic topologies is used; for example, a bus topology may be combined with a star topology to achieve a particular configuration, as demonstrated in Figure 3.7.

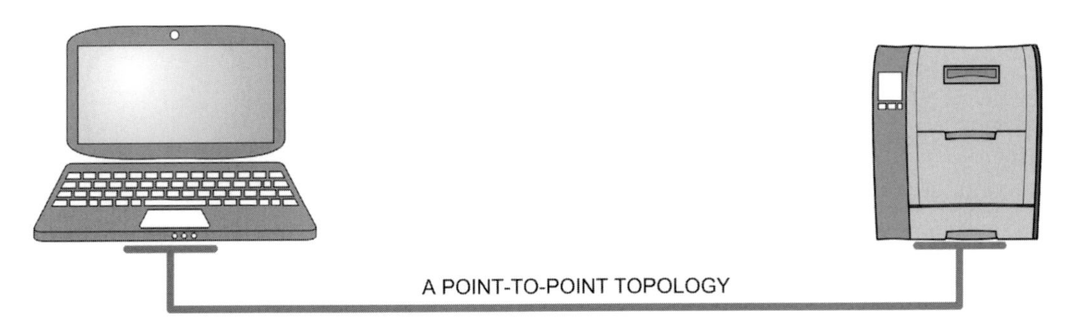

Figure 3.6. A point-to-point topology uniquely characterizes the one-to-one endpoint connection of two devices, such as the computer and printer.

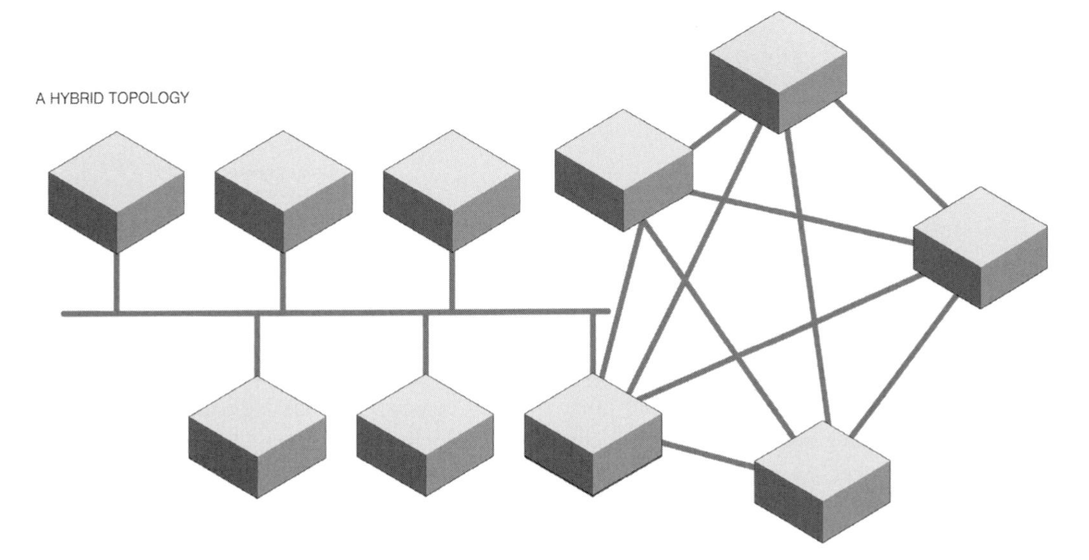

Figure 3.7. The combination of a bus topology with a mesh topology in turn creates a hybrid topology.

In another example, a *Local Area Network* (LAN) is characteristic of a number of computers within limited proximity that are interconnected and, in turn, capable of intercommunication over a relatively short distance, as illustrated by the bus topology in Figure 3.3. Likewise, a personal area network uniquely characterizes a number of devices, again within a certain proximity, that are personal to an individual, as illustrated in Figure 3.8. Mesh networking is typically used with technologies such as EnOcean, ZigBee, Z-Wave, Bluetooth low energy, and Wi-Fi, for example. A mesh topology can extend over several hundreds of meters, wherein data can be carried across the network *nodes* like a baton in a relay race.

In short, a disruptive topology occurs when a technology permits connectivity through a scheme that is not uniquely conceptualized through one of these distinct topologies. For example, a 3G/4G capable smartphone permits connectivity to the wide area network

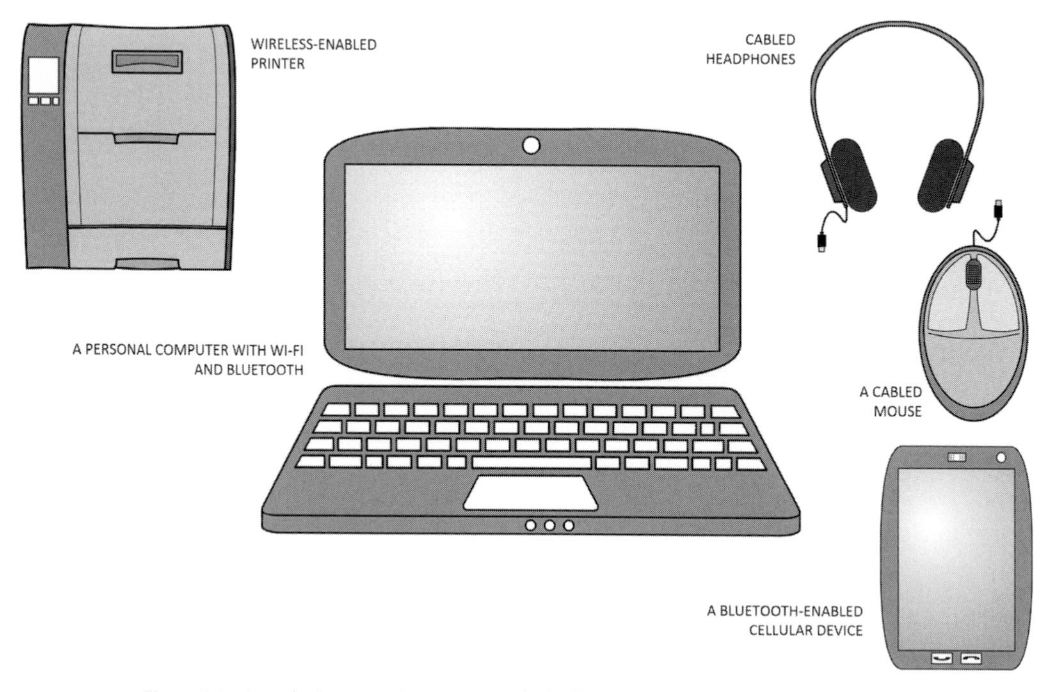

Figure 3.8. A typical personal area network that's personal to you.

in a personal area context; in turn, the smartphone typically disrupts our notion of area networking and dilutes the topology sufficiently such that we are no longer sure where one topology begins and the other ends. In essence, a technology that is capable of disrupting our traditional area network conceptualization is deemed *disruptive*.

In fact, we could tenuously argue that Wi-Fi is also a disruptive technology, as it technically cannot be considered to be a wireless personal area networking technology; in fact, it is a *Wireless Local Area Network* (WLAN) technology (we discussed this earlier in Chapter 2, "What is a Personal Area Network?"). Nevertheless, Wi-Fi permits connectivity to the wider area network or to a neighboring LAN, but your Wi-Fi router or access point remains inherently integral to your PAN. Similarly, as we have already mentioned, your smartphone is disruptive as it can enable connectivity to the wider area network or Internet, whilst remaining an integral component within your PAN space.

3.1.2 Dissolving the Boundaries

The supposition of disruptive topologies will inevitably destabilize the conceptual boundaries of the PAN and other area networks. What's more, if we take another look at Figure 3.2, we can clearly see that the WAN topology engulfs the entire range of conceptualized area networks. This is a reflection of networking topologies today, as WAN technologies permit pervasive connectivity irrespective of location. We touched upon the Lawnmower Man Effect earlier in this chapter, and suggested that it characterizes a consumer's ubiquitous ability to traverse digital systems across the globe (the one-network). Furthermore, it represents the consumer's ability to connect to what may be perceived as a virtual community, devoid of boundaries and limitations. We conjectured that this one-network enables consumers to share, with empowered censorship, their *Private Network* (PN). In other words, a consumer has the ability to decide who they let in to their personal area "bubble." It seems as if everyone now shares this one space, and it is difficult to understand where this space begins and where it might end. More and more, the PAN space is infused with technology *convergence*, that all-important ubiquitous ability to connect and traverse the digital space irrespective of location.

3.2 Technology Convergence

In *Convergence Culture*,[1] Henry Jenkins conjectures four facets of convergence:[2] *economic, technological, social*, and *cultural*. In short, Jenkins describes the role of digitized media and the part consumers have to play in forming and creating new media, as well as the relationship that consumers have with such media. Naturally, an additional factor to consider in the role of convergence culture is *social media*, and consequently the interaction and involvement of the consumer across all social media platforms. The label "media" has nowadays become so generic that it encompasses a myriad of digital platforms and has become a significant factor in today's consumer electronics and other associated applications.

So where, and how, does technology convergence fit in, and what relevance does convergence culture have? The transition of media to a digitized form, its widespread accessibility and availability, and the delivery of digitized media, whether over a cabled or wireless connection, have all largely contributed to multiple technologies duplicating the same function. In other words, there are multiple technologies providing similar mechanisms enabling the delivery of media – for example, the phone (voice), data, and video, all once independent applications, now operate together synergistically. The evolution of technology has consequentially converged media into multiple channels operating on a single platform across our one-network.

> Convergence does not depend on any specific delivery mechanism. Rather, convergence represents a paradigm shift – a move from media-specific content toward content that flows across multiple media channels, toward the increased interdependence of communication systems, toward multiple ways of accessing media content, and toward ever more complex relations between top-down corporate media and bottom-up participatory culture. (Henry Jenkins, *Convergence Culture*)

Essentially, we are witnessing new intelligent and pervasive connectivity models and strategies that are enabling consumers to communicate with disruptive topologies through technology convergence, which, in turn, straddles our more traditional topologies such as the PAN, LAN, *Metropolitan Area Networks* (MANs), and so on. In the sections to follow, we briefly touch upon some of the technologies that are said to be

[1] Jenkins, H., *Convergence Culture*, 2008.
[2] Tripp, M., "Henry Jenkins, *Convergence Culture*," ~2006.

responsible for disrupting our topologies; technologies such as white space radio, femtocells, 3G/4G, and satellite broadband (for the most isolated areas), which are all allowing us to sustain an online presence regardless of our location.

3.2.1 White Space Radio

We discuss white space radio further in Chapter 16, "Future and Emerging Technologies," but, for now, we'll briefly cover the subject as it has the potential to become a technology that provides holistic connectivity across rural homes (*wireless broadband*) as well as to provide a host of *Machine-to-Machine* (M2M) applications to include *smart metering*. The name "white space radio" emerged as a consequence of a liberated analog television spectrum, due to the analog to digital switchover. The focus of the group supporting the technology, the Weightless *Special Interest Group* (SIG) surrounds the 600 MHz frequency. The propagation characteristics of this frequency lend themselves quite well to distance, along with the ability to travel through walls and other objects with ease. White space's innocuous ability is planned to offer consumers the freedom of connectivity in the remotest of areas.

3.2.2 3G, LTE, and LTE Advanced

3G has been available to consumers for many years now. The race toward 4G is an ambition to serve an increasing demand upon cellular network operators – that of *data-hunger* (see the discussion of *packet-hungry* in Chapter 1, "It's a Small Wireless World"). 4G, and in particular *Long-term Evolution* (LTE) *Advanced*, which is a significant enhancement to the *3rd Generation Partnership Project* (3GPP) LTE standard, aims to offer true wireless broadband, which the *International Telecommunications Union* (ITU) governing body touts as reaching data rates of up to 1 Gbit/s. The members of an insatiable, data-hungry consumer-base regularly use their cell phones to connect to the wider area network to ensure that they remain part of their virtual community (see Chapter 4). More and more, smartphones are becoming connectivity-dependent, since the multiple "apps" require constant updates, and your social media network is normally integrated within your smart device. Typically, data rates are an inclusive part of your subscription, but, with an increasing demand on data services, network operators are struggling, both to keep up with this demand and to provide adequate coverage for consumers in rural areas. A number of network operators have devised a "band-aid"[3] approach to resolving over-capacity of their cells and are turning to alternative techniques to ensure consumers utilize their data packages.

3.2.2.1 Femtocells

As we have already intimated, femtocells are considered to be a "band-aid" approach to resolving both the over-capacity of cellular network operators' infrastructures and the coverage issues in rural areas. Femtocell products are touted as an extension of the

[3] Gratton, D. A., "Femtocells: You're Either Right or Impatient!," June 2011.

cellular infrastructure, insofar as they offer a cell data connection within the home or business through a broadband backend, in turn enabling consumers in poor cell coverage areas to continue to utilize their data package (as part of their regular subscription with their cell operator). The technology is not limited to rural areas, as the product can be utilized in urban areas to provide better cell phone coverage. The growth of this technology has been restricted by a largely effective broadband backend, and it has received a mixed reaction from the market overall. It is anticipated that femtocells will serve a niche consumer-base, offering a cell service within the home or business for up to five concurrent users in areas that are affected by poor cell coverage or for consumers wanting to receive an improved cell service.

3.2.3 Satellite Broadband

In the remotest of areas, or where rural broadband is intolerably stagnant, satellite broadband can be used to provide connectivity within the home or business. The technology is often expensive, but nonetheless does enable consumers and business owners to sustain a network presence. There are several providers offering a variety of data packages for both home and business users.

4 Introducing the Lawnmower Man Effect

We are all connected.

The need for consumers to sustain a permanent connection has been driven by a deep-seated need, fueled by peers, the furor of social media, and just simply by a trend to have immediate access to anything, anytime, anywhere. The technology supporting this trend has kept up with its ever-evolving pace and is only guilty of stumbling on occasions. Nevertheless, the availability of Wi-Fi[1] in bars, shops, restaurants, and so on, along with favorable data packages offered by both fixed and cellular providers, have allowed us all to sustain that all-important "IP-fix." In fact, most (if not all) *Mobile Network Operators* (MNOs) and *Mobile Virtual Network Operators* (MVNOs), along with multiple *Telecommunication Service Providers* (TSPs) and *Internet Service Providers* (ISPs), have provided their consumer-base with multiple opportunities to remain connected to the wider area network, or the Internet as it is commonly known.

> The Lawnmower Man Effect (LME) represents the consumer's ability to traverse digital systems across the globe, all captained from their personal area networking space utilizing pervasive WAN technologies. (Dean Anthony Gratton)

4.1 Overview

So, we have penned the *Lawnmower Man Effect* (LME) as a definition; a term that classifies the ability afforded to a new generation of consumers who seek to have a permanent connection to anything, anytime, anywhere. A generation of consumers who are equipped with numerous electronic devices, which have an inherent ability to connect. In Figure 4.1 we further conceptualize the LME supposition that we, as individuals, have this unique ability to cross the globe virtually, in an instant. The term is derived from the 1992 film starring Jeff Fahey and Pierce Brosnan, *The Lawnmower Man*. In the film, we witness Brosnan conducting numerous experiments on Fahey using *virtual reality* in an attempt to increase his intelligence, but, of course, this is not the focus of our definition here. Rather, in short, despite Brosnan's character's malevolent motivation, Fahey becomes physically holistic within the wider area network, having an ability to traverse computers, technology and telephony systems, and a range of

[1] Wi-Fi has become synonymous with connecting to the Internet, something which we discuss later, in Chapter 13, "The 802.11 Generation and Wi-Fi."

Figure 4.1. The Lawnmower Man Effect typifies a consumer's ability to traverse digital systems across the globe, all captained from their personal area network through pervasive WAN technologies.

other digital applications and services across the globe. Essentially, the Lawnmower Man Effect typifies the consumer's ability to traverse similar systems across the globe, whether sitting at a computer, waiting for a train, or cheering at a sporting event, hopefully devoid of the malevolence portrayed in the movie, and all captained from their personal area network. The Lawnmower Man Effect differs from the *Internet of Things*[2] (IoT) insofar as the IoT refers to the interconnection of distinguishable and smart/intelligent objects or "things" and their virtual manifestation within the Internet or similar IP structure. Likewise, as consumers, we are equally identifiable and physically

[2] The term "Internet of Things" was penned by Kevin Ashton in 1999; the term refers to the interconnection of distinguishable objects or "things" within the Internet or similar IP structure.

traceable – our IP-address might already be sufficient for us to become holistic participants in an extended IoT. We discuss IoT and how everyday objects have become smarter in Chapter 6, "Enabling the Internet of Things."

What's more, the LME refers to a supposition that consumers, irrespective of their location and despite their conceptual boundaries, are all connected to this *one-network* – if you like, one IP-enabled global community, where each consumer has every opportunity to remain connected, no matter where they are in the world – and this ethos applies equally well to the IoT. Nowadays, the immediacy of content and access, and perhaps that of social media, is also responsible for this need to be ever-present. Put simply, we want access to information "Now"; we can no longer wait for that all-important email, tweet, Facebook, or Google+ message, and, of course, there are those eager to triumph through the adornment of a new badge awarded to them by Foursquare, as they oust their colleague to become *Mayor* of a popular establishment or venue. All of these web-based communities have a well-established presence with available access through fixed or cellular-enabled devices, ensuring your attendance is (emphatically) required, either sitting at or away from your computer, or even while you are traveling mid-flight over the Atlantic.

4.2 The Social Media Phenomenon

In fact, the dynamic social media shift in communication has given rise to a new revolution of instant knowledge acquisition and *curation*, supporting a culture of virtual communities mapped across a virtual space and echoed through platforms such as Twitter, Facebook, Google+, and LinkedIn, something we have already touched upon. Gone are the days of monolog-based promotion from traditional forms of media broadcasting. Today we are our own promoters, campaigners, and cynics – the influencers on whose content both lifestyle and purchasing decisions are made. Social media has instilled a deeper sense of belonging and community, which goes far beyond just a tweet here or a tweet there! It has given birth, in effect, to a growing army of *brand ambassadors*. Consumers who previously yearned to feel part of the latest social platform trends are now key influencers of where new technology is leading – all supported by an omnipresent fixed or wireless connection.

Big Brother is indeed watching – Big Brother is us!

4.2.1 Cross Platform Promotion

With holistic tools such as *Cross Platform Promotion*[3] (CPP), we now have the ability to provide a 24-hour global presence, incorporating our thoughts into daily blogs, tweets, and posts – our likes and dislikes have the power to influence thousands. We are masters and commanders of our virtual space, and we are perhaps only one step away from allowing narcissism to overpower our sense of fair play. There is no denying the

[3] Gratton, S. and Gratton, D. A., *Zero to 100,000*, 2011.

double-edged sword analogy when we consider the power of social media and where it may lead us in terms of future connectivity. A brave new world of truth, or a narcissistic monopoly of power-hungry evangelists?

Only time will tell.

4.3 Intelligent Personal Area Networking

With our 24-hour connectivity paradigm, in turn sustaining our presence in the one-network, consumer electronic and other devices are becoming consciously aware – devices are becoming smarter! A humanistic behavior, if you like, which permits an intrinsic ability to auto-configure. The provision of intelligent personal area networks is the premise of one example of smart personal devices that are becoming savvy enough to auto-configure and offer users a conversational approach to completing their set-up. In Chapter 1, "It's a Small Wireless World," we discussed the possibility of a wireless utopia where consumer electronic products, when brought into proximity, would instantly recognize features and capabilities of other similarly enabled products and auto-configure accordingly, ultimately needing little or no user intervention. The notion of intelligent personal area networking isn't entirely new, and several companies are endeavoring to support an ecosystem in which devices can come together and interoperate seamlessly and transparently – all part of the simplicity stance that so many products previously lacked. In taking several strides forward, the *Digital Living Network Alliance* (DLNA) already provides a standard such that DLNA-enabled consumer electronic devices within the home are capable of sharing media content by recognizing other like-enabled devices on the user's home network. This is another extension of the personal area network, as we are now capable of streaming media to our TV directly from our PC. The next step is to offer these same devices within the PAN and home, to auto-configure with minimal interaction from the consumer. In another initiative, the *Universal Plug and Play* (UPnP) forum defines a number of networking-specific protocols that attempt to enable seamless and transparent connection of network devices. Both DLNA and UPnP have been around for many years and, if you like, are the unsung heroes of seamless coexistence and interoperability between today's consumer electronic products.

The premise is that our personal area network and other devices within our homes, such as the computer, TV, *Set-top-box* (STB), and DVD, are all network-capable and, as such, should auto-configure gracefully. The ability to discover other like-enabled devices should be innate, which brings us neatly to the *extended* personal area network.

4.4 The Extended Personal Area Network

As our personal area networking space has diverged, multiple naming conventions have been adopted to characterize uniquely its scope and function. In Chapter 2, "What is a Personal Area Network?," we offered several colloquial references that are used to portray a specialized, ad hoc, or unique form of area network. We suggested such terms

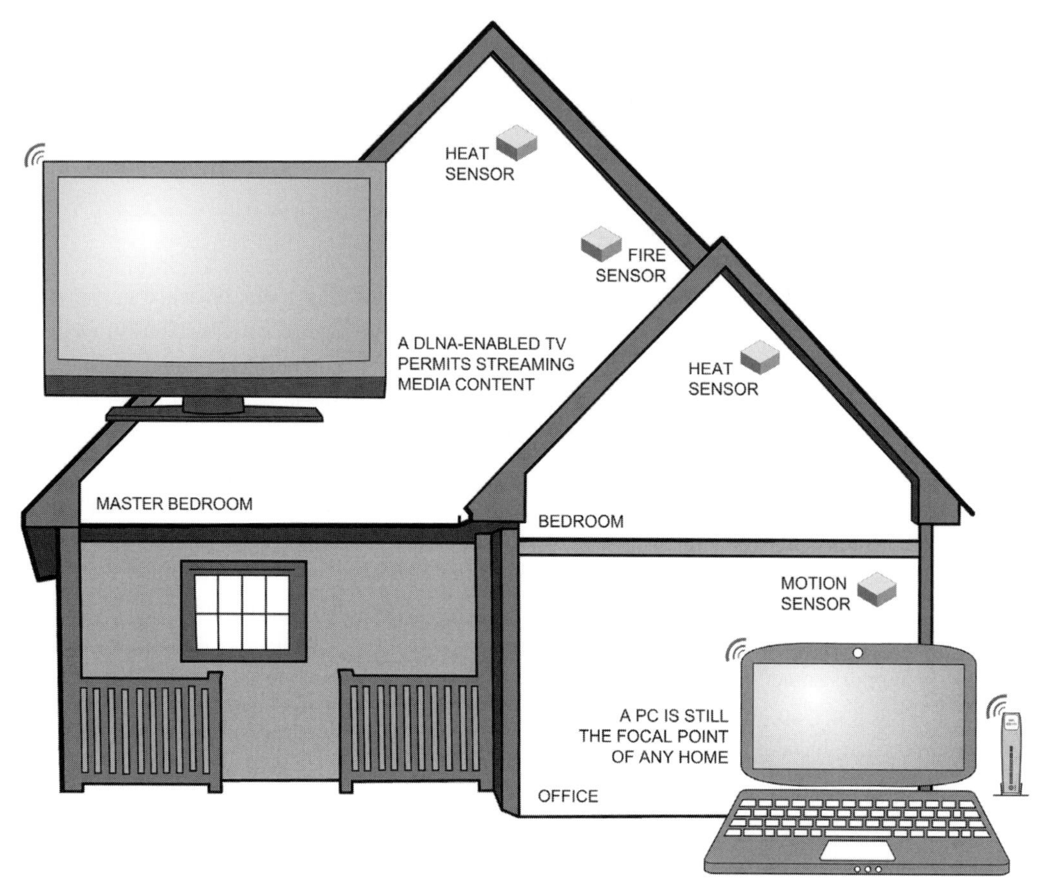

Figure 4.2. The home area network may comprise a PAN, along with a sensor network. The area network within the home comprises a multitude of fixed and wireless devices, in turn creating a blurred conceptual area.

as *Vehicle Area Network* (VAN) and *Wireless Video Area Network* (WVAN) as a couple of examples. However, increasingly, these once colloquial terms are being used in everyday language. The *Body Area Network* (BAN), *Home Area Network* (HAN), and VAN are becoming part of mainstream conventionalism. Again, these conventions that we've touched upon are used to characterize the function, scope, and purpose of the devices that occupy such a network. In Figure 4.2, we illustrate an example where the home uses several low energy sensors to monitor environmental influencers; the computer in the office can stream media to a TV located in the bedroom, and the Wi-Fi access point and router provide that all-important Internet connection to the *Wider Area Network* (WAN).

4.4.1 The Home Area Network

The home area network is still very much a PC-centric perspective and, despite manufacturers attempting to steer away from PC-dependent technologies and applications, the PC

Figure 4.3. The Boxee Box by D-Link (DSM-380) provides DLNA capability. (Courtesy of D-Link.)

still remains integral to the home. For example, some TV manufacturers are integrating "net"-capable features, such as LoveFilm,[4] Netflix,[5] YouTube,[6] and other dedicated TV services, offering TV content for pay-per-view channels through the Internet-capable TV. However, these rental service providers also offer the ability to view channels on a number of Internet-capable devices, such as smartphones and tablets, as long as they are logged in to their respective account. In fact, some manufacturers provide an all-in-one solution, such as D-Link (dlink.com) with their Boxee Box product, which harnesses local PC-content and the Internet to stream media to your TV. We illustrate D-Link's Boxee Box in Figure 4.3 and discuss the influence of such products in Section 4.6, "Media Convergence," where we look at how the media industry has converged its mainstream content over the Internet. Some TV manufacturers are also looking to integrate Skype (skype.com), a video and telephony service which is widely used on the PC – another attempt to steer away from the PC as the primary "command" center. In fact, some reports[7] suggest that the TV has surpassed the PC, tablet, smartphone, and other devices as the preferred device choice to watch online content, but portability and ease of access will ultimately become the deciding factors as to where consumers digest their content. Nonetheless, the PC still remains (and arguably will remain) the focal point of most homes and, as such, it will inherit the responsibility of managing area networks within the home.

[4] A British provider and an Amazon subsidiary (lovefilm.com, circa 2002) that offers streaming video on demand in Northern Europe.

[5] An American provider (netflix.com, circa 2007) of streaming video on demand services across the United States, Canada, and Latin America. Netflix launched their services in the United Kingdom and Ireland, competing against British provider LoveFilm, in 2012.

[6] YouTube also provides rental services for the United Kingdom, Canada, and the United States.

[7] Knight, S., "Televisions Overtake Computers for Online Video Streaming," 2012.

4.4.1.1 Domotics and the Smart Home

Domotics,[8] a subject which we discuss later on in the book, broadly describes technology and/or automated devices and applications within the intelligent home; for example, intelligent home heating, intruder/motion sensors, and so on, may all report back to a central unit, which can then be viewed at a console. However, a PC would permit greater flexibility in the manipulation and control of the collated data, from where the consumer would be able to act upon such information accordingly. Typically, media content is located on a PC, which can then be streamed to any nominated capable device on the home network, although TV manufacturers are initiating a shift away from PC-dependent content, as we have already discussed. With the popularity of tablet devices, control can easily be shifted to the portable device (irrespective of operating system), but a PC would remain integral to the personal area network. In short, the home area network, as illustrated in Figure 4.2, naturally encompasses the entire home, but there remains uncertainty as to where the PAN begins and ends. Is the PAN simply limited to the office, or does it encapsulate the extended home area network? It is safe to conclude that, indeed, the personal area network remains in charge of the domotic space.

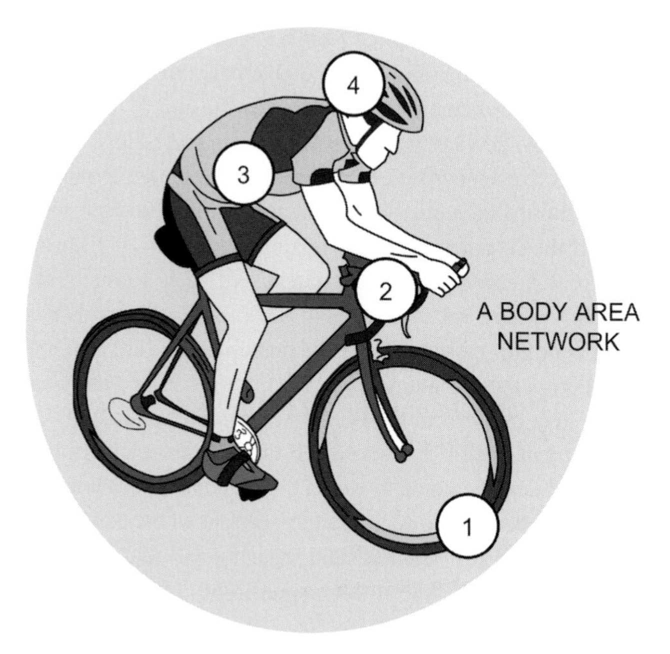

Figure 4.4. A BAN may comprise a number of fixed or wireless-enabled products that occupy your person.

[8] "Domotics" is a term penned in 1984 by French journalist Bruno de Latour to characterize the use of computer science and automated technologies to form intelligent homes (*maison intelligente* or "smart home," 1986).

4.4.2 The Body Area Network

The body area network has emerged as a topical and popular topology that typically characterizes *wearable technology* and the ability to collect data, more often in real-time, from a number of interconnected or independent devices about your person, which may be part of your clothing or simply technology that's placed in your pocket. For example, in Figure 4.4 we illustrate a cyclist using a bike power sensor (**1**) which has been mounted on a bicycle to capture data relevant to the cyclist's power output. The bike power sensor may transmit information to a mounted display (**2**) or to a wireless watch (**2**) – the cyclist is also using a heart-rate monitor (**3**). And finally, the cyclist is listening to music using a wireless stereo headset (**4**). The sports, fitness, well-being and health market sectors greatly benefit from such technology, and we will expand on our understanding further in Chapter 5, "Introducing Low Power and Wireless Sensor Technologies," and in Chapter 10, "The Power of Less: ANT."

So, in short, the number of devices emerging onto the market is prolific. The range of MP3 players and headset products, for example, all begin to form your unique BAN. The prevalence of low energy technologies, in turn creating products for wireless watches, heart monitors, and other medical devices, are used to collate data onto a PC for performance, trending, and behavioral statistics. Again, all your performance data can be viewed from a device that's capable of displaying such information, but, of course, your PC (PAN) would offer you greater flexibility.

4.4.3 The Vehicle Area Network

Another popular topology emerging into everyday use is the vehicle area network, as illustrated in Figure 4.5. A VAN may incorporate an MP3 player, SD-Card, USB-stick, and a Wi-Fi or Bluetooth-capable device, such as a cellular phone, which becomes

Figure 4.5. A VAN may comprise a number of wireless-enabled products that occupy your vehicle, and it is extended to your PAN and WAN.

integral to the vehicle's entertainment and telephony system (**1**). The satellite *Global Positioning System* (GPS) (**2**) uses wide area networking technology to gather information regarding your location, and you may then extend this to Foursquare, for example, to allude to your location, which you may share with other participants.

Other use cases may permit streaming over Wi-Fi (**3**) (music playlists, for example) to your vehicle in preparation for your journey into work the following day – the PAN will, of course, manage your vehicle area network playlist. And finally, a *Radio Frequency Identification* (RFID) tag (**4**) is used to automate passage through freeway/motorway tolls, security barriers at your place of work, and so on.

4.5 Wireless Convergence

The economic downturn has witnessed numerous companies converge in their effort to continue to deliver their products and services, as the world's optimism is quashed by escalating costs and empty wallets. This consolidation of effort has inadvertently streamlined and economized industries who may have sat back and wondered: "Why didn't we do this before?"

If we extend this philosophy to wireless technology and play devil's advocate, is there any need to have multiple technologies all competing for a similar, if not the same, market sector? We could possibly conjecture whether it is necessary for a consumer to deliberate how they might choose to transfer a picture from their mobile phone to their PC. The consumer might simply be thinking: "I just want to do it." The consumer technically needn't worry that they would have to use a particular technology to enable the transfer; rather, they should have the capability of simply making a transfer, irrespective of the technology underlying their product. In a similar manner, as consumers we pick up a mobile phone and make a telephone call – we are blissfully ignorant as to how this is achieved – we just become connected and then we're speaking with our friends, family members, or even our colleagues.

Figure 4.6. The Texas Instruments WiLink™ 7.0 Solution chipset (WL1281/WL1283). (Courtesy of Texas Instruments.)

4.5.1 A One-Size-Fits-All Technology

Naturally, playing devil's advocate allows us to consider willfully whether the wireless industry will witness a "one-size-fits-all" technology, in turn banishing the plethora of standards and the incumbent competitive "mine's better than yours because..." theology. This may be perceived as a little harsh, but we have already witnessed a number of manufacturers consolidating multiple radios onto a single chipset. For example, Texas Instruments (ti.com) offer their WiLink™ 7.0 Solution platform (WL1281/WL1283), which consolidates GPS, Wi-Fi, Bluetooth wireless technology (classic), Bluetooth low energy, ANT technology, and an FM radio transceiver into a single chip, as we show in Figure 4.6 and conceptualize in Figure 4.7. Likewise, other manufacturers have also consolidated multiple radios into a single chipset offering, which can lead us only to speculate as to a timeframe in which this evolutionary step toward wireless convergence will take place – a "one-size-fits-all" silicon chip.

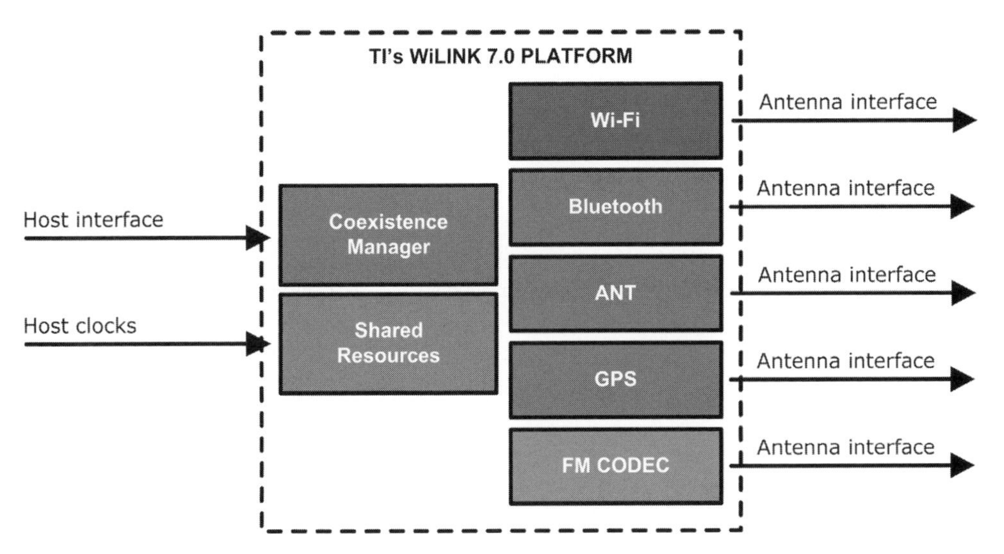

Figure 4.7. Conceptualized image of TI's WiLink™ 7.0 Solution platform, comprising a number of wireless technologies all consolidated into a single chip. (Source: ti.com.)

4.6 Media Convergence

The media industry is not oblivious to the convergence phase either, with many Hollywood, American, and British media giants consolidating their efforts and diversifying their content delivery. Further fueling the LME hypothesis, with consumers permanently connected to the wide area space, audio/video content can today be sourced from almost anywhere. We live in a continually evolving world of expanded media choice that has both empowered us as consumers and shortened our concentration spans dramatically, as we'll discover in a moment.

With the arrival of the home video recorder in the 1980s, a dramatic shift in control occurred; one that saw the move from a TV network-controlled programming perspective to a consumer-controlled one. The ability to watch one television program whilst recording another, to fast-forward, rewind, and freeze-frame our favorite moments, became an intoxicating need that has become increasingly insatiable. "We were empowered by a new ability to watch what we wanted *when* we wanted."[9]

What's more, we have moved to a media-converged state of existence, in which we are continuing to seek out new ways to stay permanently connected to its flow – wherever we are these days, we can connect. Traditional broadcasting corporations are aware of this shift and have adapted to it by further empowering their viewers through online applications such as BBC iPlayer (bbc.co.uk/iplayer) in the United Kingdom and Hulu (hulu.com) in the United States, for example. In an increasingly competitive viewer market, however, these applications must compete with a growing number of online services, such as LoveFilm, Netflix, and YouTube, which offer pay-per-view TV content direct to Internet-capable devices, something we touched upon earlier in Section 4.4.1, "The Home Area Network."

Figure 4.8. Zeebox cleverly exploits the power of the ever pervasive LME to provide a connected, interactive experience of television viewers around the world. (Courtesy of Zeebox.)

4.6.1 Social TV

But we have become more than mere observers. Nowadays, we are able to interact with the information and entertainment we receive, through our ability to capture and deliver content as it happens. For example, *The Voice UK* (bbcthevoiceuk.co.uk) and *Britain's Got Talent* (itv.com/BritainsGotTalent) both offer interaction through Twitter via their clever use of Twitter *hashtags*,[10] namely #TheVoiceUK and #BGT, respectively. During live and key moments of the aired show, the hashtags will appear on screen encouraging audience participation and promotion through their Twitter interaction, using the hashtags to bring together the conversation and, in turn, create a *trending topic*.[11]

We carry our content with us and it's become pocketsize! We can flit between taking a call, writing an email, reading a text, and watching our favorite band play live on stage at Radio

[9] Gratton, S., *Follow Me!*, 2012.

[10] Hashtags are words or phrases prefixed with the # (hash) symbol to denote a group or topic, used to facilitate searches using Twitter.

[11] A hashtag that has been populated at a higher rate than other hashtags on Twitter is said to be a *trending* topic. More often than not, these trends are generated by a specific or topical current event, or simply by a large community of people talking about the subject.

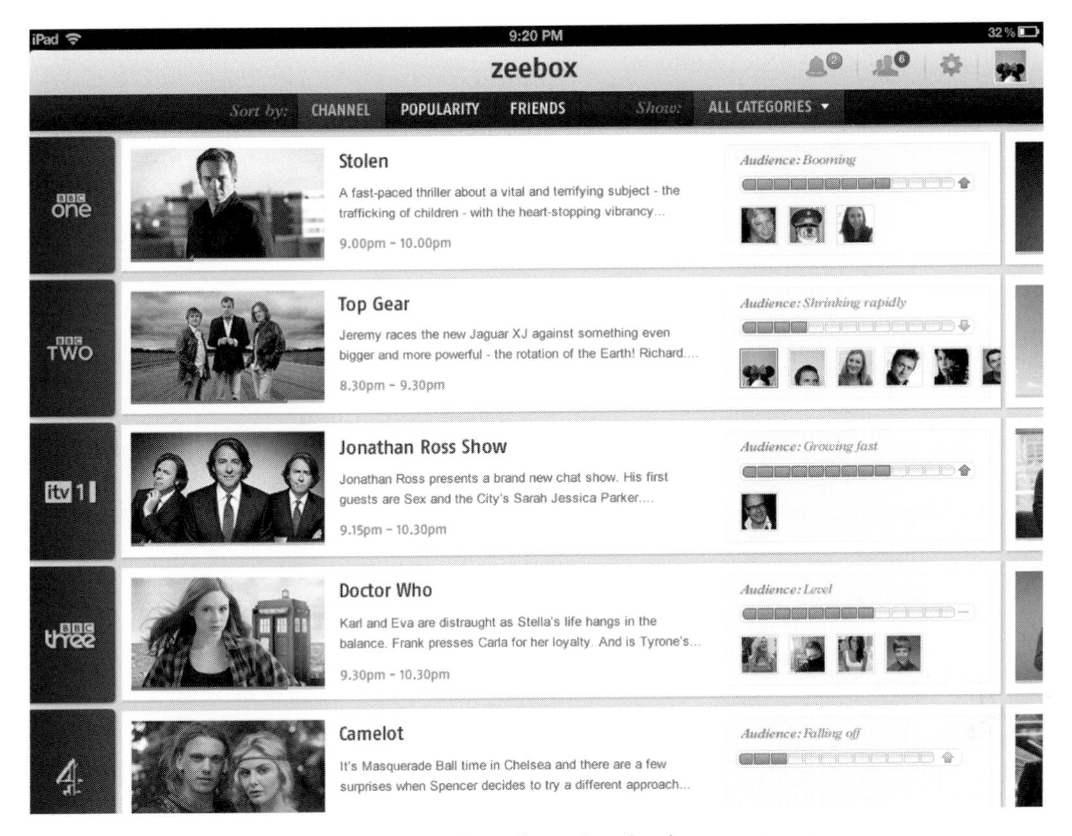

Figure 4.9. We are no longer just the audience, but also the content creators.

> City Music Hall. With all these choices constantly available to us, it's no wonder that our synapses have adapted to filter out any unwanted noise. (Dr. Sarah-Jayne Gratton, *Follow Me*!, 2012)

This interaction has expanded into even greater ways to connect and share traditional media moments, with applications such as Zeebox (see Figure 4.8) tapping into that all-important need for a permanent connection. Zeebox is a London-based company founded by Ernesto Schmitt, a former President of EMI, and Anthony Rose, the man responsible for the design and delivery of the BBC iPlayer. It is a company that's cleverly exploiting the power of the ever-pervasive LME to provide a connected, interactive experience for television viewers around the world. Its growing popularity among users echoes the fact that we are no longer just the audience, but also the content creators (see Figure 4.9). As such, a psychological shift has also taken place; one that has resulted in our attention spans shrinking as we try to condense our media mix into bite-size pieces that can be quickly absorbed and shared. We know that our content stream is continually updating, and we live with a "What's coming next?" attitude toward the way that we choose to process it, always ready to jump ahead to the next and best new item in what's become known as a *fast-forward* society.

Part II

The Wireless Sensor Network

5 Introducing Low Power and Wireless Sensor Technologies

"It's trendy to be green" is a mantra chanted by many companies, manufacturers, and consumers alike. If you're not whistling an eco-friendly tune, you're deemed to be pillaging the planet of its unsustainable resources. Increasingly, we are encouraged to think of the bigger picture and to understand the repercussions of our decisions today, which ultimately may have an impact on our future. Inevitably, manufacturers are seeking alternative energy solutions that either extend battery life or harness ambient resources to energize new products. In this chapter, we touch upon some of the techniques that have been used to extend battery life alongside those used to capture natural resources as an alternative to using batteries or a fixed power source. The chapter also looks at the diversified market, in which low energy, low power sensor technology is being used; then, in the upcoming chapters, we will explore in greater detail specific areas of technology, market, and scope.

Figure 5.1. Consumer electronics and appliance manufacturers have a variety of energy saving programs available to them that encourage reduced energy consumption and product efficiency, in turn placing the planet into safer hands.

5.1 Energy Efficient Labeling

Let's look first briefly at how manufacturers inform their consumers about products that are labeled and classified in terms of energy efficiency. Numerous programs are available that encourage manufacturers to build energy efficient products, in turn placing the future of our planet into safer hands (see Figure 5.1).

The United States, Europe, and Asia utilize a number of schemes and logos that offer the intended consumer the ability to assess what kind of impact a chosen product may have on their energy bills, as well as a potential longer term consideration of the planet's unsustainable energy resources. The *European Union energy labeling* scheme for white goods (or household appliances), such as washing machines, dryers, refrigerators, and so on, means such items now display an *energy label*, an example of which is shown in Figure 5.2. This scheme is scoped in several EU directives and has been widely adopted across Europe. In consumer electronics, an *Energy Star* logo, as shown in Figure 5.3, is used to identify energy efficient electronic products. This voluntary international standard, which originated in the United States, has been adopted across the world and is used to identify the energy efficiency of computer equipment and its peripherals as well as kitchen equipment, in addition to that of homes.

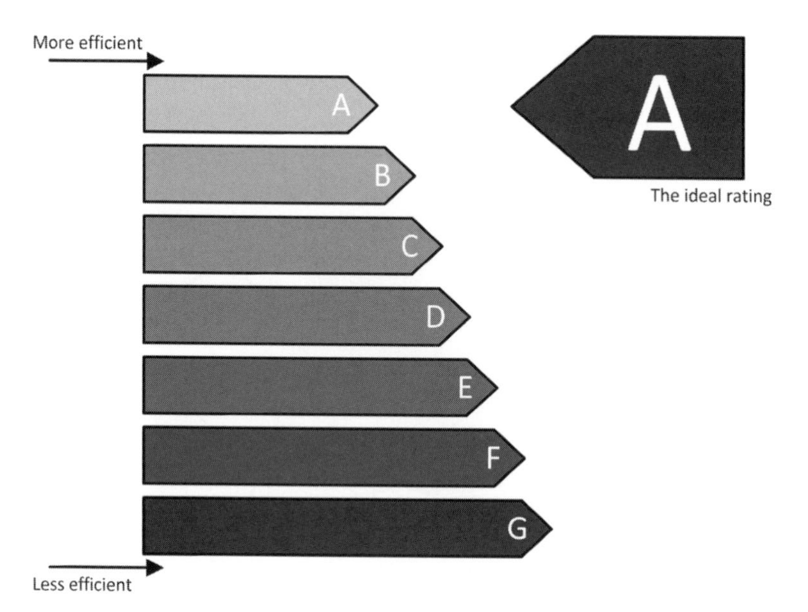

Figure 5.2. The European Union energy labeling scheme used for white goods, vehicles, housing, and window installations assigns a classification from A to G.

5.1.1 The European Union Energy Label

The energy labeling scheme isn't limited to white goods – it also applies to vehicles, buildings, houses, and window installations. Naturally, the scheme varies, to some extent, from country to country, and, in the examples we use here, we refer to the European

Figure 5.3. The Energy Star program and its associated logo (sources: energystar.gov and eu-energystar.org) results from a joint collaboration between the EPA and the EU. The voluntary program encourages manufacturers to use the energy logo on their products.

Union energy labeling program. Nevertheless, the energy efficiency class is used to inform potential consumers of the energy efficiency of a particular product, and its class rating, as shown in Figure 5.2, is used to inform the purchaser of how efficient that product is. The energy efficiency class shown on an appliance (see Figure 5.4) is a rating given from G (less efficient) to A (most efficient). The classification is provided as an overall summary following exhaustive energy efficiency testing and verification. If we use the example of a washing machine, a consumer can quickly ascertain parameters that were used to classify the appliance. The energy label will include information relating to factors such as how much water and energy are consumed during a washing cycle and how efficient the spin cycle is during the drying process. The label may also include the overall capacity of the appliance and information relating to noise pollution. In short, a consumer is provided with exhaustive information relating to how eco-friendly and green the product/appliance is.

5.1.1.1 A+ and A++ Energy Classes

Naturally, manufacturers are constantly evolving techniques to conserve power and provide more efficient energy-aware mechanisms. As a consequence of these improved energy efficient techniques, which are nowadays increasingly becoming inherent in a new generation of appliances and electronic products, the European Union energy label has been accommodated to include top rated A classes. The new A+ and A++ classes are now included on the energy label to offer consumers greater choice when selecting appliances, homes, vehicles, and so on.

5.1.2 The International Energy Star Logo

The Energy Star program (originally defined in the United States, see energystar.gov) has become a joint collaboration between the *Environmental Protection Agency* (EPA), an American governmental body, and the *European Community* (EC). The specific agenda

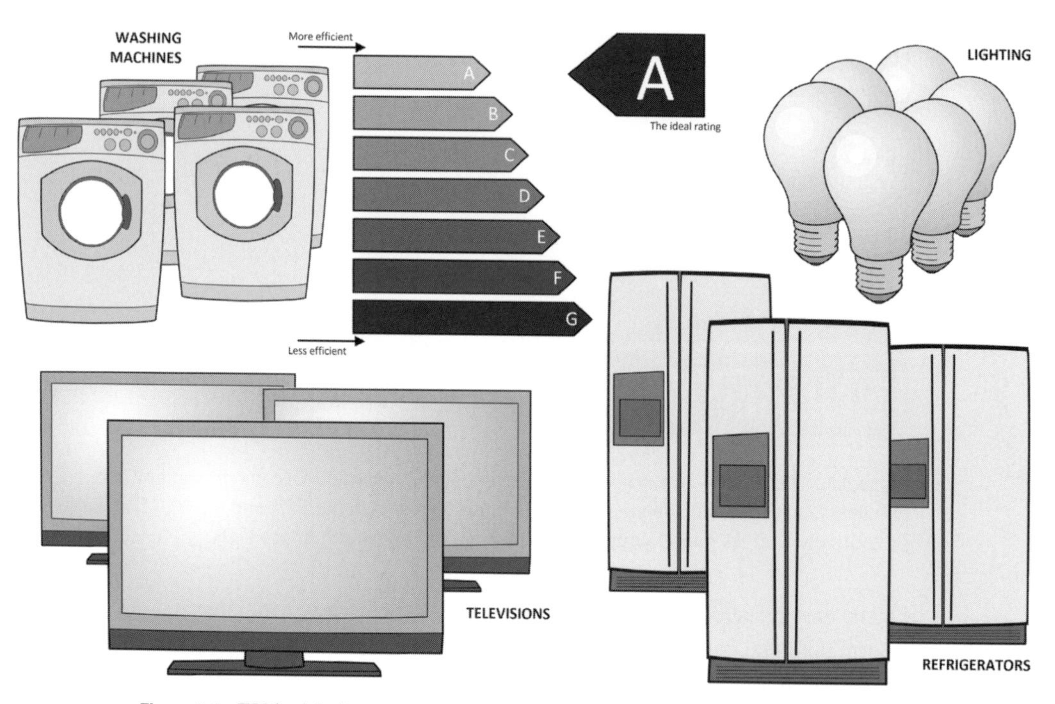

Figure 5.4. EU legislation ensures that all white goods are labeled suitably, informing the consumer of the energy efficiency class.

of both governmental agencies is to encourage manufacturers to achieve Energy Star ratings of their appliances and products, with the objective of reducing greenhouse emissions and overall energy consumption. The program isn't limited to the United States or Europe (eu-energystar.org); other countries, to include North America (oee.nrcan.gc.ca), Australia (energystar.gov.au), and some parts of Asia, have all adopted the program. The program and its numerous technical specifications cover a broad range of consumer electronic products, such as computers and monitors; other electronic products, such as televisions, cordless phones, audio/video equipment, and so on; lighting for the home and commercial buildings; appliances such as refrigerators, dryers, and washing machines; and home and commercial temperature control systems (see Figure 5.5).

The Energy Star program is also extended to the home. In particular, Energy Star tools enable homeowners to assess their existing energy use, in turn encouraging them ultimately to improve energy efficiency. Similarly, the program's portfolio includes new homes, in that new builds can qualify to receive the Energy Star logo (source: energystar.gov). In comparison, the European Union energy label provides its own program for appliances and homes, as we discussed previously.

5.2 Energy Efficient Techniques in Wireless Technology

Manufacturers are encouraged to provide energy efficiency within their products. The programs we have already discussed have their own criteria that a manufacturer must

Figure 5.5. The Energy Star program provides a comprehensive agenda including numerous electronic products and home appliances.

adhere to in order to receive an eco-friendly product label. In the low power and sensor technology market, devices or *nodes* that are installed within homes and industrial buildings must demonstrate energy efficiency, in turn reducing maintenance overheads. In the sections that follow, we look at how energy efficient techniques are being used within wireless technology ultimately to extend battery life or even negate the need to use batteries. However, in the rest of this section, we initially look at *renewable energy* and how some techniques derived from renewable sources have uncanny parallels with some mechanisms used within wireless technology products.

5.2.1 Deriving Energy from Alternative Resources

We are inevitably aware that the cost of energy over the last few decades has risen exponentially, burdening the majority of consumers and industries with increasingly large bills for electricity, gas, and oil. Across the world, we have seen the demand for resources increase as the population grows, and the incessant demand on resources remains unabated. The growing knowledge that there is not an infinite supply of energy has left many industries and manufacturers seeking alternative energy solutions, such as those found with renewable energy.

Figure 5.6. A wind farm and its collection of wind turbines are an excellent example of renewable energy. The turbines are often interconnected to produce sufficient voltage through a transformer before the power is finally delivered to the grid.

5.2.2 What Is Renewable Energy?

Renewable energy is the acquisition of energy derived from natural resources, such as wind, sunlight, rain, and so on; it describes an energy source that is considered inexhaustible. In one such example, a wind farm and its collection of wind turbines are used to harvest energy from the wind, as we illustrate in Figure 5.6. The wireless technologies that form the *Personal Area Networking* (PAN) domain are not immune to the need to find alternative energy resources or at least to offer sustainability in terms of extended battery life. Increasingly, consumers are reminded of the planet's depleting resources, and, in turn, they place an unrelenting pressure on innovators to ensure that our electronic-enabled products are energy efficient and eco-friendly. Unquestionably, in the consumer electronics industry, innovators continually strive to provide alternative sources or improved transmission techniques in which an economized send/receive regime, for example, is architected, in turn optimizing or extending battery life. In fact, some innovators have devised other ingenious techniques that rely upon the notion of renewable energy for wireless-enabled products. For example, *energy harvesting* (like our wind farm example) relies upon solar and wind energy as well as kinetic energy in the ambient environment to provide sufficient power for a sensor device.

5.2.3 No Batteries Required

We now take a more detailed look at some of the techniques used within wireless technology to extend battery life or to derive energy from alternative sources. Let's look at EnOcean GmbH (enocean.com), a company that has increasingly become synonymous

Table 5.1. The EnOcean application portfolio, illustrating specific areas of expertise

Application	Areas of expertise
Building automation	Lighting, heating, cooling, ventilation, security, metering, and so on.
Industrial automation	Condition monitoring, process optimization, control, metering, and switching.
Automotive and aviation	Condition monitoring and switching.
Medical	Temperature, blood pressure, heart beat, monitoring, and so on.

with energy harvesting. EnOcean prides itself on its notion of *no wires* and *no batteries*. With this in mind, EnOcean provides maintenance free *Wireless Sensor Networks* (WSNs) and has deployed its technology into over 100 000 buildings. In Table 5.1, we highlight EnOcean's application portfolio whilst providing specific areas of expertise.

In Chapter 9, "Green, Smart, and Wireless: EnOcean," we will explore the technology in greater detail, but, for now, we simply want to focus on EnOcean's technology and its ability to use ambient changes in its environment to self-power sensors within its network. Incredibly, minor fluctuations in the environment are adequate enough to power a node sufficiently for it to transmit data (a uni-directional sensor). For example, small mechanical, temperature, solar, vibrational, and thermal fluctuations are among some of the renewable sources which feed EnOcean's energy convertor, as we see in Figure 5.7.

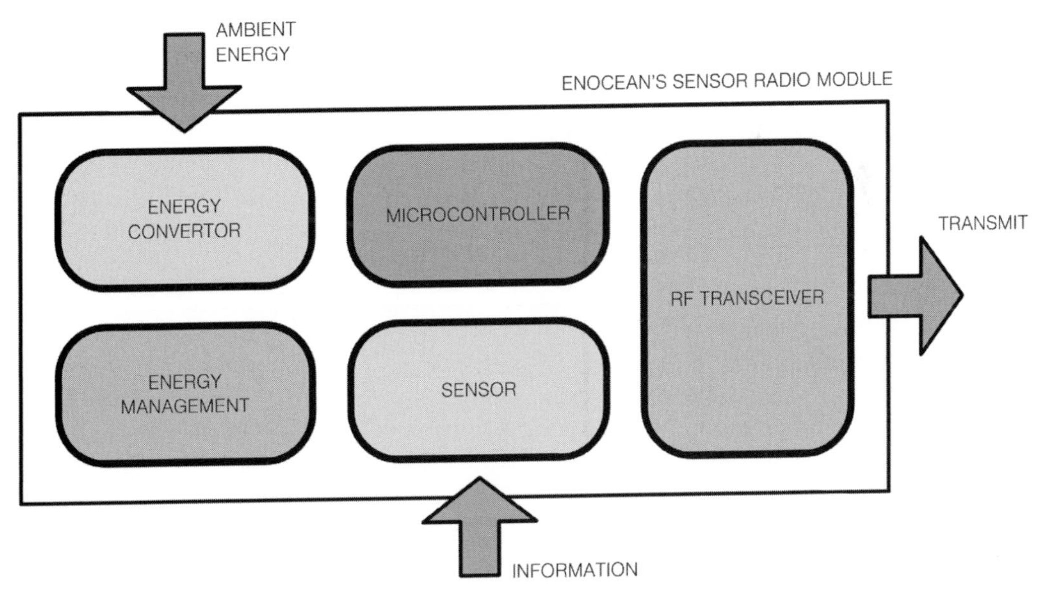

Figure 5.7. The system blocks showing EnOcean's uni-directional self-powered wireless sensor technology. (Source: EnOcean Alliance.)

5.2.3.1 More About EnOcean's Energy Conversion

EnOcean's energy convertor technology is capable of capturing varied fluctuations to help self-power sensors within its network. For example, a mechanical energy convertor may simply use a single button press like that found in a rocker light switch to harness sufficient energy. What's more, small celled solar-enabled sensors can obtain energy from internal or external (natural) lighting sources. However, the EnOcean technology also provides effective energy management, ensuring that, once the energy is harvested, it isn't unnecessarily squandered. As such, further techniques are used to ensure that data is transmitted in the shortest amount of time, in turn consuming less power. Likewise, effective management of standby and sleep timers further ensures the longevity of the network sensor, which brings us neatly to ANT Wireless.

5.2.4 Optimizing Data Transmission

EnOcean technology is somewhat unique in terms of its batteryless way of working, yet the topology still utilizes optimized data transmission paradigms to ensure the shortest duration possible when sending data. Likewise, ANT Wireless (thisisant.com) technology adopts a similar principle in terms of optimizing data transmission, which, in turn, extends battery life. ANT Wireless is widely used within the sports, fitness, and well-being markets, something which we will touch upon later, in Chapter 10, "The Power of Less: ANT."

ANT Wireless devices require batteries for both sides of the wireless connection, and the technology aims to offer one year of battery operation. Furthermore, ANT Wireless has created a software protocol stack that efficiently controls the radio, as shown in Figure 5.8. Similarly, its operation within the 2.4 GHz unlicensed *Industrial, Scientific and Medical* (ISM) band ensures the radio is used for the shortest amount of time.

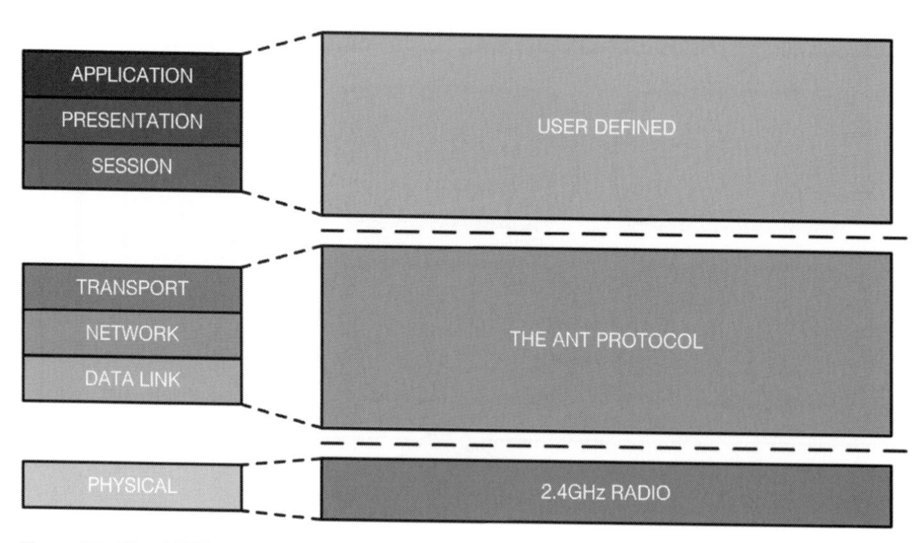

Figure 5.8. The ANT+ stack architecture has been developed to support an effective transmission scheme that, in turn, optimizes energy consumption.

What's more, wireless technologies such as Wi-Fi utilize a "doze" period technique between the transmission of frames. The *Wi-Fi Multimedia* (WMM) *Power Save* (PS) or WMM-PS program, introduced by the Wi-Fi Alliance, encourages consumer electronic devices to reduce energy consumption and, in turn, extend battery life. We discuss this program further in Chapter 13, "The 802.11 Generation and Wi-Fi."

5.3 What Do Low Power and Sensor Technology Provide?

The low power and wireless sensor technology sector has become increasingly popular due to a new generation of *green energy* conscious consumers and manufacturers. Manufacturers, innovators, and consumers alike are becoming increasingly aware of the energy impact and the footprint they leave on the planet. With the cost of energy escalating exponentially over the last decade or so, the associated maintenance and expense for homes and industry have become immeasurably expensive. As such, innovators continually seek energy efficient technologies as a solution to optimize energy consumption within homes, manufacturing, and industry. Low power and wireless sensor technologies empower many buildings' infrastructures with an intelligent network of sensors (or *nodes*), which are capable of monitoring the ambient environment; for an example, see EnOcean's TCM 320C transceiver, shown in Figure 5.9. Typically, the information sourced by these sensors is collated by a central/console unit or monitoring station that reacts to environmental factors, in turn offering optimal utilization of energy consumption. With this mesh of intelligent low power nodes, lighting and heating/cooling can all be regulated and controlled. For example, the temperature of an office or commercial building can be regulated at a lower temperature if no one is in the building or if it's the

Figure 5.9. A typical programmable wireless transceiver module, the EnOcean TCM 320C. (Courtesy of EnOcean GmbH.)

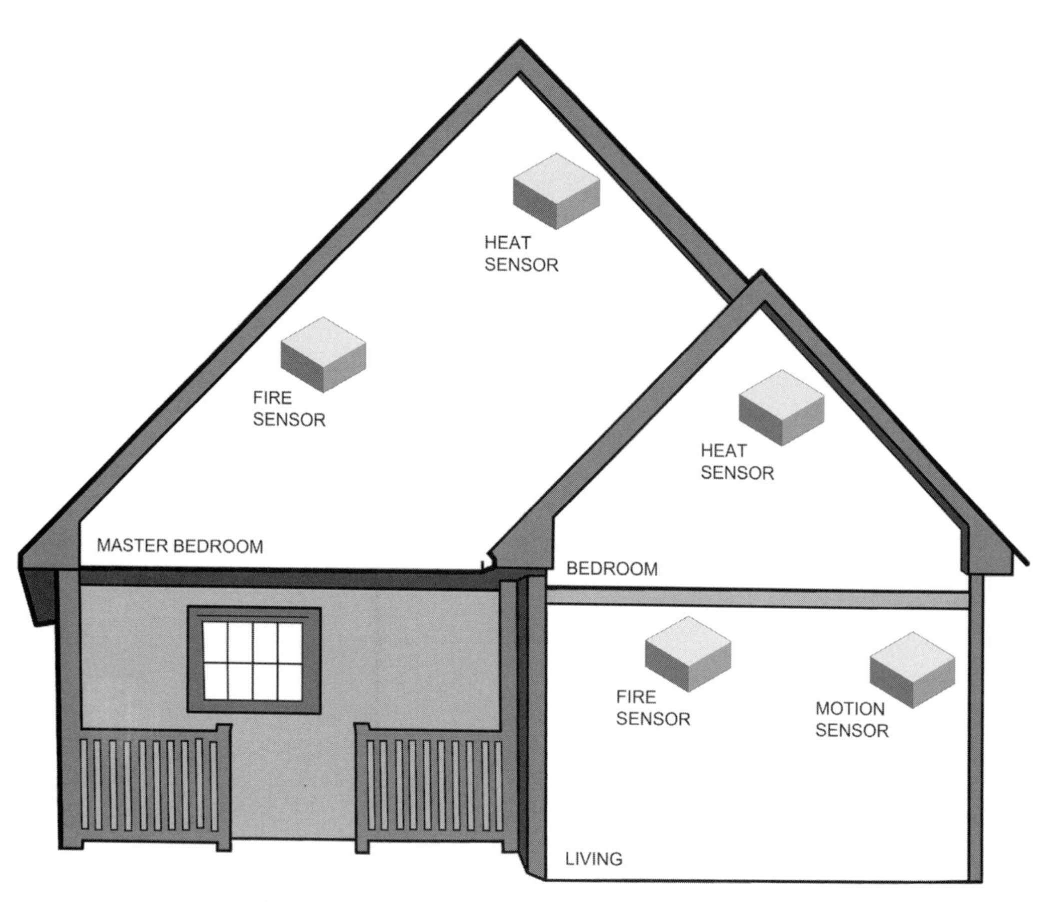

Figure 5.10. The home itself forms a HAN, which may be monitored and maintained by a console unit or through a personal area network.

end of the working day; conversely, the temperature can be increased when the normal working day starts. Similarly, lighting can be automatically switched off if the sensor doesn't detect motion and switched back on once movement is detected. EnOcean provides some excellent use cases, which we will pick up on in Chapter 9, "Green, Smart, and Wireless: EnOcean." Naturally, some homes already use similar devices, ensuring that the home owner's utilization of energy remains optimal. In Figure 5.10, we illustrate a typical room in which a number of sensors are used to regulate the environment.

Within the home, a central or console unit, as we mentioned already, may be provided to help the user regulate, monitor, and maintain the *Home Area Network* (HAN), although further management of the network may be extended to the PC (as part of the personal area network). In Chapter 4, "Introducing the Lawnmower Man Effect," we discussed how the PC is still integral to the home and, as such, that it is still very much a PC-centric model. We also touched upon *domotics* (penned by French journalist Bruno de Latour), something we will see manifest itself in Chapter 8, "Control Your World with ZigBee," and Chapter 9. Essentially, within the *smart home*, or *maison intelligente*,

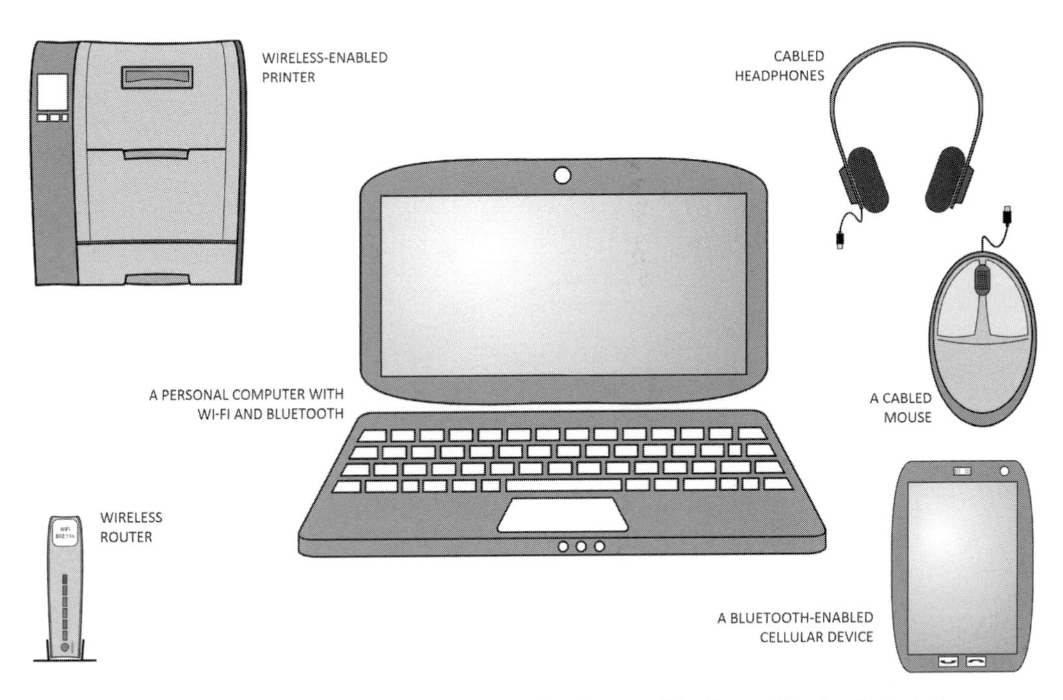

Figure 5.11. A typical personal area network, where combinations of fixed and/or wireless devices occupy your personal space.

various products, such as home heating sensors, intruder/motion sensors, and so on, all monitor the environment intelligently and often report back to a localized or central unit.

You may recall in Chapter 2, "What is a Personal Area Network?," that we characterized a personal area network as a collection of fixed or wireless devices, where these devices are located in proximity, for example on a desk at home or in a small office, as illustrated in Figure 5.11. Typically, these devices will not have a transmission range greater (on average) than 30 meters or so. You may also recall from Chapter 2 that we mentioned that the devices which form such a network are personal to you. However, with low power and wireless sensor technology, a *mesh* network (see Figure 5.12), combining a series of wireless nodes, may not necessarily be integral to your personal area network. We discussed networking topologies in greater detail in Chapter 3, "Disruptive Topologies through Technology Convergence."

Nevertheless, these nodes still form an holistic network, which you may or may not have control over; moreover, a building maintenance engineer, for example, may have access to this *private* network. In a home environment, you may have access to your own *extended* personal area network; for example, if your home has an integral alarm system controlling numerous sensors within the home, or, similarly, if your home system automates heating and cooling, then this personal area network may typically be controlled through a dedicated web-interface from within the home or through a web-interface enabling you to access the home remotely via your PC.

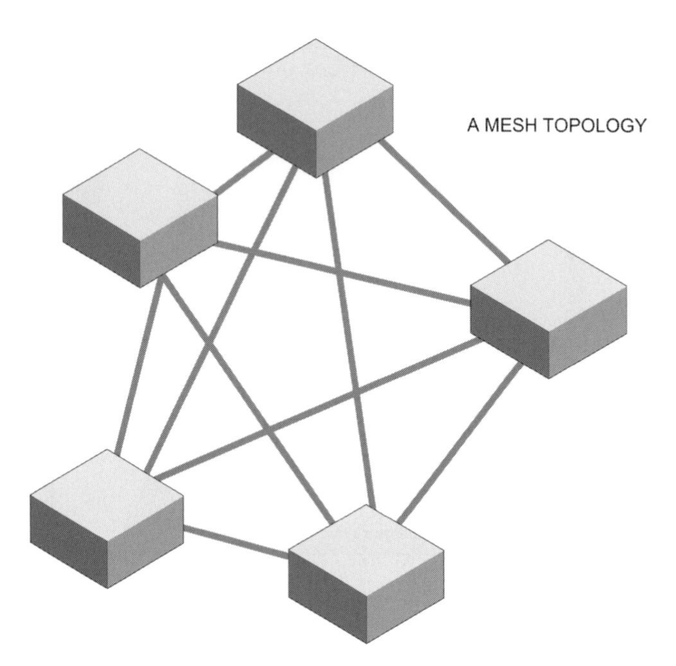

Figure 5.12. A mesh topology is characterized by an ability to traverse data across nodes in a relay race fashion.

5.4 What Should We Expect from Part II?

One of the early chapters in this second part of the book looks at Bluetooth low energy (Chapter 7, "Bluetooth low energy: The *Smart* Choice"); a technology that has only recently been introduced to the market and is aggressively attempting to secure a significant market share. In Part II, "The Wireless Sensor Network," we present a variety of low energy and wireless sensor technologies, each purporting low energy characteristics and low cost. Naturally, these technologies individually possess their own strengths and weaknesses, which may suit many applications and products. Each chapter provides a structured approach in explaining the technology whilst providing insight into specific technology alliances and membership structures. As such the benefits of each alliance may be determined. Furthermore, each chapter offers a comprehensive review of the market sector that a given technology targets and endeavors to explain the overlap and redundancy between competing technologies. Likewise, the technology makeup is covered exhaustively, ensuring that aspects of the physical medium and software building blocks are presented, in addition to explaining the attributes of the expected networking topologies.

6 Enabling the Internet of Things

We briefly touched upon the *Internet of Things* (IoT) earlier, in Chapter 4, "Introducing the Lawnmower Man Effect." The *Lawnmower Man Effect* (LME) is a supposition used to classify the ability afforded to a new generation of consumers who seek to have a permanent connection to access anything, anytime, anywhere. In short, it represents the consumer's ability to traverse digital systems across the globe, all captained from their personal area networking space utilizing pervasive *Wide Area Networking* (WAN) technologies. Let's take a moment to form a better understanding of the Internet of Things. Firstly, the IoT needs to be distinguished from the *Internet*. The Internet, of course, represents a globally connected number of networks, irrespective of a wired or wireless interconnection. IoT, on the other hand, specifically draws your attention to the ability of a device to be tracked or identified within an IP structure – similarly, we (humans, consumers, and so on) are nowadays capable of being tracked *and* we can all be identified, as we intimated earlier in Chapter 4, perhaps via our IP address. The IoT is often misconstrued, invariably carte blanche, to accommodate anything that resembles Internet capability, and this is certainly not the case. More specifically, the IoT is solely used to characterize the ability to track and identify objects or things within the Internet or similar IP structure. However, from this perspective, we have now empowered these objects to collate new information, in turn extending the initial IoT concept. The IoT refers to the interconnection of distinguishable smart/intelligent objects or "things" and their virtual manifestation within the Internet or similar IP structure. The IoT also recognizes the empowerment[1] of these objects or things to gather new information and, in turn, portray a representation of their "world." After all, the Internet is wholly reliant on information that has been/is typically derived from human input. In essence, this collated and vast amount of data and/or information, available through your permanent connection, was captured at some point by humans. Nonetheless, human input can on occasions be inaccurate and, increasingly nowadays, people no longer have the inclination or time to undertake seemingly endless data entry. Kevin Ashton[2] describes this simply by stating that computers and, for that matter, the Internet, "are almost wholly dependent" on humans for information. In essence, people are responsible for feeding the Internet petabytes of data, which may have been derived from typing at a keyboard,

[1] Ashton, K., "That 'Internet of Things' Thing," 2009.

[2] The term "Internet of Things" was penned by Kevin Ashton in 1999 to characterize identifiable objects or things and their virtual representation within the Internet or similar IP structure.

Figure 6.1. The number of people using the Internet is expected to grow exponentially, whilst the interconnection of objects or "things" is anticipated to exceed human use and population.

through oral entry, or simply by taking a picture or posting a video. Ashton suggests[3] that this "input" is constrained by time and ultimately accuracy because "people are not very good at capturing data about things in the real world." So, with the IoT, we can empower objects with a modicum of intelligence, and these objects can become instrumental in collating new and accurate data to continue to feed the incumbent knowledge-base of the Internet, alongside the dedicated human "resourcers." The technologies that feature in this section, namely Bluetooth low energy, ZigBee and EnOcean, are a few wireless technologies that are significant in supporting an IoT ecosystem of intelligent objects or things by gathering data and/or information governed by our (human) remit of what should, or can, be relevantly collated and, in turn, consumed. IoT takes away from humans the need to collate mediocre information (what can only be described as trivia) and, in turn, shares that information effectively and accurately to allow recipients to make informed decisions. As such, we have empowered technology to feed us and source content diligently, more often with a greater sense of reliability and, ultimately, accuracy. In Figure 6.1, we conceive the reality of the IoT in the context where smart objects are globally interconnected.

What's more, since its inception, the IoT has broadly been used to cover a variety of trends, subjects, and phenomena within mainstream computing and, of course, Internet

[3] Ashton, K., "That 'Internet of Things' Thing," 2009.

popularity, growth, and usage. Notwithstanding, people have taken the term and have molded it to analogize their own supposition, shaping it to accommodate a specific industry trend or phenomenon. In fact, the population of users (that is, people) using the Internet is expected to grow to two billion in just a few years, although the actual interconnection of objects or "things" within the Internet is anticipated to exceed[4] human use and population – some other reports suggest this has already happened, as we conceptualize in Figure 6.1. It seems that new and emerging *smarter* objects or things are becoming increasingly intelligent,[5] with the provision of processors, memory, storage, and so on, and, of course, they have the ability to sustain that all-important IP connection whilst maintaining the IoT's founding concept of "tracking."

In this chapter, we briefly cover how the interconnection of smart devices forming the IoT uses the *Internet Protocol* (IP) holistically to provide a "one-network" of intelligent devices. We also touch upon the use of *IPv6 over Low power Wireless Area Networks* (6loWPAN), a protocol that will empower objects with the next generation IP. In the section that follows, we look at how organizations such as the IPSO Alliance are becoming proponents of the use of IP to empower smart objects.

Figure 6.2. The IPSO Alliance trademark logo. (Courtesy of the IPSO Alliance.)

6.1 Shaping an IP-enabled World

The IPSO Alliance (ipso-alliance.org) is a non-profit organization with currently over 60 member companies that circle the globe, all from a diverse and rich market sector. The Alliance works to establish IP as the technology empowering a new generation of smart objects, and encourages growth within various industries through marketing, awareness, and education. We illustrate in Figure 6.2 the IPSO Alliance trademark logo. The Alliance's primary objective is to advocate IP networked devices for use within the energy industry and a host of other commercial and consumer applications. With IP at its core, smart objects can form this holistic network of intelligent devices, as we have already discussed. The prevalence and success of wireless technology have spawned a plethora of *nodes*, or low powered devices, which have been integrated into residential, business, and other commercial environments, all providing eco-friendly aware and

4 IBM Social Media, "The Internet of Things," 2010.
5 *EE Times*, "Photos from the Frontier: The Internet of Things," 2012.

intelligent monitoring of our energy consumption, and offering security and control – as discussed in Chapter 5, "Introducing Low Power and Wireless Sensor Technologies."

With pervasive connectivity, primarily empowered with IP, we have conceived this innate belief that all devices, irrespective of size, now participate within the IoT. As we move through this section of the book, we will learn more about Bluetooth low energy, in Chapter 7, "Bluetooth low energy: The *Smart* Choice," ZigBee, in Chapter 8, "Control Your World with ZigBee," and EnOcean in Chapter 9, "Green, Smart, and Wireless: EnOcean." In these chapters, we discuss the underlying software building blocks that enable the respective technologies to converse with each other; and let's not forget Wi-Fi, which has become synonymous with connecting to the Internet, something we discuss later, in Chapter 13, "The 802.11 Generation and Wi-Fi." The IPSO Alliance suggests that IP has been well established, is incredibly stable, and can support a varied and wide range of applications – we will undoubtedly witness this as we move forward through the remaining chapters in Part II and head into Part III, "The Classic Personal Area Network."

6.1.1 What Are Smart Objects?

Earlier, we discussed how the IoT refers to the interconnection of distinguishable smart/intelligent objects or "things" and their virtual manifestation within the Internet or other IP structure. In essence, the IPSO Alliance describes smart objects as *small computers* at a cost of only a few dollars and with a minimal footprint. This means that, with the provision of processors (8-, 16- or 32-bit microcontrollers), memory (several tens of kilobytes), storage, and so on, along with an inherent ability to sustain an IP connection (perhaps limited to sending/receiving a few hundred kilobits per second), almost anything and everything can become connected at an affordable cost. These devices would typically be battery operated, although this is not an absolute requirement. As such, it is envisaged that these smart objects will be embedded in everyday products such as cars, light switches, and industrial machinery, enabling a myriad of applications for home and commercial automation, in factories, throughout industry, and even in healthcare situations. What's more, smart objects can become responsible for enabling intelligent smart grids, cities, and transportation. So, the extended IoT concept incorporates the new generation of smart/intelligent objects, and we are now entrusting this technology to educate, notify, and inform us of events that we have deemed important – this part is inescapable, as there still needs to be human interaction. The scalability and flexibility of IP are suitable for the diverse range and potential number of applications, and it is its well-established architecture, with existing applications for email, *Voice over IP* (VoIP), video streaming, and so on, that affords the protocol its robust and adaptable usage. The IP stack has been regarded as large and perhaps cumbersome, requiring high amounts of processing power and memory. However, several lightweight revisions have emerged to accommodate smaller devices and lower energy footprints. Furthermore, the lightweight nature of the new generation IP stack has allowed itself to be used in conjunction with other protocols, such as low power Wi-Fi, ZigBee, and Bluetooth low energy. Since IP can run over almost anything, adapting the stack to run over ZigBee or Bluetooth low

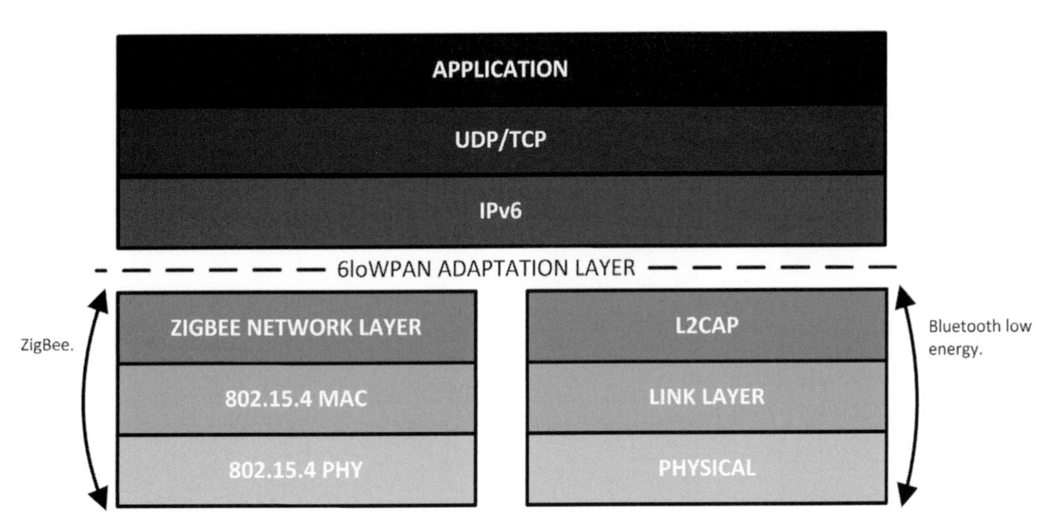

Figure 6.3. 6loWPAN over ZigBee and Bluetooth low energy, respectively.

energy, for example, is a relatively straightforward task, as we highlight in Figure 6.3. As can be seen in the figure, a 6loWPAN adaptation layer is used cohesively to allow both ZigBee and Bluetooth low energy to coexist with the upper layers of the IP stack, namely IPv6 and *User Datagram Protocol* (UDP) /*Transmission Control Protocol* (TCP).

6.1.2 Smart Agents

As already mentioned, human interaction with devices is required at some point, whether that's from the onset, when a smart object needs to be configured and defined with our initial criteria, or as a human acting on or reacting to information such as the arrival of new data. The next generation of smart objects may comprise the *smart agent*, a hybrid device that's capable of acting autonomously, possibly undertaking decisions on our behalf. It's simply about control, putting us in charge, and making us more aware of our energy footprint – just one of the many aspects of the new IoT ecosystem. Likewise, it alleviates the mediocre aspects of our routine whilst appeasing and simplifying regular based tasks, placing us in charge of energy consumption and use. And finally, do we really need *everything* to be interconnected? At what point does the IoT become "overkill" for what was once perhaps just a simple and mundane activity?

6.2 The IoT Architecture

The *Internet Protocol Suite*, commonly known as the *TCP/IP*[6] *model*, is a set of protocols used for the Internet and other similar networks, such as the *Personal Area Network*

[6] The *Transmission Control Protocol* (TCP)/*Internet Protocol* (IP) should not ordinarily need expanding as it is nowadays so commonplace within mainstream computing.

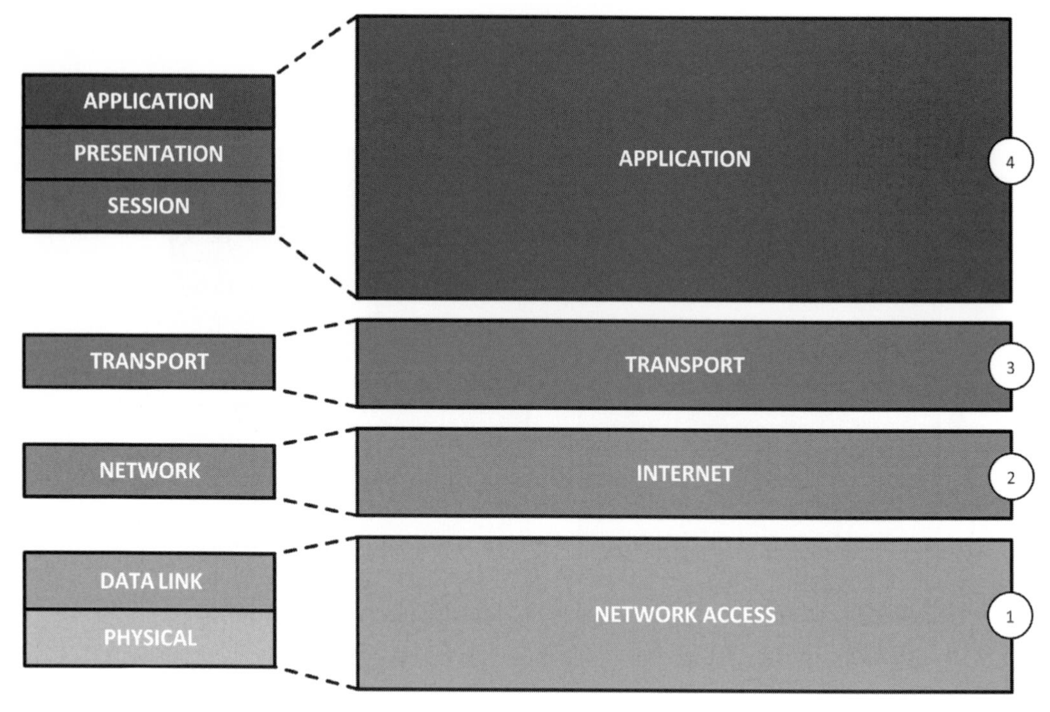

Figure 6.4. The four layers of the TCP/IP model alongside the traditional OSI model.

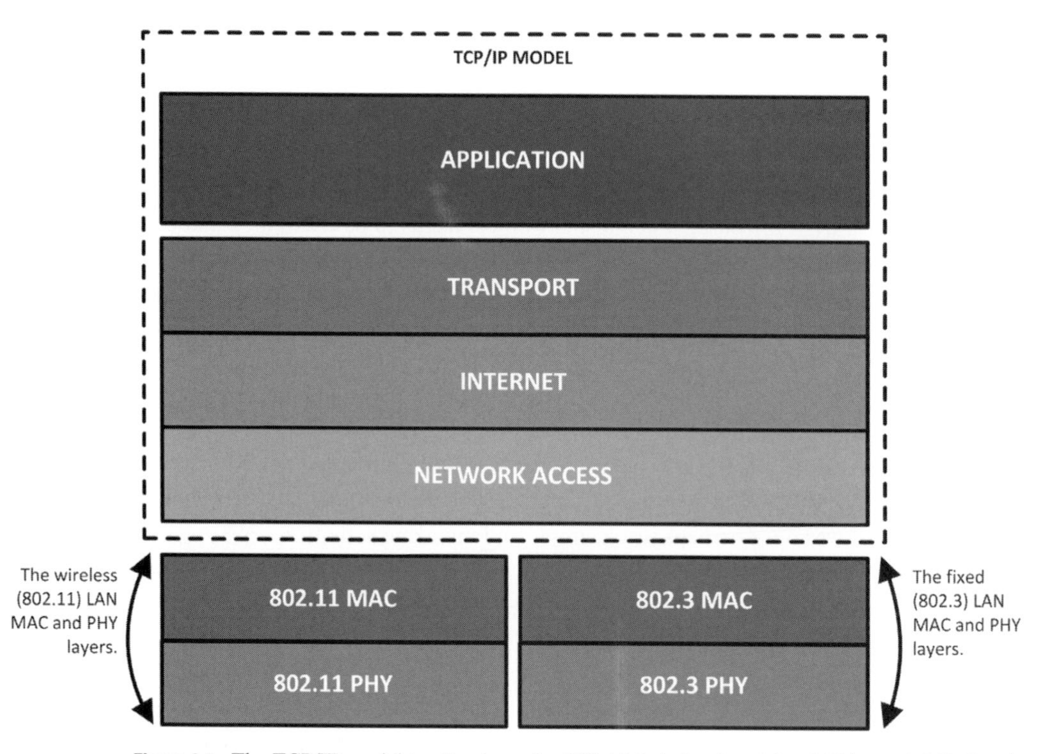

Figure 6.5. The TCP/IP model *in situ* above the 802.11 (wireless) and fixed (Ethernet) MAC and PHY layers, respectively.

(PAN), *Local Area Network* (LAN), *Wide Area Network* (WAN), or Internet, and so on. The IP suite has a four-layered reference paradigm used to conceptualize and portray the layered components that occupy the Internet architecture, as shown in Figure 6.4. However, the IP suite's "migrational" fit to the *Open Systems Interconnection* (OSI) model has caused some confusion within mainstream computing. Naturally, the TCP/IP model is used to facilitate communication over both fixed and wireless connections. Looking at Figure 6.4, we can identify four layers within the model, namely *Network Access* (**1**), *Internet* (**2**), *Transport* (**3**), and *Application* (**4**). Wi-Fi technology uses the *Institute of Electrical and Electronic Engineers* (IEEE) 802.11 standard, and it provides the technology with both its *Media Access Control* (MAC) and *Physical* (PHY) layers; these layers provide the wireless interface equivalent for Ethernet (802.3). (We cover Ethernet in Chapter 13.) In Figure 6.5, we place the TCP/IP model stack above both the 802.11 and 802.3 MAC and PHY layers. In fact, we illustrate in Figure 6.5 the complete protocol stack as used within a fixed or wireless environment. Earlier, in Figure 6.3, we illustrated how ZigBee and Bluetooth low energy might accommodate an IP stack.

7 Bluetooth low energy

The *Smart* Choice

Bluetooth low energy is an addition to the Specification of the Bluetooth System: Core, v4.0. The inclusion of low energy capabilities enables a new generation of Bluetooth *Smart* devices, which are capable of operating for months, and possibly years, just utilizing coin-cell batteries. Essentially, to bear *Bluetooth Smart* or *Bluetooth Smart Ready* logos on your product (see Figure 7.1, left to right, respectively), they must include Bluetooth v4.0 functionality, along with other criteria which we come back to later in this chapter. Bluetooth low energy, like Bluetooth classic and high speed, which we discuss later in Chapter 14, "Bluetooth Classic and High speed: More Than Cable Replacement," uses the unlicensed 2.4 GHz *Industrial, Scientific, and Medical* (ISM) frequency band.

Figure 7.1. The Bluetooth Smart and Smart Ready logos used on a product and/or its packaging to indicate v4.0 (ready) capabilities. (Courtesy of the Bluetooth SIG.)

7.1 Overview

Bluetooth wireless technology now comprises two systems, namely *Low Energy* (LE) and *Basic Rate* (BR); both systems provide capabilities that enable device discovery, connection establishment, and other mechanisms. The BR system, which we discuss in more detail in Chapter 14, provides optional *Enhanced Data Rate* (EDR), *Alternative Media Access Control* (MAC), and *Physical* (PHY) or *Alternative MAC/PHY* (AMP) extensions, in turn offering synchronous and asynchronous connections, with data rates of up to 721 kbit/s as a minimum, whilst 2.1 Mbit/s can be achieved with EDR and 24 Mbit/s for high speed operation using 802.11 AMP. The LE system, on the other hand, which is derived to some extent from the BR/EDR system, has several capabilities that ultimately offer a lower cost and low power solution; it provides reduced complexity when compared with the BR/EDR system and is designed for products and applications which utilize low data rates. The LE system provides data throughput of around 1 Mbit/s, ut it is not designed to provide file transfer, for example; rather, it's ideally suited to applications where only small packets of data are exchanged. Bluetooth devices that implement both systems are, of course, capable of communicating with other like-enabled devices and have the opportunity to provide a greater range of applications. What's more, a product implementing both systems may also communicate with other devices that implement either system.

7.1.1 The History of Bluetooth low energy

In 2001, Nokia's Research Center[1] commenced development of *Wibree* technology (see Figure 7.2); this was publically announced later, in 2006, when Nokia planned to encourage the wider adoption of its technology and initiate an open industry forum for the *Personal Area Networking* (PAN) space. As an alternative, and seen as a competitor, to Bluetooth wireless technology, Wibree technology purported better energy efficiency when compared with Bluetooth and likewise had the ability to communicate over distances of 10 m or so. What's more, Nokia intended to see their technology integrated into smaller devices such as clothing, watches, cell phones, and so on. Nonetheless, the Bluetooth *Special Interest Group* (SIG) was in active discussion with Nokia to merge and expand its technology portfolio with, what was coined at the time, an ultra-low power technology. As such, in August 2007, the Nokia Research Center announced that "the Wibree specification will become part of the Bluetooth specification as an ultra-low power Bluetooth technology." The Bluetooth SIG formally announced the adoption of the Specification of the Bluetooth System: Core, v4.0, in July 2010, which now included *Bluetooth low energy* (BLE) derived from Nokia's Wibree technology.

Figure 7.2. Nokia's Wibree logo. (Courtesy of Nokia.)

7.1.2 Opening up a New Market for Bluetooth Technology

The new generation of Bluetooth wireless technology witnessed rapid adoption as manufacturers were eager to embrace the new low energy variant. In fact, by mid 2011, just one year after the SIG's announcement, many products were already under development and verging on entering their market space; and, emerging fairly immediately, in late 2011, several cellular or smartphone manufacturers announced Bluetooth Smart Ready products. Manufacturers from several market sectors, including Nike, Garmin, Motorola, Apple, and Samsung, have all adopted and utilized the new standard, with products primed for market. The scope of Bluetooth low energy opens up an enormous

[1] Nokia Research Center, "Wibree Forum Merges with Bluetooth SIG," 2007.

and new market for Bluetooth, utilizing the low cost and low power connectivity, and in Section 7.2, "The Bluetooth low energy Market," and Section 7.3, "The Bluetooth low energy Application Portfolio," we discuss the market scope and application opportunities for Bluetooth low energy.

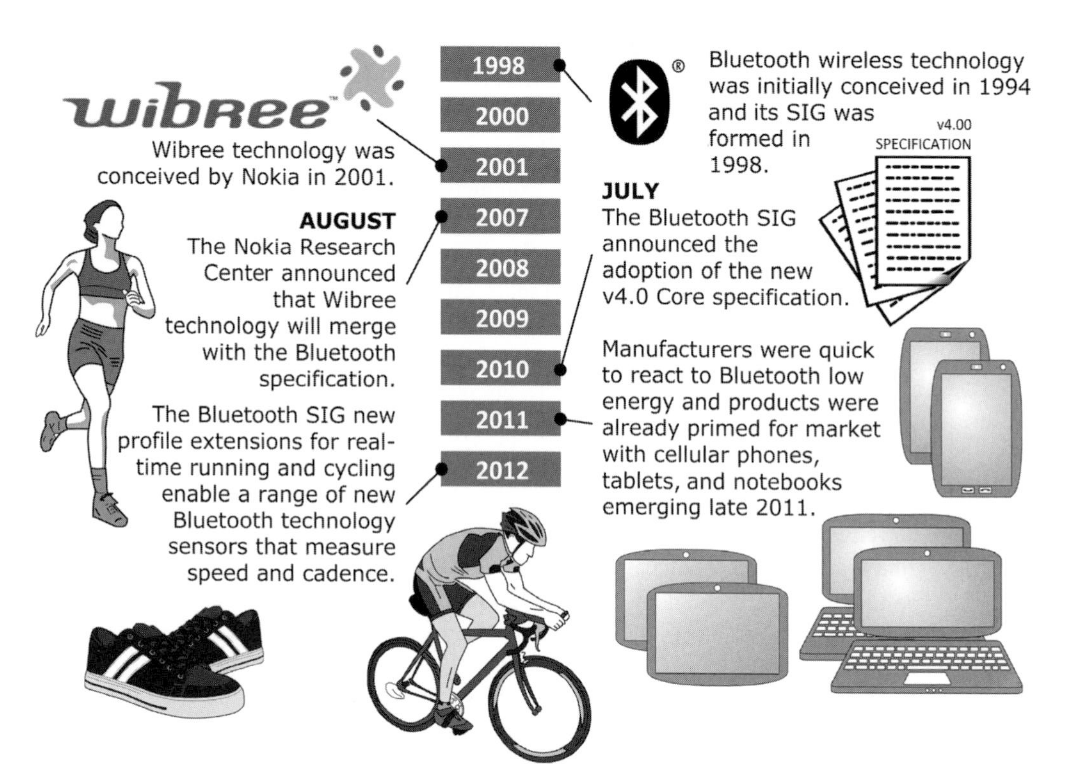

Figure 7.3. Bluetooth low energy's timeline, from its inception to where the technology is today, with its Specification of the Bluetooth System: Core, v4.0. (Wibree logo courtesy of Nokia; Bluetooth logo courtesy of the Bluetooth SIG.)

7.1.3 The Bluetooth low energy Timeline

The infographic shown in Figure 7.3 provides a snapshot of Bluetooth low energy's technology timeline, from its inception to where the technology is today. BLE has quickly become adopted by a number of leading manufacturers over a relatively short lifespan since the Bluetooth SIG announced its adoption in 2010. Bluetooth wireless technology was first conceived as far back as 1994 and the Bluetooth SIG was formalized in 1998. Nokia and the Bluetooth SIG were in active communication about merging their respective technologies, and in August 2007 Nokia announced that Wibree technology will merge with the Bluetooth specification. Later, in July 2010, the Bluetooth SIG formally announced the adoption of the Specification of the Bluetooth System: Core, v4.0, and in 2011 the Bluetooth SIG announced the Bluetooth Smart and Smart Ready brand extensions. Microsoft announced support for v4.0 in its new Windows 8 operating system and Apple likewise announced that its iPhone 4S will support v4.0, and consequently the Apple iPhone 4S became the first Bluetooth Smart Ready device. In 2012, the Bluetooth SIG announced that its member companies exceeded 17 000 members, and first-generation tablets and music players reached the market branded as Bluetooth Smart Ready. The Bluetooth SIG additionally announced in 2012 several new profile enhancements, which would witness a new generation of Bluetooth products for the sports, fitness, and well-being markets.

7.1.4 The Bluetooth Special Interest Group

In this chapter, we cover Bluetooth low energy and its associated architecture. Typically, as you will soon learn from other chapters in this book, we include a perspective from the associated technology's alliance or member group governing the technology's future, development, and evolution; however, since we cover Bluetooth *classic* and *high speed* later in Chapter 14, "Bluetooth Classic and High speed: More than Cable Replacement," we'll cover this topic later in this chapter.

7.2 The Bluetooth low energy Market

The number of wireless technologies that are shared in Part III, "The Classic Personal Area Network," typically serve high-end, high data rate products best suited for audio-, voice-, and data-heavy-centric applications. The Bluetooth SIG was eager to enter a profitable market space and identified an opportunity to provide a low cost, low energy control and sensor technology by merging with Wibree technology; as such, the SIG extended its existing technology portfolio. The new BLE ecosystem purports small or low latency data rates and has low bandwidth, which, in turn, aptly lends itself to longer battery life. Both Bluetooth classic and high speed have matured and captured an existing market for a wealth of consumer electronic products, including cellular or smartphones, PCs, car kits, and so on. BLE has been architected from the ground up and has a new radio, software protocol stack, and several new specific profiles, all designed to operate from a coin-cell battery. With the prerequisite of ensuring extended battery life firmly in place, BLE could empower its devices with an inherent consciousness about energy consumption and essentially become instrumental in supporting the *Internet of Things* (IoT). What's more, BLE, with its inherent low cost, has the opportunity to empower the *Wireless Sensor Networking* (WSN) market in providing a wealth of applications that enable it to provide a healthy, robust application-base. As we have already mentioned, a wealth of products have already started to emerge using the new Core v4.0 specification, and in Figure 7.4 we illustrate Texas Instruments' (ti.com) CC2540 chip, a new generation of BLE silicon specifically optimized for the new generation of products and applications.

Figure 7.4. Texas Instruments' *Bluetooth*® low energy CC2540 chipset empowering a new generation of Bluetooth low energy products. (Courtesy of Texas Instruments.)

7.3 The Bluetooth low energy Application Portfolio

Bluetooth wireless technology has been in use for over a decade and has matured to accommodate a diverse and competitive consumer electronics market sector. Other technologies that are described in this section, along with energy and its associated management to reduce consumption (and, in turn, cost) effectively, have become increasing concerns for consumers and industries alike. It seems a natural evolution for a popular and prevalent technology to provide a low power and low cost solution and so enable a new generation of products. The application portfolio discussed in the sections to follow simply demonstrates BLE's overwhelming popularity and what the Bluetooth SIG has described as "The second wave of Bluetooth technology" (bluetooth.com). As such, the Bluetooth SIG has developed a number of *profiles*, which initially target the sports, fitness, and well-being markets with sensors, heart-rate monitors, athletic shoes, and so on, as well as mainstream consumer electronics, such as cell or smartphones, tablets, notebooks, and TVs. The new version of Microsoft Windows 8 also includes full support for Bluetooth v4.0.

7.3.1 BLE-specific Profiles

A Bluetooth application achieves interoperability through the use of *Bluetooth profiles* – these simply define how an application should manifest itself from the physical layer through to the *Logical Link Contrail and Adaptation Protocol* (L2CAP), along with any other associated protocols that support the profile. Likewise, a profile will prescribe specific interaction from layer to layer and vice versa, as well as peer-to-peer interaction. The Bluetooth system uses a core profile, which all Bluetooth-enabled devices must implement – this includes both Bluetooth systems, that is, BR/EDR and LE. The *Generic Access Profile* (GAP) specifically describes behaviors and procedures for device discovery, connection establishment, security, authentication, and service discovery, and this is something we discuss further in Section 7.6.8, "Generic Access Profile."

Table 7.1. GATT-based specifications for the LE system

GATT profile and service specifications	
ANP	Alert Notification Profile.
ANS	Alert Notification Service.
BAS	Battery Service.
BLP	Blood Pressure Profile.
BLS	Blood Pressure Service.
CSCP	Cycling Speed and Cadence Profile.
CSCS	Cycling Speed and Cadence Service.
CTS	Current Time Service.
DIS	Device Information Service.
FMP	Find Me Profile.
GLP	Glucose Profile.
GLS	Glucose Service.
HIDS	HID Service.
HOGP	HID over GATT Profile.
HTP	Health Thermometer Profile.
HTS	Health Thermometer Service.
HRP	Heart-rate Profile.
HRS	Heart-rate Service.
IAS	Immediate Alert Service.
LLS	Link Loss Service.
NDCS	Next DST Change Service.
PASP	Phone Alert Status Profile.
PASS	Phone Alert Status Service.
PXP	Proximity Profile.
RSCP	Running Speed and Cadence Profile.
RSCS	Running Speed and Cadence Service.
RTUS	Reference Time Update Service.
ScPP	Scan Parameters Profile.
ScPS	Scan Parameters Service.
TIP	Time Profile.
TPS	Tx Power Service.

In Table 7.1 we present a comprehensive list of profiles and services that have already been adopted (v1.0), and in the sections to follow we provide an overview of some of the potential applications that have enabled a new generation of Bluetooth products and service/applications. The profiles listed here are specific to the LE system and are dependent on the *Generic Attribute Profile* (GATT), something we discuss later in Section 7.6.7, "The Generic Attribute Profile." What's more, an LE-specific profile is typically underpinned by a *service*, which exposes, to the given application, core functionality and/or data services that are relevant in enabling that profile.

7.3.1.1 Alert Notification Profile

The *Alert Notification Profile* (ANP) defines two roles, namely the alert notification *client* and *server*, and may coexist with other profiles; the ANP uses the *Alert Notification Service* (ANS). The ANP allows a client device, such as a watch, to retrieve alert or event data, and to receive information regarding the number of new alerts or unread items that

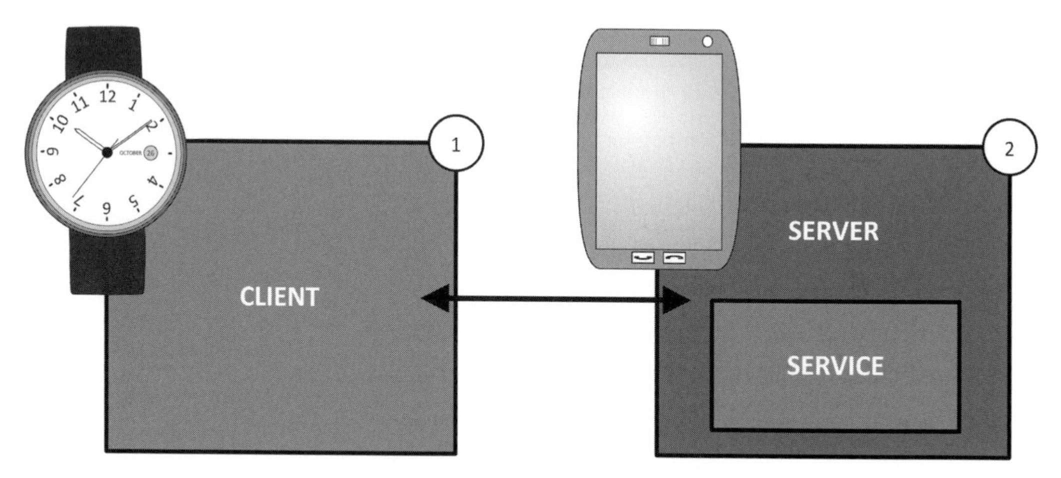

Figure 7.5. The relationship between the profile roles and service. A watch (**1**) undertakes the role of client, and a smartphone (**2**) takes on the role of server.

are retained by the server. More specifically, the service may offer information about the type of alert, the number of new alerts, and the number of unread alert items. For example, in Figure 7.5, a watch (**1**) receives information about the number of missed calls and/or text messages from a smartphone (**2**), which may include ancillary information about the caller's ID and the sender's ID of the text message (the watch would only receive the notification of a new text message and not the message itself). The alert notification server has the responsibility of instantiating the ANS.

7.3.1.2 Blood Pressure Profile

The *Blood Pressure Profile* (BLP) defines two roles, namely a blood pressure *collector* and *sensor*, and may coexist with other profiles; the BLP uses the *Blood Pressure Service* (BLS) and the *Device Information Service* (DIS). The BLP allows a device, such as a notebook, smartphone, or tablet device, to retrieve blood pressure measurement and other related data from a non-invasive blood pressure sensor. The profile can be implemented for both consumer and professional healthcare applications. The BLS may offer information about the blood pressure measurement, along with a timestamp, a user's pulse rate, user profile, and identification and measurement status. In Figure 7.6, we illustrate a notebook (**1**) "collecting" information from the blood pressure sensor (**2**). The blood pressure sensor has the responsibility of instantiating both the BLS and DIS.

7.3.1.3 Glucose Profile

The *Glucose Profile* (GLP) defines the same roles as the BLP and can also coexist with other profiles; the GLP uses the *Glucose Service* (GLS) and the DIS. The GLP allows a device, such as a notebook, smartphone, or tablet device, to retrieve glucose measurement and other related data from a glucose sensor. The profile can be implemented for both

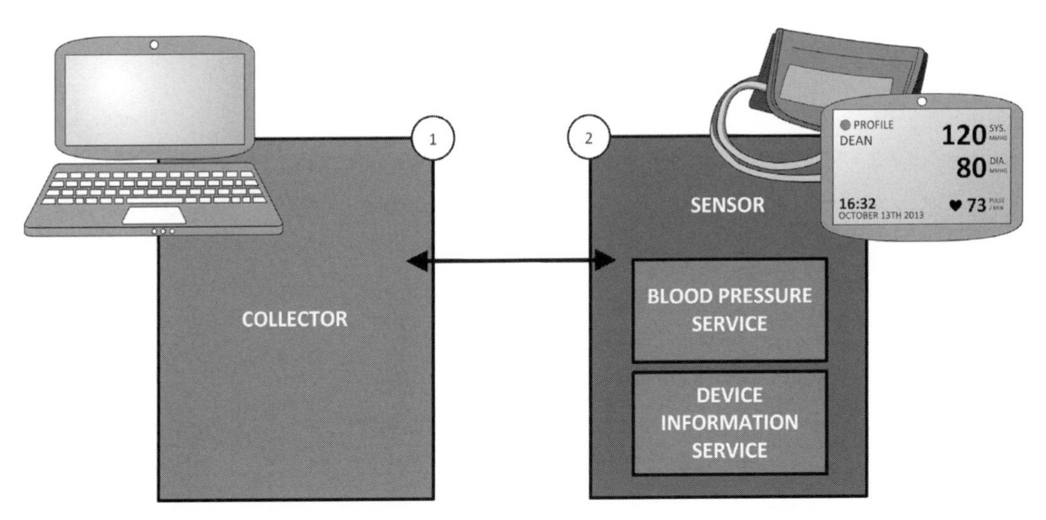

Figure 7.6. The relationship between the profile roles and services. A notebook (**1**) undertakes the role of collector receiving information from the blood pressure sensor (**2**).

consumer and professional healthcare and fitness applications. The glucose sensor has the responsibility of instantiating both the GLS and DIS.

7.3.1.4 Heart-rate Profile

The *Heart-rate Profile* (HRP) also defines the same roles as the BLP and can coexist with other profiles; the HRP uses the *Heart-rate Service* (HRS) and DIS. The HRP allows a device, such as a notebook, smartphone, or tablet device, to retrieve heart-rate measurement and other related data from a heart-rate sensor. The profile can be implemented for both consumer and professional healthcare and fitness applications. The heart-rate sensor has the responsibility of instantiating both the HRS and DIS.

7.3.1.5 Cycling Speed and Cadence Profile

The *Cycling Speed and Cadence Profile* (CSCP) defines two roles, namely a *collector* and a cycling speed and cadence *sensor*, and may coexist with other profiles; the CSCP uses the *Cycling Speed and Cadence Service* (CSCS) and the DIS. The CSCP allows a device, such as a smartphone or tablet device, to retrieve wheel revolution data and/or crank revolution data from the cycling speed and cadence sensor. Data retrieved from the sensor are calculated by the collector device. The profile can be implemented for sports and fitness applications. In Figure 7.7, we illustrate a smartphone (**1**) "collecting" information from the cycling speed and cadence sensor (**2**). The cycling speed and cadence sensor has the responsibility of instantiating both the CSCS and DIS.

7.3.1.6 Running Speed and Cadence Profile

The *Running Speed and Cadence Profile* (RSCP), like the CSCP, defines two roles, a *collector* and a running speed and cadence *sensor*, and may coexist with other profiles; the RSCP uses the *Running Speed and Cadence Service* (RSCS) and DIS. The RSCP

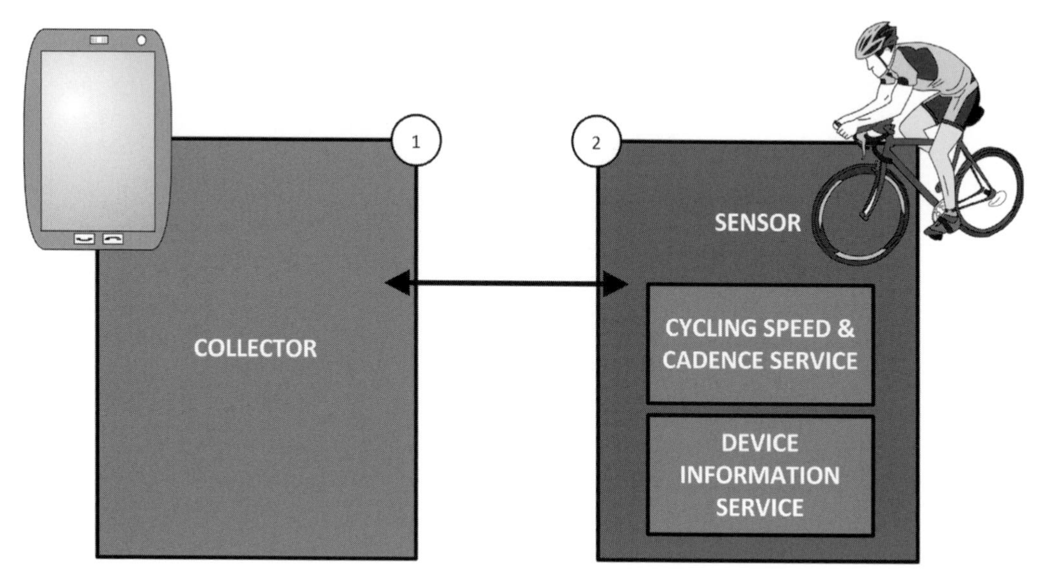

Figure 7.7. The relationship between the profile roles and services. A smartphone (**1**) collects data from the cycling speed and cadence sensor (**2**).

allows a device, such as a notebook or tablet device, to retrieve speed, cadence, and other information from the running speed and cadence sensor. Data retrieved from the sensor are calculated by the collector device. The profile can be implemented for sports and fitness applications. In Figure 7.8, we illustrate a smartphone (**1**) "collecting" information from the running speed and cadence sensor (**2**). The running speed and cadence sensor has the responsibility of instantiating both the RSCS and DIS.

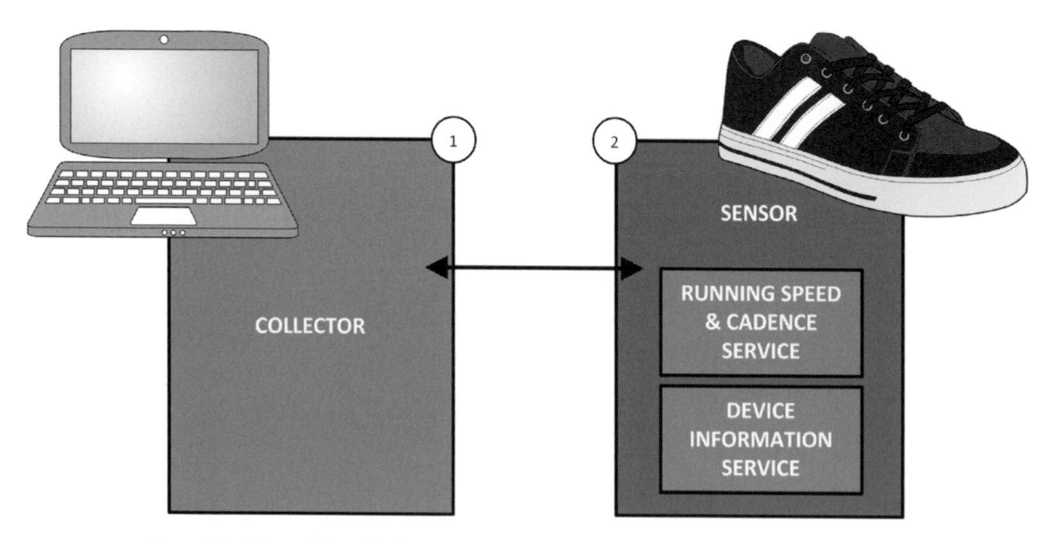

Figure 7.8. The relationship between the profile roles and services. A notebook (**1**) collects data from the running speed and cadence sensor (**2**).

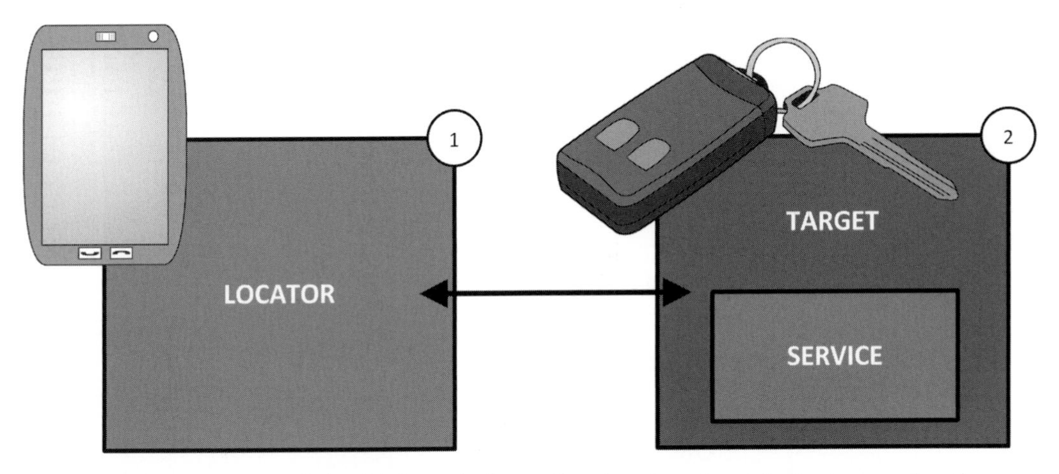

Figure 7.9. The relationship between the profile role and service. A button is pressed on the smartphone (**1**) which, in turn, alerts the target device (**2**).

7.3.1.7 Find Me Profile

The *Find Me Profile* (FMP) defines two roles, namely a *locator* and a *target*, and may coexist with other profiles; the FMP uses the *Immediate Alert Service* (IAS). When a button is pressed on a locator device, it simply initiates an alerting signal on the target device, which may be a watch, smartphone, tablet, or key-fob, and essentially allows users to locate devices that may have been misplaced. In Figure 7.9, we illustrate a smartphone (**1**) "locating" a misplaced key-fob (**2**), which will be alerted (audibly) allowing the user to locate his/her keys. The target device has the responsibility of instantiating IAS.

7.3.1.8 Health Thermometer Profile

The *Health Thermometer Profile* (HTP) defines two roles, namely a *collector* and a *thermometer*, and may coexist with other profiles; the HTP uses the *Health Thermometer Service* (HTS) and DIS. The HTP allows a device, such as a notebook, smartphone, or tablet device, to retrieve temperature measurement and other related data from a thermometer sensor. The profile can be implemented for both consumer and professional healthcare and fitness applications. In Figure 7.10, we illustrate a notebook (**1**) "collecting" information from the thermometer sensor (**2**). The thermometer device has the responsibility of instantiating both the HTS and DIS.

7.3.1.9 Proximity Profile

The *Proximity Profile* (PXP) defines two roles: a proximity *monitor* and a *reporter*, and may coexist with other profiles; the PXP uses the *Link Loss Service* (LLS), IAS, and the *Tx Power Service* (TPS). When a device moves out of radio range from a reporter device and the connection is lost or dropped, an "alert" is triggered on the monitoring device, effectively informing a user that the devices have become separated. The user may configure a device, if separated, to take additional action – for example, the reporter

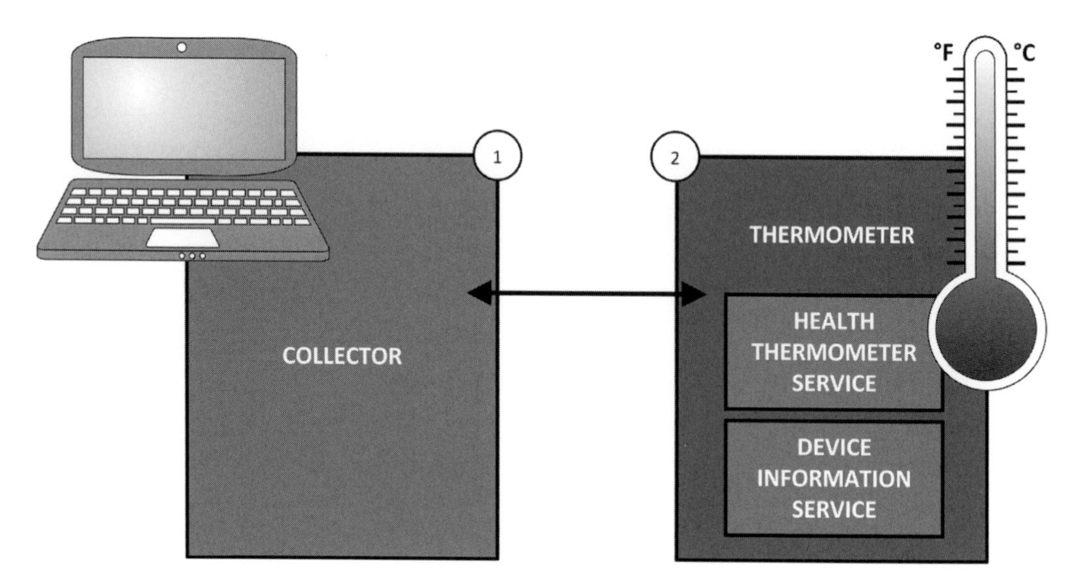

Figure 7.10. The relationship between the profile roles and services. A notebook (**1**) collects data from the thermometer sensor (**2**).

device may become locked or unusable until such time as the devices are brought back into proximity with each other. Likewise, the user can define how a device should behave once in proximity, where, for example, the devices can automatically establish a connection. The reporter device has the responsibility of instantiating the LLS, IAS, and TPS.

7.3.1.10 Time Profile

The *Time Profile* (TIP) defines two roles, namely a time *client* and a *server*, and may coexist with other profiles; the TIP uses the *Current Time Service* (CTS), *Next DST Change Service* (NDCS), and the *Reference Time Update Service* (RTUS). The profile is used to retrieve or request current date and time using the RTUS, along with data related to time zone, which is exposed by the CTS of the peer device. Likewise, the peer device will impart information related to when the next *Daylight Savings Time* (DST) will occur using the NDCS. The server device has the responsibility of instantiating the CTS, NDCS, and RTUS.

7.4 Bluetooth low energy and its Competitors

BLE faces competition from all the technologies that feature in this section, and, whilst ZigBee, EnOcean, and ANT Wireless have already succeeded in this market space, BLE has some catching up to do. Nevertheless, with the technology entering a broader market space, and with the overwhelming uptake of BLE by a number of manufacturers, several industry reports suggest that Bluetooth-enabled equipment will explode and witness sales of 4.2 billion units by 2015.[2] Bluetooth technology has an advantage over its competitors in that the technology has already been integrated into the majority, if not all, of cellular or smartphones, tablet devices, PCs, and other electronic equipment. However, Nordic Semiconductor (nordicsemi.com) has developed[3] a new converged chipset, namely the nRF51 Series, a multi-protocol "combo" solution in which ANT Wireless and BLE are both integrated onto a small *System-on-Chip* (SoC), affording ANT Wireless technology an opportunity to dominate what will become an aggressive and lucrative market sector. We discuss ANT Wireless later in Chapter 10, "The Power of Less: ANT."

7.4.1 The New 3-in-1 Specification

The Bluetooth SIG has essentially reinvented Bluetooth wireless technology with its introduction of the new Specification of the Bluetooth System: Core, v4.0. The new specification has been coined by some as *3-in-1*, a specification which includes BLE along with the existing classic and high speed variants. Naturally, BLE wishes to secure markets within the sports, fitness, well-being, and home automation markets, but curiously some industry pundits have suggested that BLE could directly compete with *Near Field Communications* (NFC). Bluetooth low energy has been touted as a viable technology candidate to offer mobile payments and other similar use cases, as the technology has already been integrated into a number of consumer electronic products, such as Apple's iPhone 4S and iPad HD.[4] Nevertheless, the industry seems very much geared toward NFC as the technology to deliver the all-important user scenarios and infrastructure, as already demonstrated by Barclaycard and other industry leaders such as Google. We discuss NFC and its associated market sector later, in Chapter 12, "Just Touch with NFC." Nonetheless, whilst many industry analysts originally regarded NFC as a competitor to Bluetooth, and vice versa, BLE has many pundits wondering if Bluetooth will venture into NFC's domain. Nonetheless, it isn't in BLE's remit to offer such use cases that directly compete with NFC; and, despite rumors of Bluetooth low energy encroaching on NFC's application portfolio, it still isn't included within its profiles to provide such applications.

7.5 Networking Topology

Bluetooth low energy supports several topologies, namely a *point-to-point* (or *peer-to-peer*), *point-to-multipoint*, and a *star*; these are illustrated in Figure 7.11. Likewise, there are also two types of BLE roles, namely a *master* and a *slave*, which we discuss later in Section 7.6.1, "The Physical Layer."

[2] CBR Mobility News, "Bluetooth-enabled Equipment Sales Will Soar to 4.2 billion by 2015," 2012.
[3] ANT Wireless, "ANT+ and Bluetooth low energy Concurrent Combo Chip Solution," 2012.
[4] Gabriel, C., "Apple iWallet May Favor Bluetooth Over NFC," 2012.

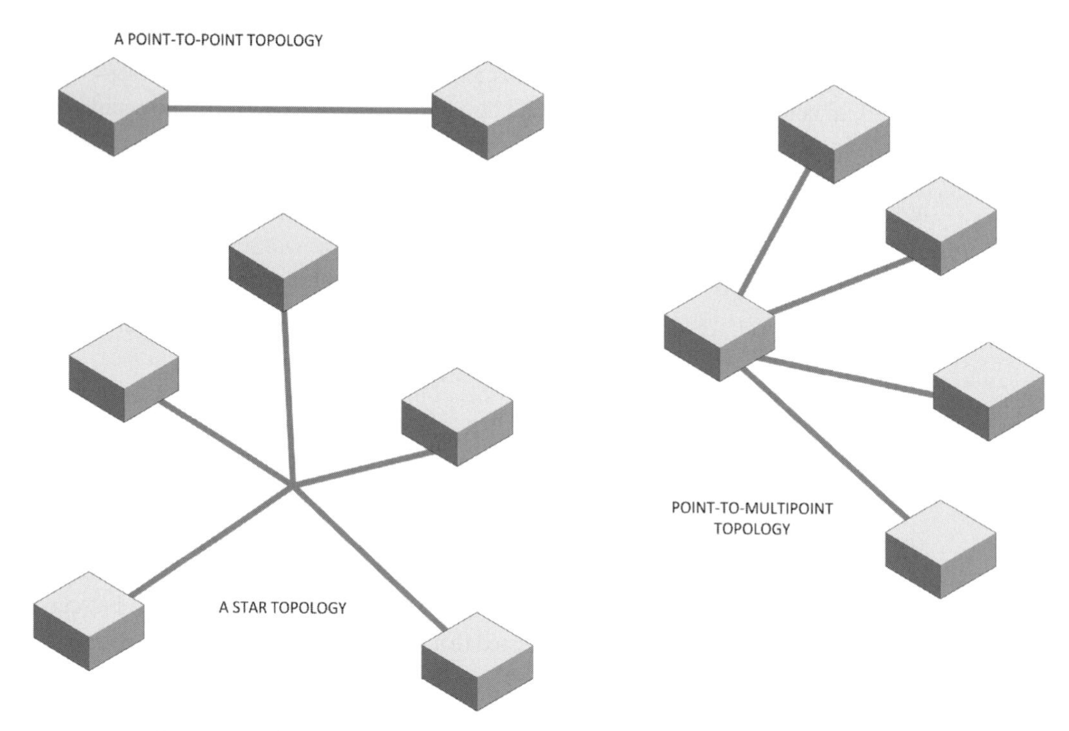

Figure 7.11. The Bluetooth low energy point-to-point (or peer-to-peer), point-to-multipoint, and star topologies.

7.5.1 Piconets

The fundamental topology used within a Bluetooth ecosystem is called a *piconet* – a *master* device communicating and interacting with a maximum of seven *slaves*. Typically, in a BR/EDR system a master/slave would share a common physical channel; however, in the LE system, each slave communicates with the master device on its own physical channel. In Figure 7.12 we illustrate a master device (**1**) interacting with its two slaves, (**2**) and (**3**). A *scatternet* is a topology in which two or more piconets overlap and interconnect, as shown in Figure 7.13. In this figure, we show two overlapping piconets (**3**), with master devices (**1**) and (**4**), and their respective slave devices, (**2**), (**3**), (**5**), (**6**), and (**7**). The LE system does not support a scatternet topology, nor does it support the master/slave role switch typically supported by the BR/EDR system.

7.6 The Bluetooth low energy Architecture

In Section 7.1, "Overview," we touched upon the fact that Bluetooth wireless technology now comprises two systems, namely BR/EDR and LE. We discuss in this section the LE system, but we may often refer to the BR/EDR system just to offer an explanation on the differences between the systems. The LE system has been architected from the ground up to enable a new generation of BLE-enabled products that have low cost, low data

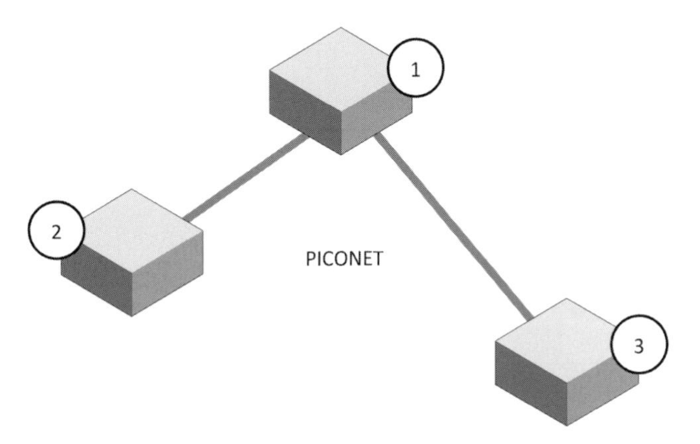

Figure 7.12. The piconet is the basic topology used within a Bluetooth ecosystem, where a master device interconnects with a maximum of seven slaves.

rates, consume less power, and have reduced complexity. In Figure 7.14 we map the *Open Systems Interconnection* (OSI) model against the Bluetooth low energy software stack. What's more, a Bluetooth system will include a *host*, an entity that typically sits above the *Host Controller Interface* (HCI), and one or more *host controllers* – entities that sit below the HCI. In Figure 7.22 we illustrate the host and host controller components of the LE stack, where the HCI permits communication between the host and controller using a serial interface. We provide a review of the HCI transport layers later in Section 7.6.3, "The Host Controller Interface," and offer a more thorough explanation in Chapter 14. In the sections that follow, we discuss the LE software stack from the bottom up, and, looking at Figure 7.14, we can see that our LE stack layers have been labeled numerically to aid in identification of the software building blocks that make up the software stack.

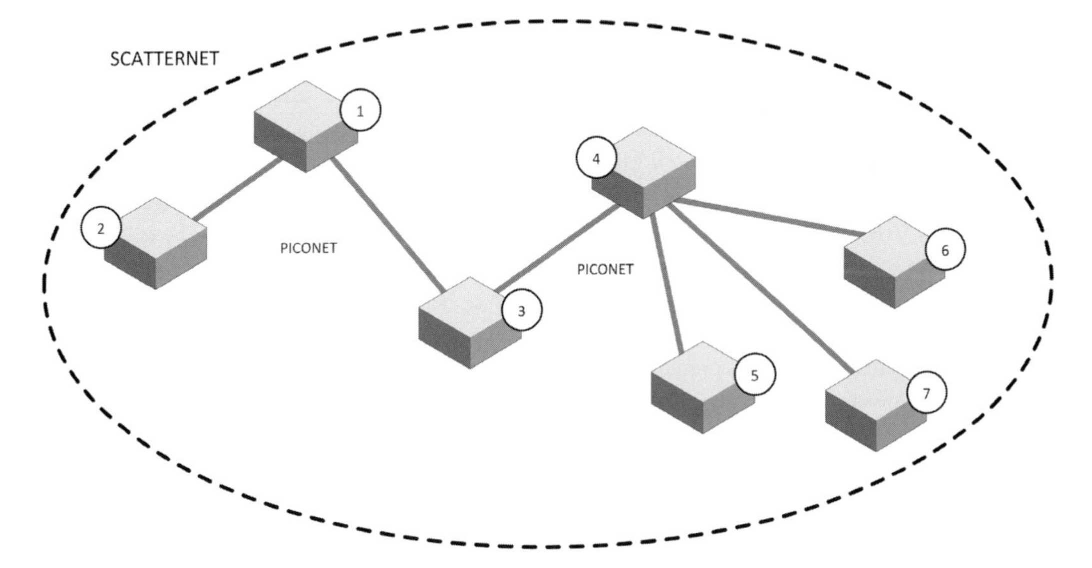

Figure 7.13. A scatternet topology is where two or more piconets overlap and interconnect.

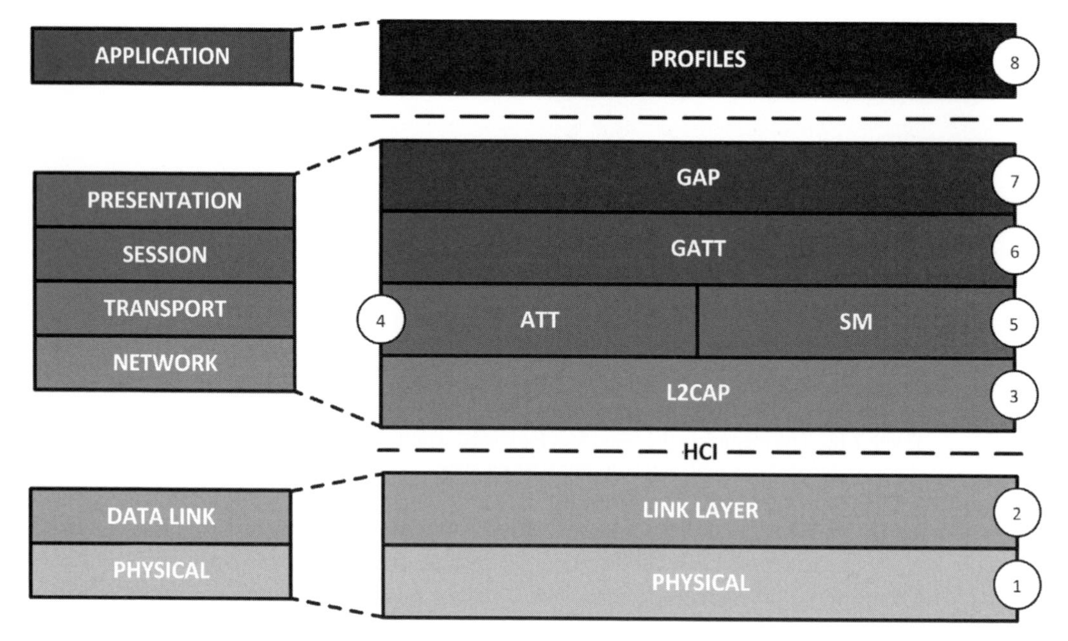

Figure 7.14. The Bluetooth low energy software stack architecture mapped against the OSI model.

7.6.1 The Physical Layer

We start with the *Physical* layer (**1**), which has the responsibility of managing the LE transceiver and operating the radio medium in the 2.4 GHz (2400–2483.5 MHz) frequency band, which uses 40 RF channels. Like the BR/EDR system, LE uses a *frequency hopping* transceiver to overcome *interference* and *fading*. The frequency hop is enabled during an active connection, in which the hop-length is pseudo-random. The *Frequency Hopping Spread Spectrum* (FHSS) technique spreads the RF power across the spectrum, aids in the reduction of interference, and divides the spectrum into 37 1 MHz (wide) data channels. We discuss advertising and data channels later in Section 7.6.2, "The Link Layer." The LE system uses two multiple access schemes, namely *Frequency Division Multiple Access* (FDMA) and *Time Division Multiple Access* (TDMA); 40 physical channels are used within the FDMA scheme – these channels are separated by 2 MHz. We discuss how physical channels are mapped later in Section 7.6.2.3, "Physical Channels." Moreover, three channels are used as *advertising* channels, whilst the remaining 37 are used as *data* channels. In the TDMA scheme, a device will transmit a packet at a predetermined interval, and the corresponding device also responds following a predetermined interval. What's more, the physical channel is sub-divided into time units, which are referred to as *events*. So, a data packet that is to be transmitted between LE devices is positioned in these events; there are two types of events, namely *advertising* and *connection*.

A device transmitting an advertising packet on the PHY channel is known as an *advertiser*. A *scanner* is a device that receives "advertising" from an advertising channel, when the device doesn't have an intention to connect to the advertising device. A device that needs to create a connection to another LE device, known as an *initiator*, will listen for connectable advertising packets. As such, the initiator can form a connection using the same advertising PHY channel which originated the connectable advertising packet. An advertising event is concluded once an advertiser receives and accepts the request for a connection to be established, wherein the initiator then becomes master of the piconet and the advertiser becomes the slave. At the onset of a connection event, channel hopping commences, and the connection event is also used to transmit and receive packets between the master and slave.

Table 7.2. The link layer's operating states and roles

State	Behavior
Standby	The link layer in standby does not transmit or receive packets; a device can enter standby from any other state.
Advertising	In the advertising state, a device will be transmitting advertising channel data and responding to such packets; a device can enter the advertising state from standby.
Scanning	In the scanning state, a device will be listening for advertising channel packets; a device can enter the scanning state from standby.
Initiating	In the listening state, a device will be awaiting packets from a specific advertising device and will respond to these packets to initiate a connection; a device can enter the initiating state from standby.
Connection	In the connection state, a device may become a master or slave; a device can enter the connection state either from an initiating or advertising state.

7.6.2 The Link Layer

The *Link Layer* in Figure 7.14 (**2**) operates a state machine with five states, namely *Standby*, *Advertising*, *Scanning*, *Initiating*, and *Connection*, and permits only one state to be active at any time. We'll come back to this in a moment. Essentially, the physical and link layers form our host controller, as shown in Figure 7.22. The next conceptual layer within the LE stack is HCI (**A**), as shown in Figure 7.22 – the presence of an HCI is dependent on the type of implementation. For example, a Bluetooth-enabled stereo headset would typically not use an HCI, as the entire application resides on a single system using one processor. However, a *Universal Serial Bus* (USB) dongle may be used to retro-fit your notebook with Bluetooth connectivity. The host will reside on your PC, whereas the dongle would be used to offer your notebook Bluetooth connectivity via the host controller. The host and host controller will use the HCI-USB transport to exchange data. Nowadays, some Bluetooth implementations utilize the HCI as a conceptual interface, retaining the host and controller topology. We discuss the HCI transport layer in Section 7.6.3, "The Host Controller Interface."

In Table 7.2, we list the link layer's roles and states, and in Figure 7.15 we illustrate the state machine and the transitions used by the link layer. The standby state is the default state within the link layer – from this state the link layer may enter the advertising, scanning, or initiating state. So, when a device enters the connection state from an

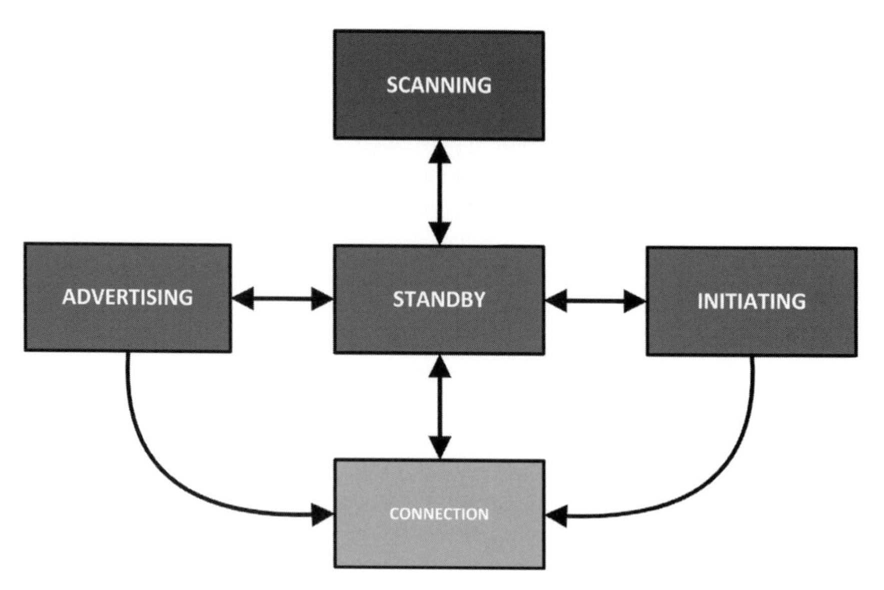

Figure 7.15. The state machine used by the link layer, showing the transition between states.

initiating state, the device will assume the role of master, whereas if the device enters the connection state from an advertising state, then the device will adopt the role of slave. The master device is responsible for defining the timings for transmissions. The link layer may optionally support multiple instances of the state machine; however, some restrictions apply. When the link layer supports multiple state machines, and whilst in the connection state, the device cannot operate as master and slave simultaneously. Likewise, a master device may support multiple connections, although the slave device will only use a single connection.

7.6.2.1 Link Layer Filtering and White Lists

The link layer may choose to utilize *device filtering*, which is based on the address of the peer device. This alleviates the number of devices to which the link layer should respond. As such, the advertising, scanning, and initiating states each have their own filter policies and, whilst the link layer is in the advertising state, the advertising filter policy is used. Likewise, whilst in the scanning state, the scanning filter policy is used, and so on. A *white list* is used to assist the link layer when applying device filtering and, as such, a *record* of the device address and its type, that is, public or random, something we discuss in the section that immediately follows, is stored within a white list record. The host is responsible for defining and configuring the white list, wherein data are only received by the host from devices that are recorded in the white list.

7.6.2.2 Device Addressing

A Bluetooth low energy device, like the BR/EDR system, uses a public and/or optional private/random 48-bit device address, which uniquely identifies a device within a piconet. The latter address type may help to assure privacy within a piconet, as it may change

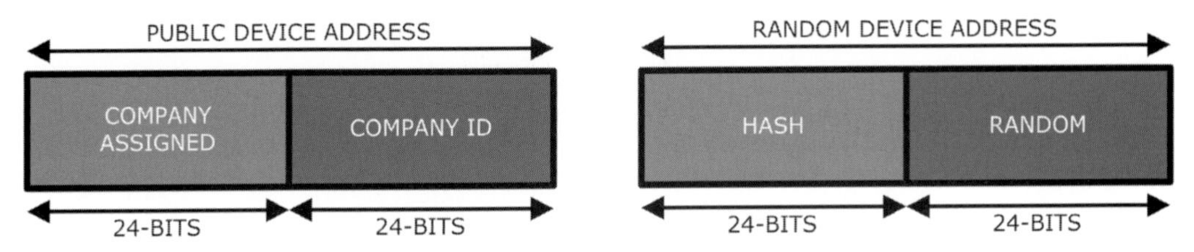

Figure 7.16. The format of the public and random device addresses.

frequently. The public address should conform with the "48-bit Universal LAN MAC addresses," as specified by the *Institute of Electrical and Electronic Engineers* (IEEE) 802-2001 standard, which will use a valid *Organizationally Unique Identifier* (OUI) that is sourced from the IEEE Registration Authority. In Figure 7.16, we illustrate the two fields that make up the public address field. We also show the random device address, which is also divided into two fields.

Table 7.3. How the 40 RF channels are mapped to advertising and data channels

Channel	Center frequency	Channel type	Data index	Advertising index
0	2402 MHz	Advertising.		37
1	2404 MHz	Data.	0	
2	2406 MHz	Data.	1	
...	...	Data.	...	
11	2424 MHz	Data.	10	
12	2426 MHz	Advertising.		38
13	2428 MHz	Data.	11	
14	2430 MHz	Data.	12	
...	...	Data.	...	
38	2478 MHz	Data.	36	
39	2480 MHz	Advertising.		39

7.6.2.3 Physical Channels

Earlier, in Section 7.6.1, "The Physical Layer," we touched upon the availability of the 40 RF channels which are used in the 2.4 GHz ISM frequency band, and in this section we discuss how these channels are allocated into two physical channels, specifically for advertising and data – something which we also discussed earlier. You may recall that the advertising physical channel actually uses three RF channels for discovering devices, and initiating and broadcasting data with devices in proximity. What's more, we also intimated that the data physical channel uses 37 RF channels to achieve communication between devices in proximity; these channels are uniquely indexed, as shown in Table 7.3.

A device wishing to communicate with another device must share a physical channel by tuning its transceiver to the same channel as the other device simultaneously. The likelihood of a collision on the same RF channel is probable, due to the limited availability

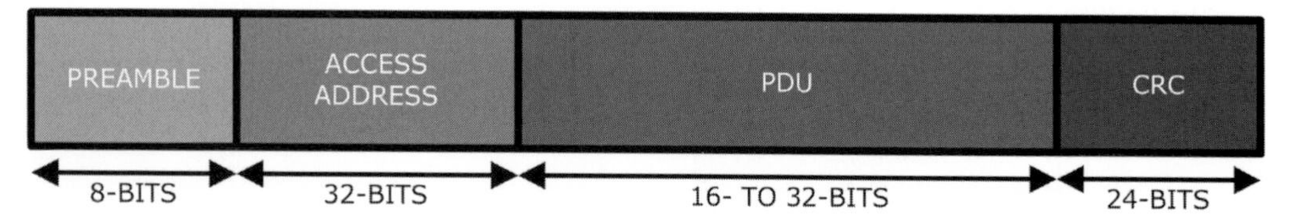

Figure 7.17. The link layer packet format.

Table 7.4. The fields that make up the link layer packet format

Name	Description	Size
Preamble	This 8-bit field is used for frequency synchronization, symbol timing estimation, and *Automatic Gain Control* (AGC). The preamble field uses 10101010b (`0xAA`) for the advertising channel packet and uses either 10101010b or 01010101b (`0x55`) when the value of the *Least Significant Bit* (LSB) of the adjoining access address is 1 or 0, respectively.	8
Access address	This 32-bit field uses the same value for all advertising channel packets, namely `0x8E89BED6`, although the value for the data channel packets is different and alternates for each link layer connection.	32
PDU	This 16- to 32-bit field contains the advertising or data channel *Protocol Data Unit* (PDU). We discuss the Advertising and Data PDUs in more detail in the following sections.	16 to 32
CRC	The 24-bit *Cyclic Redundancy Check* (CRC) field is used to validate the content of the PDU.	24

of channels. As such, for each transmission the physical channel is prefixed with an *access address*, which is used as a *correction code* alerting the link layer to the collision.

7.6.2.4 The Link Layer Packet Format

The link layer uses one common packet format, as shown in Figure 7.17, for both advertising and data channel packets. The fields that form this packet are listed in Table 7.4, along with a description and their size in bits.

As discussed in Table 7.4, there are, in fact, two types of PDU within the link layer packet. In Figure 7.18, we illustrate the format of the advertising channel PDU. The advertising channel PDU uses a 16-bit header, wherein the payload field is defined by the length field within the header, as shown in Figure 7.19. In Table 7.5, we discuss the fields that make up the advertising channel PDU header field.

The second PDU within the link layer packet, as we discussed earlier, is shown in Figure 7.20, where we illustrate the format of the data channel PDU. Like the advertising channel PDU, the data channel PDU also uses a 16-bit header, wherein the payload field is defined by the length field within the header, as shown in Figure 7.21. In Table 7.7, we discuss the fields that make up the data channel PDU header field.

The *Message Integrity Code* (MIC), as shown in the data channel PDU, see Figure 7.20 and Figure 7.21, is only used with an encrypted link layer connection, and is not

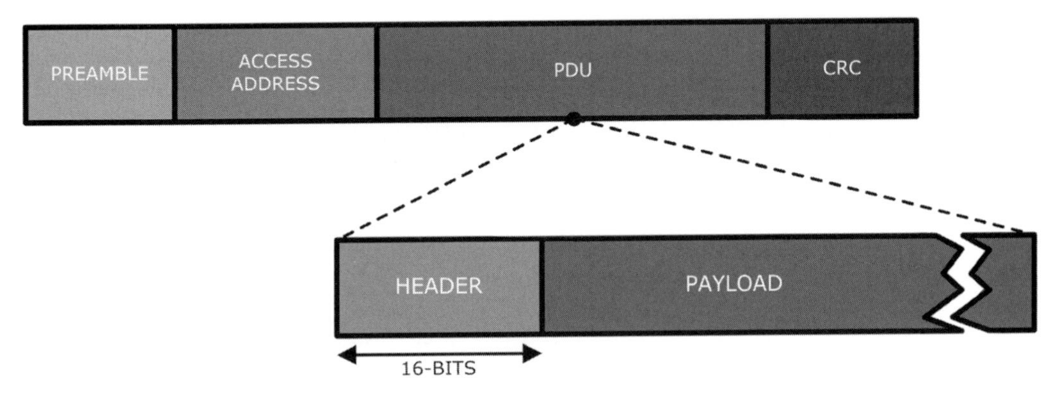

Figure 7.18. The advertising channel PDU format.

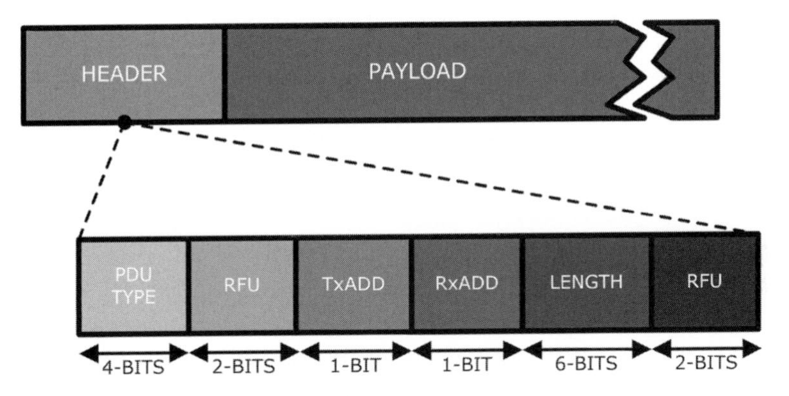

Figure 7.19. The advertising channel PDU header format.

Table 7.5. The fields that make up the advertising channel PDU header

Name	Description	Size
PDU type	This 4-bit field denotes the PDU type and is explained further in Table 7.6.	4
RFU	Reserved.	2
TxADD	The TxADD and RxADD fields are both 1-bit in length and influence the	1
RxADD	PDU type, as shown in Table 7.6. If neither of these fields is used, then they are deemed "reserved."	1
Length	The length field is 6-bits in length and is used to indicate the length of the actual payload (6 to 37 octets).	6
RFU	Reserved.	2

included when the connection is unencrypted or when there is a data channel PDU zero-length payload.

7.6.2.5 The Link Layer Control Protocol

The *Link Layer Control Protocol* (LLCP) within the link layer is primarily used to control and negotiate certain procedures when connecting between peer link layers.

Table 7.6. The PDU type field values

Type	Packet name	Description
0x00	ADV_IND	A connectable undirected advertising event.
0x01	ADV_DIRECT_IND	A connectable directed advertising event.
0x02	ADV_NONCONN_IND	A non-connectable undirected advertising event.
0x03	SCAN_REQ	This PDU type is transmitted by the link layer during a scanning state and is received by the peer link layer in the advertising state.
0x04	SCAN_RSP	This PDU type is transmitted by the link layer during an advertising state and is received by the peer link layer in the scanning state.
0x05	CONNECT_REQ	This PDU type is transmitted by the link layer whilst in the initiating state and is received by the peer link layer in the advertising state.
0x06	ADV_SCAN_IND	A scannable undirected advertising event.
0x07–0x0F		Reserved.

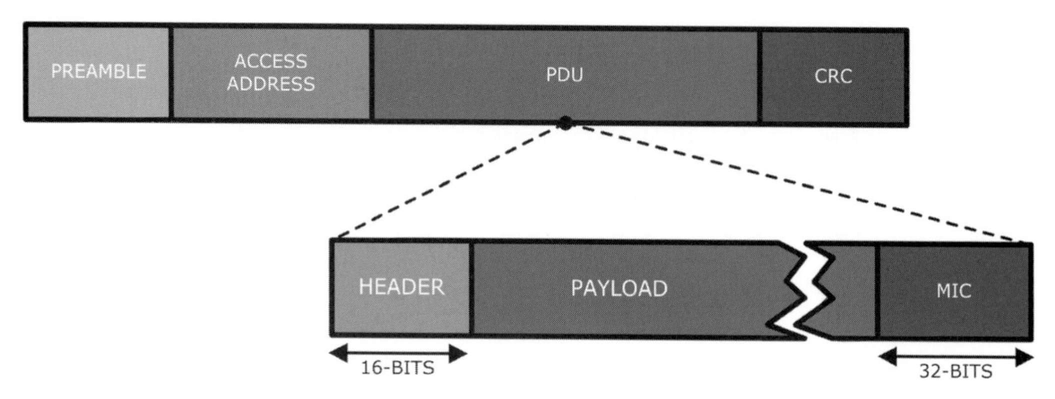

Figure 7.20. The data channel PDU format.

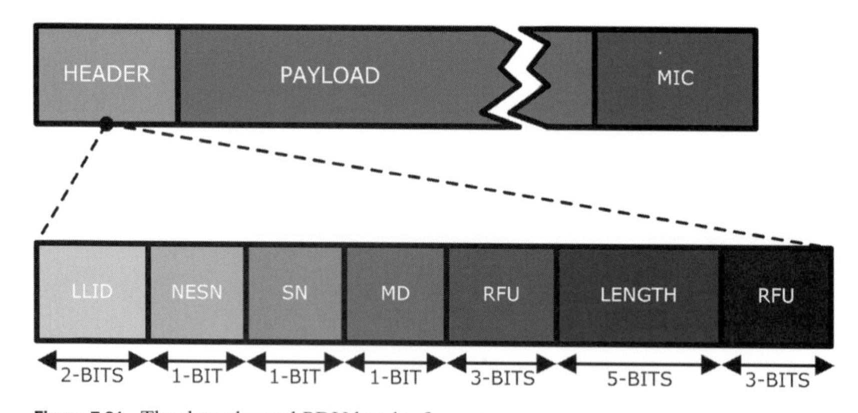

Figure 7.21. The data channel PDU header format.

Table 7.7. The list of fields that make up the data channel PDU header

Name	Description	Size
LLID	The *Logical Link Identifier* (LLID) is 2-bits in length and specifies whether the enclosed packet is an *LL Data* PDU or an *LL Control* PDU. We list the possible combinations in Table 7.8.	2
NESN	*Next Expected Sequence Number* (NESN).	1
SN	*Sequence Number* (SN).	1
MD	*More Data* (MD).	1
RFU	Reserved.	3
Length	The length field is 5-bits in length and is used to indicate the length of the actual payload in octets, which may include the MIC, if included.	5
RFU	Reserved.	3

These procedures cover aspects of connection, starting and pausing encryption, and other link-specific procedures. Naturally, these procedures have associated timeout constraints, although the *Connection Update* and *Channel Map Update* procedures have no specific rules. A master and slave use a timeout timer, T_{PRT}, to assess a non-responsive LLCP, where, on initialization of a given procedure, T_{PRT} is reset and then started. For example, each queued LL Data PDU ready for transmission resets T_{PRT}, and when completed the timer is halted. However, if the timer exceeds 40 seconds, then the connection is presumed lost and the link layer exits the connection state and enters standby; the host is duly informed of the loss of connection. We summarize these procedures in Table 7.9.

Table 7.8. The LLID type field values

Value	Description
`0x00`	Reserved.
`0x01`	An LLID Data PDU, which may denote an empty PDU, or is a continuation fragment of an L2CAP message.
`0x02`	An LL Data PDU, which may denote a complete L2CAP message with no fragmentation, or is the start of an L2CAP message.
`0x03`	An LL Control PDU.

7.6.3 The Host Controller Interface

The presence of the HCI is solely dependent on the specific Bluetooth implementation, as we have already discussed (see Section 7.6.2, "The Link Layer"). In Figure 7.23, we illustrate two possible types of implementation: **A** represents a device that utilizes one processor to execute both the host and controller components, whereas **B** represents a device that uses two processors, one to execute the host and the other to execute the controller. The HCI is present in **A** as the implementer has chosen to retain the HCI as a conceptual interface.

Table 7.9. The LLCP procedures

Procedure	Description
Connection update	The link layer parameters associated with a connection may be updated following a connection state.
Channel map update	The link layer parameters associated with a channel map may be updated following a connection state.
Encryption	As requested by the host, the link layer may enable encryption of packets once it has entered the connection state.
Feature exchange	The link layer parameters associated with the current feature set may be exchanged following a connection state.
Version exchange	The link layer parameters associated with version information may be exchanged following a connection state.
Termination	This procedure is used to terminate a connection whilst in the connection state.

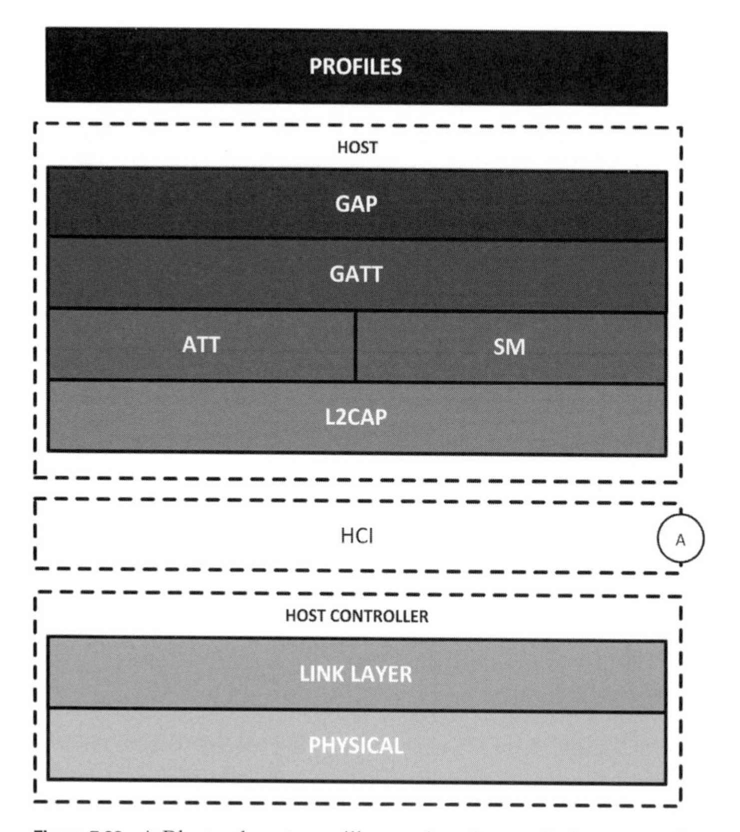

Figure 7.22. A Bluetooth system will comprise a host and a host controller, which uses the HCI to communicate.

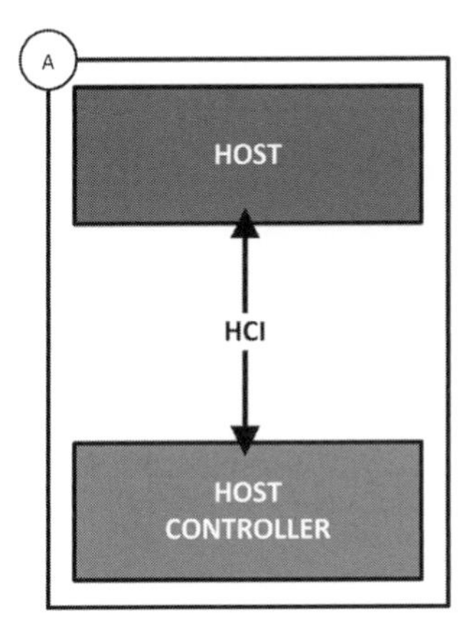

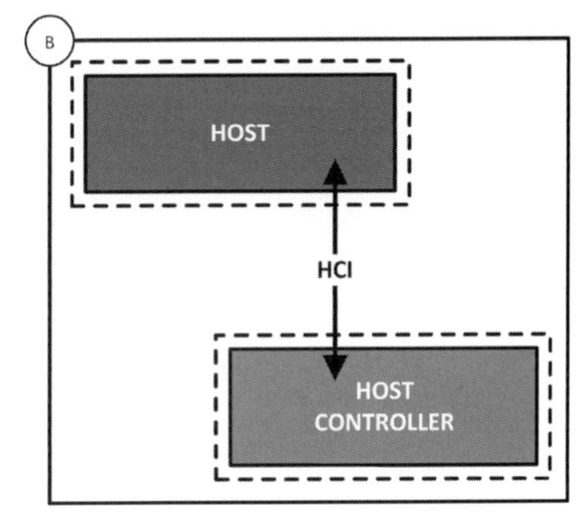

Figure 7.23. The HCI permits communication between a host and controller, but some implementations using a single processor may retain the HCI as a concept interface.

The HCI transport layer provides four interfaces to enable communication between a host and controller; we summarize these interfaces in Table 7.10. The *Universal Asynchronous Receiver/Transmitter* (UART) interface provides a basic RS232 null-modem-like connection between the host and controller entities. The 3-Wire UART interface offers an opportunity to use the HCI between two UARTs. The USB interface would permit a USB dongle, something we touched upon earlier when discussing retro-fitting a notebook; it may also be used when, for example, an independent USB module may be integrated on to a *Printed Circuit Board* (PCB). Finally, the *Secure Digital* (SD) interface creates a pathway between a Bluetooth-enabled SD device (controller) and a Bluetooth host.

Table 7.10. The HCI has four types of interface

Interface	Description
UART	An RS232 null-modem-like interface.
USB	The HCI-USB interface allows a USB dongle or an independent USB module, which may be integrated on to a PCB, to communicate.
SD	Creates an interface between a Bluetooth-enabled SD device (controller) and a host.
3-Wire UART	The 3-Wire interface provides an opportunity for a system to use the HCI between two UARTs.

The HCI uses a *command/event* paradigm and exchanges certain messages over the interface. What's more, the HCI provides a consistent method for accessing a controller's capabilities when an LE controller uses a reduced set of commands and events. In Figure 7.24, we illustrate the lower layer components of the HCI for both the host and controller; these layers are unaware of the content of the data being exchanged.

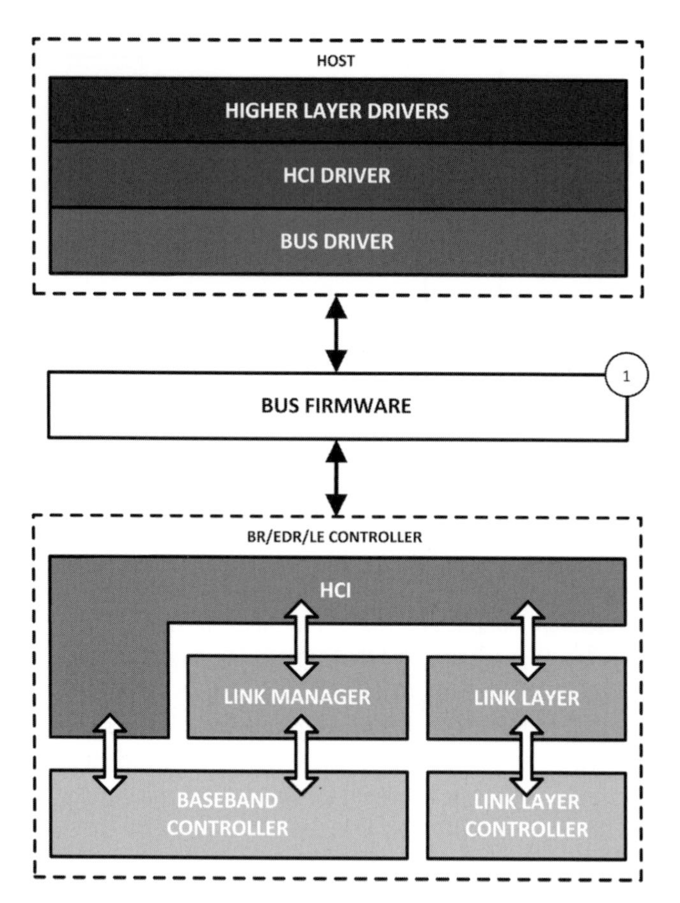

Figure 7.24. The lower layers of the HCI at both the host and host controller.

7.6.4 L2CAP

The *Logical Link Control and Adaptation Protocol* (L2CAP) layer, as shown in Figure 7.14 (**3**), has the overall responsibility to provide *connectionless* data services to the layers above, and it does so using protocol multiplexing and *Segmentation And Reassembly* (SAR) capabilities. The L2CAP is common to both BR/EDR and LE systems and therefore we should only take a moment here to note some minor differences. L2CAP functionality within the LE system operates in a *Basic Mode*, offering *fixed channel* type data services, as shown in Table 7.11, and does not support *connection-oriented* data services.

The philosophy of the L2CAP layer is based on the notion of *channels*, and each endpoint of an L2CAP channel is referred to as a *Channel Identifier* (CID). In Table 7.11, we list the CIDs for a number of fixed channels for LE operation. In fact, CID `0x0004` is used to transmit *Attribute* (ATT) protocol PDUs, where the flow is designated as "best effort." We discuss the ATT protocol in the following Section 7.6.5, "The Attribute Protocol." Each fixed channel possesses certain attributes, such as the ability to change the following parameters: configuration parameters for reliability, *Maximum Transmission*

Table 7.11. The fixed channel type values for the LE system

CID	Channel type
`0x0000`	Null (not used).
`0x0001`	Signaling (BR/EDR only).
`0x0004`	Attribute protocol.
`0x0005`	Signaling (LE only).
`0x0006`	Security manager protocol.

Unit (MTU) size, *Quality of Service* (QoS), and security. We'll discuss L2CAP later in Chapter 14, "Bluetooth Classic and High speed: More Than Cable Replacement." In the following sections, we describe two new protocols, specifically designed for the LE system, that sit on top of L2CAP, as shown in Figure 7.14. These are the *Security Manager Protocol* (SMP), which uses a fixed L2CAP channel, specific for security between LE devices, and the ATT protocol, which is merely used to provide the exchange of relatively small amounts of data over a fixed L2CAP channel.

Figure 7.25. The ATT PDU format.

7.6.5 The Attribute Protocol

The ATT layer, as shown in Figure 7.14 (**4**), is specifically optimized for LE use, but may also be used within a BR/EDR system; the ATT is comparable to the BR/EDR system's *Service Discovery Protocol* (SDP) in terms of learning more about a neighboring device's capabilities. The protocol specifies two roles, namely a *client* and *server*, and is used by a device to discover the services and capabilities of other devices in proximity. The GATT profile is dependent on the ATT implementation and, as such, GATT utilizes ATT to transport one of six protocol PDU types, namely *commands*, *requests*, *responses*, *indications*, *notifications*, and *confirmations*. In Figure 7.25, we illustrate the ATT PDU, and in Table 7.12 we discuss the fields that make up the ATT PDU along with their size in bits.

The attribute parameters field comprises four properties, namely an attribute *handle*, *type*, *value*, and *permissions* – we illustrate the attribute parameter format in Figure 7.26 and summarize these fields in Table 7.13.

Furthermore, an ATT PDU has one of six protocol PDU types, namely a *request*, *response*, *command*, *notification*, *indication*, and *confirmation*, which we summarize in Table 7.14. A server device can receive and respond to both "find information" and read requests. However, if a client sends a request, then the client will likewise support all

Table 7.12. The fields that make up the ATT PDU

Concept	Description	Size
OpCode	This 8-bit field contains the PDU type, as described in Table 7.14.	8
Attribute parameters	This variable length field contains data which is specific to the PDU type; it comprises four further fields.	
Authentication signature	This optional 96-bit field may provide authentication of the originating OpCode.	96

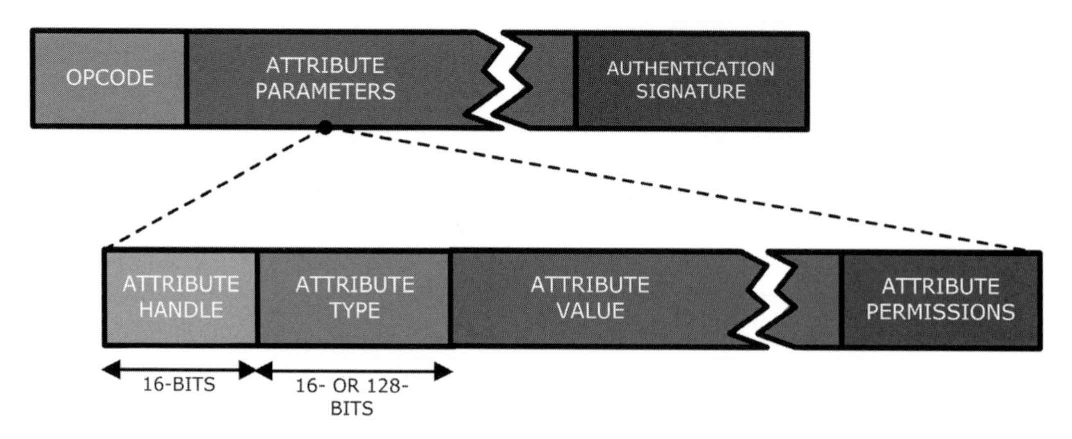

Figure 7.26. The format of the attribute parameters.

Table 7.13. The four properties that comprise the attribute parameters

Attribute	Description	Size
Handle	This 16-bit field is used to identify uniquely an attribute located on a server, in turn allowing a client to address it during read or write requests.	16
Type	This 16- or 128-bit field defines what an attribute actually represents. The Bluetooth SIG defines many attributes, but non-Bluetooth SIG attribute types are also possible. A *Universal Unique Identifier* (UUID) is a 128-bit value,[a] which is used to identify each attribute type.	16 or 128
Value	This variable length field contains information which describes the attribute type.	
Permissions	The permissions field is not accessible by the ATT layer, but instead is used by the GATT profile and by the server to determine if read and/or write permissions are permitted.	

[a] Some common or frequently used UUIDs may be shortened to 16-bits to ensure efficiency.

Table 7.14. An ATT PDU has one of six protocol PDU types

PDU type	Description
Request	This PDU is sent by the client device and it expects a *response*.
Response	Sent by the server in response to a request PDU.
Command	A client device sends a server a command, but the client doesn't expect a response.
Notification	A server device notifies a client of new information and doesn't expect a confirmation.
Indication	A server indicates to a client new information and expects to receive a confirmation.
Confirmation	Sent by the client in response to an indication PDU.

responses for that PDU. A client will send an attribute protocol request to learn more about specific attributes located on a server device where the server will always respond. In fact, a Bluetooth LE device can concurrently execute and implement both client and server roles, although only one instance of a server can be maintained by an LE device. What's more, a server will retain a single set of attributes and offer multiple services whilst providing support for multiple clients.

Table 7.15. One of three phases is used to establish keys

Phase	Description
1	Pairing feature exchange.
2	*Short-term Key* (STK) generation.
3	Transport specific key distribution.

7.6.6 The Security Manager Protocol

The SMP layer, as shown in Figure 7.14 (**5**), is specifically optimized for LE devices and is designed to complete responsibility for pairing, authentication, and encryption. More specifically, the security manager provides a key distribution method to undertake identity and encryption procedures. Each LE device has the sole responsibility of key generation and distribution. Pairing between devices is a three-phase process and is used to establish keys, which may then be used for encryption.

In Table 7.15, we list the three phases where phases *one* and *two* are mandatory whilst phase *three* is optionally performed – we illustrate in Figure 7.27 the LE pairing phases in context. In the first instance, I/O capabilities are exchanged during the *pairing feature exchange* process between devices to determine where one of the methods, as shown in Table 7.16, will be used during phase two. Phases one and two can be undertaken on either an encrypted or an unencrypted connection and, as we have already discussed, phase three is optionally used to distribute specific keys. It is also only executed on an encrypted link using STK, which was generated at phase two.

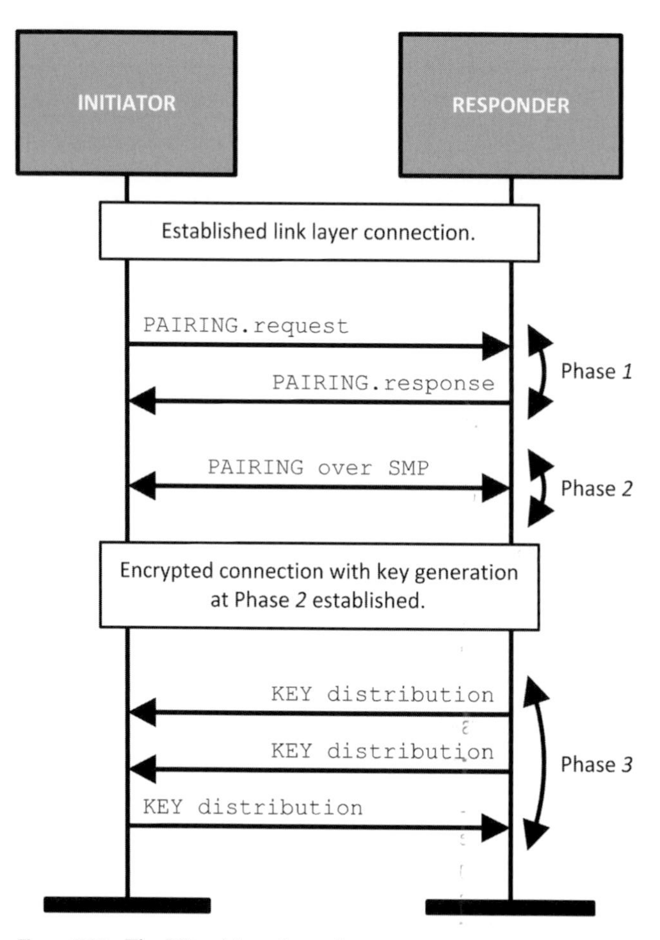

Figure 7.27. The LE pairing phases in context.

Table 7.16. Phase two methods

Method	Description
1	Just works.
2	Passkey entry.
3	*Out of Band* (OOB).

A cryptographic toolbox is used to support random addressing, pairing, and other operations within the security manager and, as such, several functions are defined, as we identify in Table 7.17. The primary cryptographic function supporting all functions as described in Table 7.17 is security function *e*. The purpose of the security function[5] *e* is to generate a 128-bit `encryptedData` from a 128-bit key and 128-bit `plaintextData` using the *Advanced Encryption Standard* (AES) 128-bit block cypher.

[5] So, `encryptedData` = *e*(key, `plaintextData`).

Table 7.17. The cryptographic toolbox functions

Function	Description
ah	The random *address hash* (ah) generates a value used for resolvable private addresses.
c1	A confirm value (c1) is exchanged and generated during the pairing process.
s1	A key function (s1) is used to generate the STK during the pairing process.

Table 7.18. The SM channel configuration parameters

Function	Description
MTU	The MTU has a fixed value of 23.
Flush timeout	The timeout value set here is infinite (`0xFFFF`).
QoS	QoS is defined as "best effort".
Mode	As we discussed earlier in Section 7.6.4, "L2CAP," the basic mode is used for LE devices.

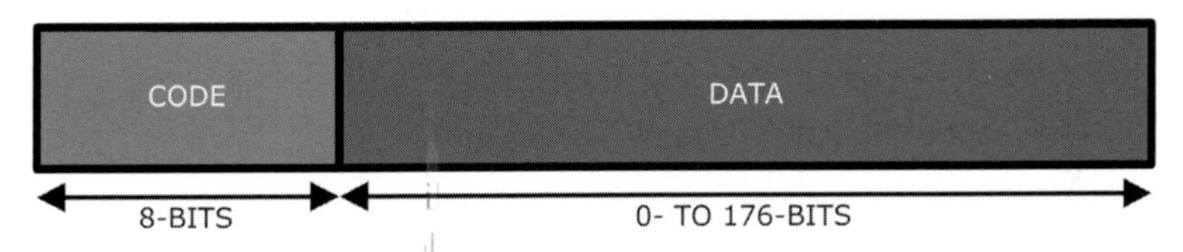

Figure 7.28. The SM protocol command format.

Table 7.19. The two fields that make up the SM protocol format

Name	Description	Size
Code	This field is 8-bits in length and identifies the command being used; we list the possible codes used in Table 7.20.	8
Data	The data field is variable and its content is determined by the code field.	

At the onset of pairing, the Pairing Feature Exchange sends a pairing *request command*, as shown in Table 7.20. The pairing process is started by the initiating device, and if the "responding" device neither supports pairing nor can it be performed, then the pairing process terminates with a corresponding error message. As we have already mentioned, I/O capabilities are exchanged during the Pairing Feature Exchange, but other authentication requirements are also exchanged, together providing the STK generation method for phase two. A 128-bit *Temporary Key* (TK) is used during the pairing process and in generating the STK. The authentication requirements are defined by GATT, which we discuss in a moment (Section 7.6.7, "The Generic Attribute Profile").

All SMP communication is sent over L2CAP, which uses a dedicated fixed channel (`0x0006`), as we showed earlier in Table 7.11; in Table 7.18 we show the configuration

Table 7.20. The SM command codes

Code	Description
0x00	Reserved.
0x01	Pairing request.
0x02	Pairing response.
0x03	Pairing confirm.
0x04	Pairing random.
0x05	Pairing failed.
0x06	Encryption information.
0x07	Master identification.
0x08	Identity information.
0x09	Identity address information.
0x0A	Signing information.
0x0B	Security request.
0x0C--0xFF	Reserved.

parameters that are used to set up the SM channel. In Figure 7.28, we illustrate the SM protocol command format, and in Table 7.19 we discuss the fields that form this packet, along with a description and their size in bits.

7.6.7 The Generic Attribute Profile

The GATT profile provides a framework for applications and other profiles which, in turn, use the underlying ATT protocol in which profile data can be exchanged. GATT defines basic elements, such as procedures, the format of services, and characteristics. These are managed and contained by attributes used in the ATT protocol, which transports this profile data. The procedures defined in the GATT profile in combination with the ATT protocol include *discovery*, *reading*, *writing*, *notify*, and *indication*, as well as the broadcasting of characteristics. Like the ATT protocol, the profile defines two roles, namely a *client* and *server*. The GATT profile ultimately specifies how a client communicates with a server device. In Figure 7.29, we illustrate the peer-to-peer model used by the GATT profile.

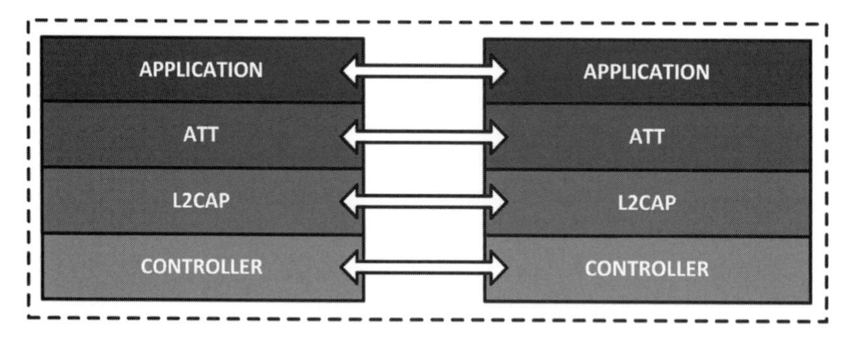

Figure 7.29. The peer-to-peer model used by the GATT profile.

In essence, a client device directs a command or request toward a server, where the client device then receives a response, indication, or notification. Conversely, a server will

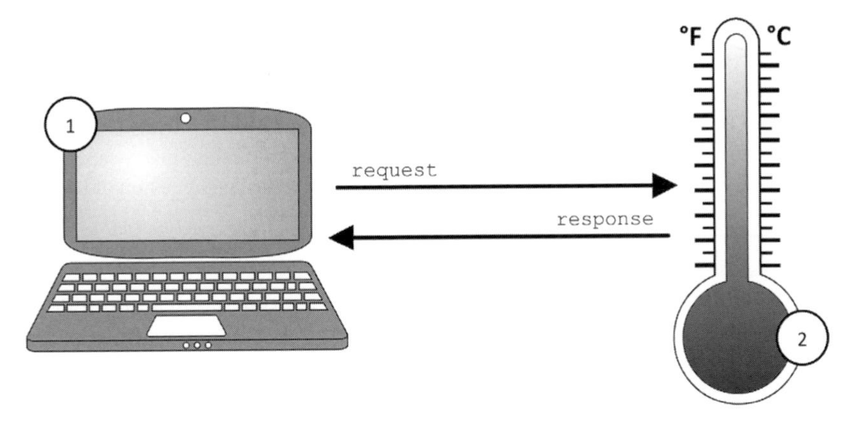

Figure 7.30. The role of client and server is typified with a notebook, by a client interrogating the thermometer sensor to learn more about its characteristics.

receive a command or request from a client and will return a response, indication, and/or notification. In Figure 7.30, we typify the role of client and server, where we witness a notebook (**1**) as a client device initiating procedures to interrogate the thermometer or sensor device (**2**), either to configure it or read its values.

The thermometer will reveal its characteristics as part of its temperature service. The thermometer may possess primary or secondary services along with several characteristics, which may include the availability of certain features and/or a data dictionary where values and other similar data may be exposed. So, the GATT profile can offer several user requirements and scenarios – for example: the exchange of configuration parameters; discovery of service and its characteristics; writing and/or reading characteristic values; and notification and indication of characteristic values.

7.6.7.1 GATT Fundamentals

A profile is the highest layer within the Bluetooth LE stack, as we illustrated earlier in Figure 7.14 (**8**). It may specify one or more services that are used to complete an application, where these services contain specific characteristics, which may then reference other services. So, a service is a collection of data or information that has associated behaviors, which together are used to fulfill a function or feature of the given profile. A characteristic is used within a service, along with any other associated properties and configuration data as to how such data can be manipulated. Each characteristic contains a value, which may refer to further information about that value and perhaps link to other peripheral information about that value.

7.6.8 Generic Access Profile

As we touched upon earlier in Section 7.3.1, "BLE-specific Profiles," Bluetooth profiles achieve interoperability for a host of manufacturers with the use of *capabilities*, as prescribed in numerous SIG-profile specifications that, in turn, define such capabilities from the physical layer or air-interface through to L2CAP. A profile will also prescribe

Table 7.21. GAP is supported by all device types

Type	Description
BR/EDR	Bluetooth-enabled devices that support the basic rate operation.
LE only	Bluetooth-enabled devices that support the low energy operation.
BR/EDR/LE	Bluetooth-enabled devices that support a combination of basic rate and low energy operations.

specific interactions at the application layer – in other words, a profile will describe naming conventions and offer visual cues by defining commonality in behavior and user experience. *The Generic Access Profile*, or GAP, is a core profile used for all device types, as shown in Table 7.21, and is something we now discuss in this section. GAP specifically describes the essential requirements for all Bluetooth-enabled devices, along with performing a single defining role. In essence, a Bluetooth device may incorporate either initiating or accepting procedures, and the peer device must support the corresponding functionality. Within an LE system, however, we define *four* roles, namely *broadcaster*, *observer*, *peripheral*, and *central*, where only one role can be supported at a given time. We'll come back to this in a moment.

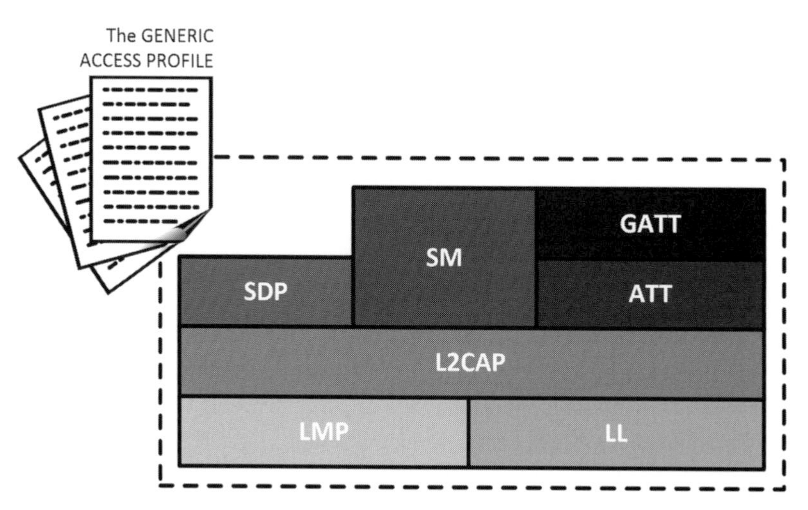

Figure 7.31. The all-encompassing Generic Access Profile provides common functionality, capabilities, and user experience for the LE system.

What's more, GAP describes specific roles that are undertaken and, further, defines discoverability, connection, and security modes and procedures. Later, in Chapter 14, we discuss roles, user expectations, modes of operation, security, and authentication. However, in this section we only cover aspects that govern LE-specific functionality. In Figure 7.31, we depict the overall relationship that governs GAP capabilities across the Bluetooth stack architecture. You will note that the stack architecture as shown covers both the BR/EDR and LE systems – we listed the supported device types in Table 7.21. So, looking at the architecture from the bottom up, LL was discussed earlier, in Section 7.6.2, "The Link Layer." We discussed L2CAP in Section 7.6.4, "L2CAP," and this

Table 7.22. LE-specific GAP roles

Role	Description
Broadcaster	A broadcaster device sends advertising events (see Section 7.6.2, "The Link Layer").
Observer	An observer device receives advertising events (see Section 7.6.2, "The Link Layer").
Peripheral	A peripheral role represents a device accepting (slave) the establishment of an LE physical link.
Central	A central role represents a device initiating (master) the establishment of an LE physical link.

is common to both Bluetooth systems. ATT and GATT were also discussed in earlier sections, along with SM. Incidentally, device types, that is, BR/EDR or BR/EDR/LE, will use SDP for service discovery; moreover, BR/EDR device types that utilize ATT will also include a GATT implementation. However, BR/EDR/LE and LE-only device types will implement GATT for service discovery, which is transported over the LE channel.

7.6.8.1 Broadcaster, Observer, Peripheral, and Central Roles for the LE System

In Table 7.22, we list and describe the GAP roles that have been defined for the LE system. Later, in Chapter 14, we cover and discuss in greater detail GAP user interface expectations, other modes of operation, and security and authentication.

8 Control Your World with ZigBee

ZigBee (zigbee.org) is a protocol-based global standard, which has been developed by a consortium of over 400 companies around the world. The protocol relies on the *Institute of Electrical and Electronics Engineers* (IEEE) 802.15.4 standard, which provides the protocol with its air-interface. ZigBee has been around for a number of years, with its initial ratified standard appearing in December 2004; later, in June 2005, the standard was made public. ZigBee specifies a software protocol that sits upon the IEEE 802.15.4 *Media Access Control* (MAC) and *Physical* (PHY) layers, which we'll discuss later on.

In this chapter, we discuss ZigBee's inception and evolution since making its first public appearance in 2005. What's more, we'll explore the ZigBee Alliance, along with its membership benefits and structure. The chapter also discusses the diverse market scope that ZigBee has already captured and is currently targeting, and explores the ZigBee product range and potential. Naturally, we'll also look at how ZigBee is set apart from other competing technologies within the same market sector. Finally, the chapter lifts the lid on ZigBee and takes a closer look at its software architecture and protocol.

Figure 8.1. The ZigBee trademark and logo. (Courtesy of the ZigBee Alliance.)

8.1 Overview

ZigBee technology is driven by a market need: a belief in building intelligence into everyday devices, where these devices are, if you like, offered a voice[1] – something which the ZigBee Alliance feels is often overlooked and has enormous market potential. Essentially, the objective of ZigBee technology is to empower devices such as light switches, thermostats, electricity meters, and other complex sensor devices typically located within residential, commercial, and industrial automation environments. The offer of a global standard-based wireless platform specifically architected toward remote monitoring and control applications, whilst ensuring simplicity, reliability, low power, and cost, is the intelligent concoction of ingredients that sets ZigBee apart from its significant competitors. With this in mind, the underlying networking topology, along with its security makeup and software protocol stack, ensure successful interoperability. Later on in this chapter, we'll explore in greater detail many of the technology's attributes.

8.1.1 The ZigBee Alliance

ZigBee technology and its associated Alliance (in Figure 8.1 and Figure 8.2, we provide the ZigBee logo and Alliance trademark, respectively) drive the future evolution and development of the protocol-based wireless standard. The Alliance, a non-profit organization founded in 2002, remains independent and vendor-neutral, which solidifies an holistic common objective across the corporation. Typically, the Alliance is responsible for public education, serving certification, and compliance programs, in addition to pursuing market development. The Alliance remains open to any participant and currently retains over 400 members across the Americas, Europe, and Asia.

As well as creating brand awareness, the ZigBee Alliance's primary focus is to build and promote the ZigBee brand across the globe.

Figure 8.2. The ZigBee Alliance trademark and logo. (Courtesy of the ZigBee Alliance.)

8.1.1.1 Membership and the ZigBee Alliance

In short, the Alliance offers potential participants three classes of membership, as shown in Table 8.1. The three classes of membership enable companies to decide on how they wish to participate within the Alliance. Like most standards development organizations, membership is structured, from *Adopter*, a basic level where the member can access completed specifications and standards before they're made public, through to *Promoter*,

[1] "ZigBee Overview," ZigBee Alliance, 2009.

Table 8.1. The ZigBee Alliance offers potential participants a three-tier membership structure to include Promoters, Participants, and Adopters

Membership level	Scope
Promoter	The highest level of membership allows its members to define and steer the Alliance along with other Promoter members. Furthermore, this membership level permits members to take board level seats and essentially have uninhibited access to "everything."
Participant	Members at this level have access to "everything," along with having early access to all Alliance activities pertaining to both specification and internal documentation. Primarily, this level has been created for companies wishing to contribute IP and to influence specifications and/or application standards.
Adopter	This basic membership level offers approved standards and other documentation once made public and access to rudimentary information. Other benefits include the use of the Alliance brand; other facets are available by special invitation only.

where board level membership entitles the member to participate in the future direction of the Alliance and the technology. Any company wishing to use ZigBee branding is obliged to join the Alliance prior to marketing or selling their products.

8.1.1.2 ZigBee Certified

The majority, if not all, of the technologies featured in this book require your product to be *certified*. The ZigBee Certified program offered by the ZigBee Alliance encourages interoperability of ZigBee products as well as ensuring the highest inherent quality. Products that demonstrate conformance with the ZigBee standard, along with successful interoperability, are awarded the "certified product" logo, as seen in Figure 8.3.

Figure 8.3. The ZigBee Certified Product logo is awarded to a ZigBee product that has successfully passed standard and interoperability testing and conformance.

Prequalification, the process required in order to achieve product certification, sometimes taking a day or less, could be achieved through successful interoperation at a ZigFest. Alternatively, test providers, such as TRaC (tracglobal.com), National Technical Systems (ntscorp.com), and TUV Rheinland (tuv.com), offer testing, regularity, and compliance schemes for ZigBee products and platforms.

8.1.2 ZigBee's Timeline

The infographic shown in Figure 8.4 provides a snapshot of ZigBee's technology timeline, from its inception to where the technology is today. ZigBee has matured over the

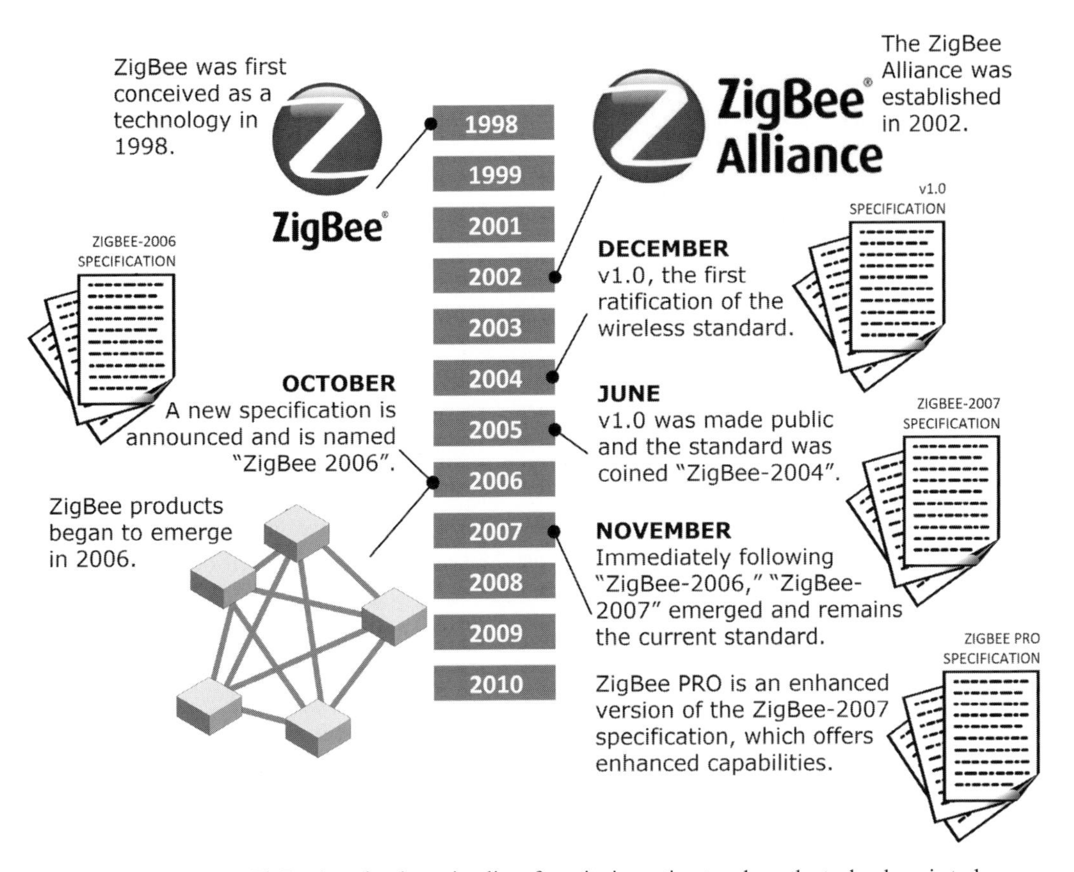

Figure 8.4. ZigBee's technology timeline, from its inception to where the technology is today, with its "ZigBee-2007" and "ZigBee PRO" specifications.

last decade or so, especially so from when it was first conceived as far back as 1998. In 2002, the ZigBee Alliance was established, and its first ratified v1.0 specification appeared in December 2004, followed by its introduction to the public in June 2005 – the standard was known as "ZigBee-2004." It wasn't until 2006 that the first generation of ZigBee products began to emerge. During the same year, in October, the "ZigBee-2006" specification emerged, immediately followed by "ZigBee-2007" in November 2007. The initial "ZigBee-2004" specification is now obsolete, but the current "ZigBee-2007" specification retains backward compatibility with "ZigBee-2006." ZigBee PRO is an enhanced version of the ZigBee-2007 specification that offers additional capabilities, such as increased network hops, increased devices within a ZigBee network, and improved security.

8.2 ZigBee's Market

A number of wireless technologies that are shared in Part III, "The Classic Personal Area Network," typically serve high-end, high data rate products best suited for

Figure 8.5. The Ember EM300 Series, the next generation ZigBee SoC family empowering the connected home including smart energy, home automation, security, and other monitoring and control applications. (Courtesy of Ember.)

audio-, voice-, and data-heavy-centric applications. ZigBee identified an opportunity within the market to provide a low cost, low energy control and sensor technology system. Such devices within the ZigBee ecosystem would require small or low data rates and have low latency, which, in turn, would aptly lend itself to longer battery life. As such, ZigBee aimed to become a global leader with its control and sensor networking standard, and was specifically architected to provide overall low power consumption, which was simple to deploy and implement. With a prerequisite of ensuring extended battery life firmly in place, ZigBee could empower its devices or *nodes* with an inherent consciousness about energy consumption. What's more, the end-cost to users was also a tangible and significant factor, as consumers didn't necessarily want to be burdened with high installation, maintenance, and energy costs.

Since ZigBee's inception, it has adapted both to market needs and to increasing competition within the *Wireless Sensor Networking* (WSN) market in providing a wealth of applications that enable it to provide a robust application-base. Its initial onset into the market sector was with *home automation* or in the *smart home*, with the technology targeted to smart lighting, advanced heating management, and security, to name a few. A wealth of products have emerged using the ZigBee standard, and in Figure 8.5 we illustrate Ember's (ember.com) EM300 series *System-on-Chip* (SoC), a new generation of ZigBee silicon specifically optimized for *Home Area Networks* (HANs) and *Smart Energy*. You may recall from Chapter 4, "Introducing the Lawnmower Man Effect," that we explored the *extended Personal Area Network* (PAN) with topologies such as HAN and the *Vehicle Area Network* (VAN). Moreover, ZigBee is a significant technology in supporting the *Internet of Things* (IoT), something we discussed earlier in Chapter 6, "Enabling the Internet of Things."

8.3 ZigBee's Application Standards Portfolio

Like most technologies within this book, ZigBee has existed for a decade and has matured to accommodate a diverse and competitive market sector – like other technologies in part of the book, energy, and its associated effective management to reduce consumption and, in turn, cost, have become increasing concerns for consumers and industries alike. The application standards discussed in the sections to follow demonstrate ZigBee's acumen and foresight within this competitive market. As such, the ZigBee Alliance has developed a number of standards, which individually target an area, that aim to satisfy consumers, businesses, industries, and governments. Overall, the greener view from the Alliance continues to help improve efficiency of homes and offices, as well as in a host of consumer electronic products.

8.3.1 Home Automation

The ZigBee *Home Automation*™ (HA) standard is provided to support an ecosystem within the home that ultimately empowers the home owner to achieve a greener, more eco-conscious, and cost-effective environment that is simple to maintain. As we have touched upon earlier, the escalating cost of energy has become a significant factor in managing energy consumption, and the ZigBee Home Automation standard affords consumers the ability to manage and maintain energy use effectively within the home. A number of manufacturers offer various ZigBee Certified products that enable the smart home, such that appliances (white goods), lighting, heating, security, and so on can all be managed through the range of interoperable products across the ZigBee mesh network.

Figure 8.6. The AlertMe Smart Energy kit comprising their SmartDisplay, SmartHub, and SmartMeter Reader. (Courtesy of AlertMe.)

In Figure 8.6, we illustrate one such smart energy solution offered by AlertMe (alertme.com), the *SmartEnergy* kit comprising their *SmartDisplay*, *SmartHub*, and *SmartMeter Reader*. The reader provides consumption statistics, pricing information, and so on, which can then also be viewed on a PC and/or smartphone. The SmartEnergy kit affords its users the ability to monitor, control, and predict their energy consumption within the home. Such products can also aid in the understanding of how and where energy is being consumed.

8.3.2 Building Automation

In a similar manner to home automation, the ZigBee Building Automation™ standard is provided to enable the monitoring and control of building environments. The ZigBee Alliance has partnered with BACnet[2] (bacnet.org); ZigBee technology is the only approved wireless standard within this group. Again, the emphasis of the standard is to manage energy consumption effectively and, as a consequence, to demonstrate eco-friendly awareness. ZigBee's partnership with BACnet will help to expand the number of buildings that can be maintained and monitored, whilst eliminating the need for a cabled system.

8.3.3 Smart Energy

With a green and eco-friendly stance, the ZigBee Smart Energy™ standard offers the ability for a host of products to "monitor, control, inform and automate the delivery of energy and water".[3] The current standard, v1.1, additionally offers an ability to provide pricing schemes and prepayment features, whilst maintaining backward compatibility with the initial v1.0 standard. The ZigBee Smart Energy ecosystem helps consumers, utilities, and governments to implement a smart grid solution, thereby empowering consumers to gain a better understanding of their consumption of utilities within the home or office.

8.3.4 Remote Control

In Chapter 11, "Introducing the Classic Personal Area Networking Technologies," we briefly touch upon how ZigBee and Bluetooth low energy, along with other wireless technologies, are looking to evolve the remote control device. *Infra-red* (IR) requires a line-of-sight operation, but the ZigBee Remote Control™ standard aims to offer a variety of consumer electronic devices (to include HDTV, home theatre systems, *Set-top-boxes* (STBs), and a range of audio equipment) a new sense of freedom by removing the need to point and control devices. The standard, created by the Alliance, has been designed for use with the ZigBee RF4CE specification, which we discuss in greater detail in Section 8.13, "ZigBee RF4CE."

[2] BACnet is a global standard, a protocol designed to manage heating and air-conditioning units effectively within commercial premises.

[3] Source: zigbee.org.

8.3.5 Input Device

Additionally designed for use with the ZigBee RF4CE specification, the ZigBee Alliance has provided the ZigBee Input Device™ standard, which promotes greener and energy efficient input devices for a host of consumer electronic products, to include keyboards, mice, touchpads, and so on. The standard encourages greater battery life within products that utilize the Input Device standard. Just like the Remote Control standard, these devices do not require line-of-sight operation and can be operated from greater distances.

8.3.6 3D Sync

The ZigBee 3D Sync™ standard fits within the ZigBee RF4CE specification, and it will enable active 3D glasses to interoperate with the existing remote controlled HDTV, STBs, Blu-ray players, gaming consoles, and other similar devices. The fundamental premise of this standard is to provide 3D glasses, which are commonly used with 3D-capable products and devices, to permit simplified connectivity scenarios from a number of video sources. It is anticipated that 3D Sync glasses will also intelligently switch between display contexts, for example between 2D and 3D displays. Moreover, as is common with the ZigBee standards featured in this chapter, products will benefit from a greener operation; this, in turn, aims to extend overall battery life when compared with IR technology.

8.3.7 Telecom Services

The ZigBee Telecom Services™ standard offered by the Alliance encroaches on other technologies within this market, such as *Near Field Communications* (NFC) and Bluetooth wireless technology. The standard leverages the mobile phone for a host of new services, to include location-based services, gaming, advertising, mobile payments, and data sharing, to name but a few. The ecosystem supported by this new standard will enable consumers to pay for products and services, create ad hoc networks for gaming, and receive discounts, offers, and coupons from participating retailers. Potentially, ZigBee's telecom ecosystem may lend itself well to allegiances with retailers through social media and may tie-in their customers through loyalty schemes.

8.3.8 Health Care

The healthcare industry is battling[4] with increased cost, which is further compounded by lack of resources, in turn compromising the healthcare needs of so many. The ZigBee Health Care™ standard provides the opportunity for many manufacturers to develop

[4] Gratton, D. A., "Keeping Mum: Transforming the Healthcare Industry *Wirelessly*," 2011.

key healthcare products that can aid in non-critical management and offer independence to the infirm. The standard does not aim to replace healthcare generally, but instead offers a service to those individuals that would normally be hospitalized or placed into a care home. The technology provides independence for the elderly, whilst alleviating the burdened healthcare system by allowing some people to remain within their home to receive essential, but non-critical, treatment. The ZigBee Alliance has partnered with and has been endorsed by the Continua Health Alliance (continuaalliance.org), a non-profit organization which collaborates with healthcare organizations and technology companies to improve the overall quality of healthcare, fitness, and wellness.

8.3.9 Retail Services

And lastly in this section, the ZigBee Retail Services™ standard offers an ecosystem with a range of interoperable products that will aid the retail sector to "monitor, control, and automate the purchase and delivery of goods."[5] A diverse ecosystem will offer both consumers and retailers handsets that will permit shoppers to "shop-pay-and-go," whilst taking advantage of in-store promotional and location-based services.

8.4 ZigBee and its Competitors

With many technologies spanning and vying for the same application space, ZigBee is up against some strong competition. Since ZigBee's inception, it has had moderate success and continues to endure a turbulent journey in establishing itself as a de facto low energy standard. What's more, the Bluetooth *Special Interest Group* (SIG) has essentially reinvented Bluetooth wireless technology with its introduction of the new Specification of the Bluetooth System: Core, high speed v4.0. The new specification has been coined *3-in-1*, a specification which includes *Bluetooth low energy* (BLE), along with the existing *classic* and *high speed* variants. We discuss Bluetooth low energy in Chapter 7, "Bluetooth low energy: The *Smart* Choice," and classic and high speed Bluetooth in Chapter 14, "Bluetooth Classic and High speed: More Than Cable Replacement."

8.4.1 Overlapping Technologies

The Bluetooth SIG has mustered sufficient momentum within the low energy industry to open up numerous prospective opportunities. As we have already touched upon in Chapter 5, "Introducing Low Power and Wireless Sensor Technologies," the ability to offer energy efficient technologies to empower intelligent *green* buildings has its

[5] Source: zigbee.org.

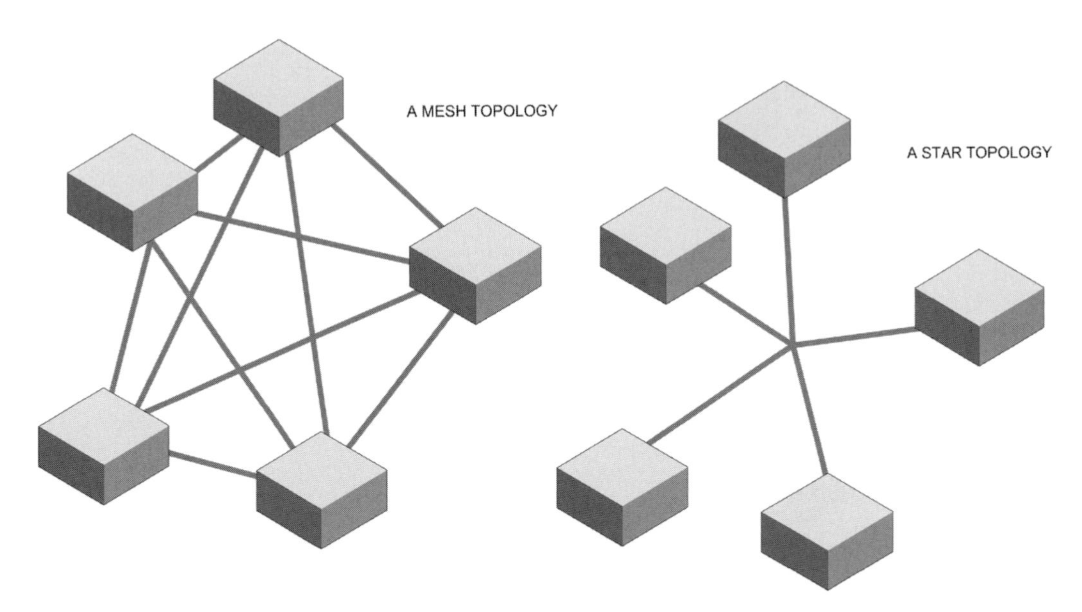

Figure 8.7. The ZigBee mesh and star topologies.

cost benefits, and ZigBee also faces competition from other technologies to include low power, that is, Wi-Fi, Z-Wave, and EnOcean, all of which target a similar, if not identical, application-base.

Table 8.2. ZigBee device types

Device	Description
Coordinator	The coordinator device initiates and maintains the ZigBee network.
Router	A router device has the responsibility of passing messages within the network.
End	An end device may be battery powered and communicates with the coordinator or router.

8.5 Networking Topology

ZigBee technology supports three topologies, namely *mesh*, *star*, and *tree*; these are illustrated in Figure 8.7 and Figure 8.8, respectively. Likewise, there are three types of ZigBee device, namely a *coordinator*, an *end device*, and a *router*, which are summarized in Table 8.2. A ZigBee network can operate in either a *beacon* or a *non-beacon* environment, although in practice non-beacon environments are used. A beacon environment is important to both a mesh and star topology, as it permits all devices within the network to remain synchronized in terms of when to communicate whilst conserving energy consumption. A beacon-enabled network is useful if all devices are battery-operated and need to optimize energy consumption in large networks. A non-beacon context

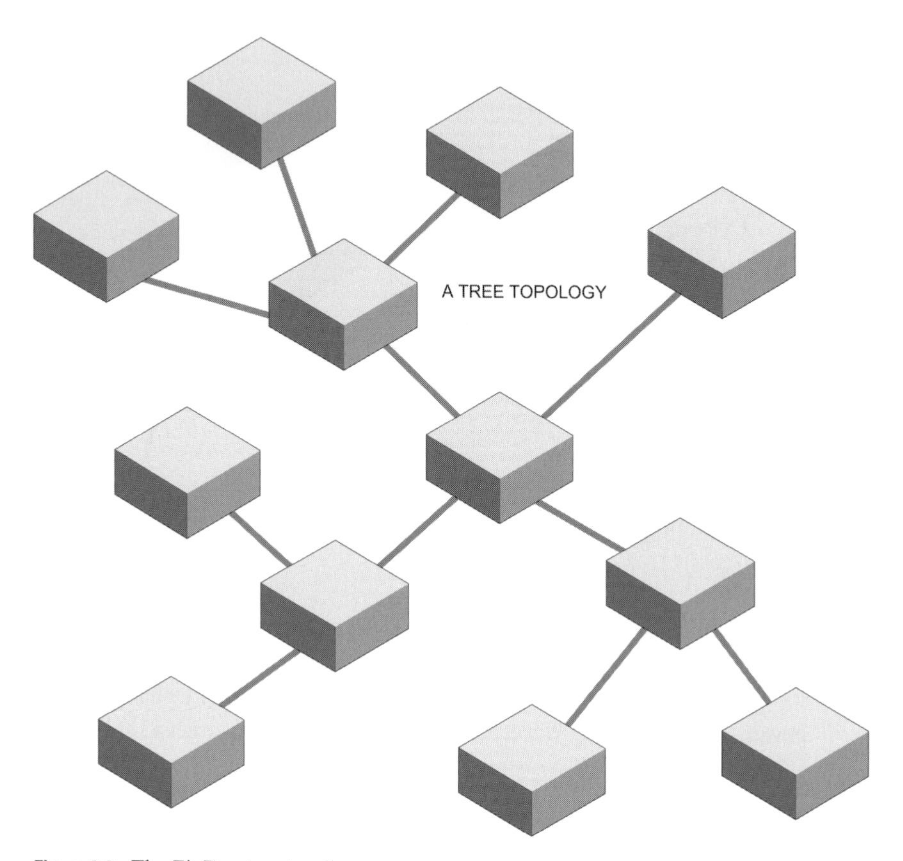

Figure 8.8. The ZigBee tree topology.

offers simplicity and enables ad hoc communication between devices, which may cause interference with other devices in proximity. Typically, non-beacon devices sleep for the majority of their life-cycle, only transmitting a notification to a coordinator device when a specific event, such as movement which has been detected by a motion sensor in a secured environment, occurs.

8.5.1 ZigBee Devices

So, a *coordinator* within a star topology has the responsibility to start and maintain the nodes within the network; likewise, a coordinator in a mesh and tree topology has the responsibility to initiate the network, but its responsibility is extended to the selection of significant network parameters and also retains the network's security keys in a *repository.* There is only one coordinator within a network, which is a *Full Functioning Device* (FFD) and can operate in any topology. A router device is also an FFD and has the ability to extend networks, such as a tree topology (as shown in Figure 8.8), and route messages (from other devices) throughout the network using a hierarchal or a peer-to-peer routing strategy. All other devices normally forming a ZigBee networking topology are the low powered end device, which communicates directly with the coordinator, or a

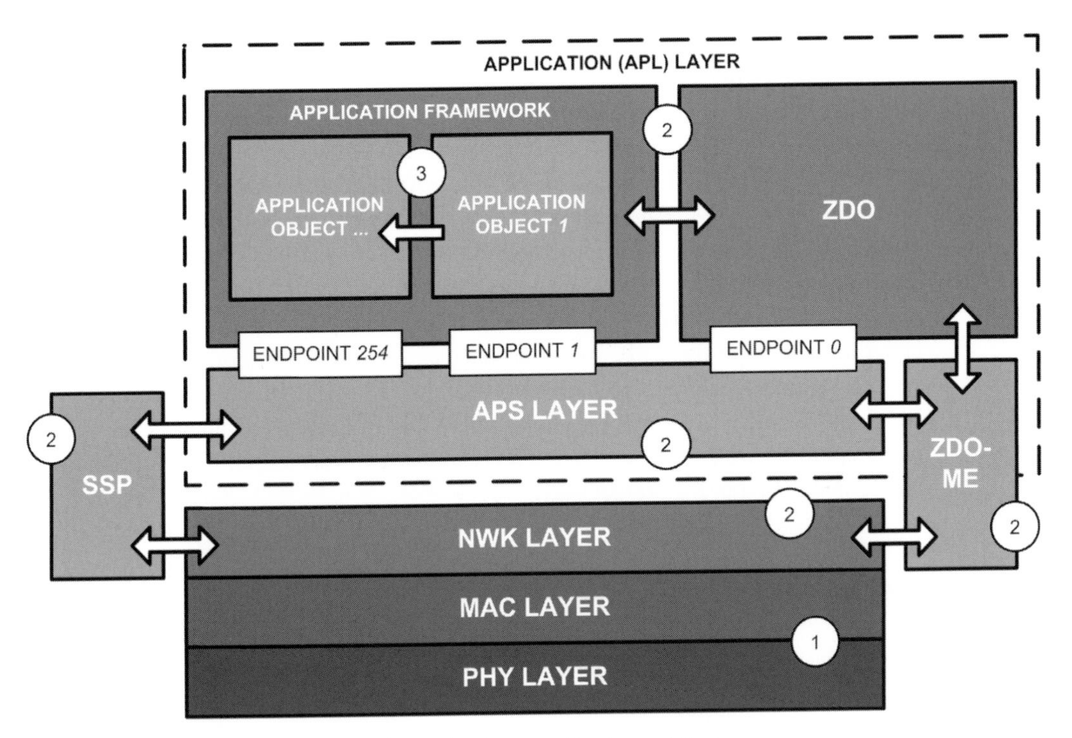

Figure 8.9. The building blocks that form the ZigBee architecture.

router device. The end device is optimized to run for many years using as little power as possible, but it can also become *orphaned* if it loses communication with the coordinator (or *parent*, as it sometimes called). The end device has the ability to scan for and join alternative networks and behaves as a *Reduced Function Device* (RFD).

8.6 The Application Layer

In the sections that follow, we take a top-down approach to the ZigBee stack architecture – essentially we take the technology from the top to the bottom and explore in greater detail the technology architecture and its software and application building blocks. In Figure 8.9, we provide the key building blocks of the ZigBee architecture – you will note that these building blocks have been labeled numerically to ease identification and classify responsibility. So, referring to Figure 8.9, blocks labeled (**1**) represent the MAC and PHY layers, which we will discuss later on in Section 8.12, "MAC and PHY Layers"; those labeled (**2**) represent the building blocks that are ZigBee Alliance responsible – in essence this is the ZigBee protocol standard; and finally those labeled (**3**) represent building blocks that are manufacturer-specific, namely the end-applications.

The ZigBee stack architecture provides a number of blocks, which are layered. Looking at it from the top down, the *Application* (APL) layer retains the *Application Framework*, along with the *ZigBee Device Object* (ZDO). Within this application framework, *Application Objects* (or ZigBee profiles/standards) with their respective endpoints are included – as we mentioned, these are manufacturer-specific, but ZigBee also provide common profiles (a *cluster library*) that are also supported (more about this later). Further down, we arrive at the *Application Support Sub-layer* (APS), which provides an interface between the APL and the ZigBee *Network* (NWK), namely the APS *Data Entity* (APSDE) via the APSDE *Service Access Point* (APSDE-SAP), but we will discuss this further in Section 8.9, "Application Support Sub-layer." You will also notice that the *Service Security Provider* (SSP) and the ZDO *Management Entity* (ME) or *management plane* are present; these are also discussed later in Section 8.9. In short, a data entity provides a service for data transmission; a service entity provides an interface to the upper layer through the SAP; and a management entity provides other miscellaneous services. Finally in the stack architecture, we arrive at the ZigBee *Network* (NWK) layer, which sits upon the MAC and PHY layers from the IEEE 802.15.4 air-interface – these building blocks are discussed in Section 8.10, "The Network Layer," and Section 8.12, "MAC and PHY Layers," respectively.

8.7 Application Framework

The application framework is a conceptual container for ZigBee devices which houses application objects or profiles. In the following sections, we discuss further the makeup of application profiles, along with their endpoints and device descriptor types.

8.7.1 Application Profiles

Just like the Bluetooth profiles, the *application profiles* for ZigBee within the application framework prescribe behavior in terms of how an application should interact with its peer. The ZigBee Alliance describes profiles as an "agreement" as to how messages and their formats, combined with their behavior, enable developers to create interoperable applications that occupy separate devices. In turn, this establishes a common way of working, enabling developers to build interoperable products confidently. In essence, data can be retrieved from another device, messages are exchanged and understood, and the processing of commands and requests can be executed accordingly. Furthermore, other behavior mechanisms that are common to all profiles, such as service and device discovery, are also needed, as this ability empowers devices to connect autonomously, to form networks, and to learn more about other devices' attributes and features. The ZigBee Alliance has authored a number of profiles, for example *Home Automation* and *Smart Energy*, which we discussed earlier in Section 8.3.1, "Home Automation," and Section 8.3.3, "Smart Energy."

The ZigBee Alliance is responsible for issuing *profile identifiers*, which are provided for two types of device classes, namely *manufacturer-specific* and *public* – the Alliance assists with establishing criteria as to how profiles should be defined. Profile identifiers are unique, and once obtained they permit the developer to define further the *device descriptors* (more about this in Section 8.7.3, "Device Descriptors") and cluster identifiers. What's more, the profile identifier is the primary enumerated feature within the ZigBee protocol. So, a profile identifier of 1 has a device descriptor with a 16-bit value,

Table 8.3. ZigBee device descriptors

Name	Description
Node	Defines the type and capabilities of a node device and is mandatory.
Node power	Defines node power characteristics and is mandatory.
Simple	Device descriptor contained within each node and is also mandatory.
Complex	Additional information relating to device descriptor and is optional.
User	User-specific descriptor and is optional.

which permits a possible 65 536 device descriptors from one profile identifier. Likewise, a cluster also has a 16-bit value, which allows a possible 65 536 cluster identifiers where, in turn, each cluster identifier has a possible 65 536 attribute identifiers.

It is the responsibility of the profile developer to define and allocate device descriptors and cluster and attribute identifiers; however, with public profiles, a cluster library is provided. A cluster library offers developers some common definitions for and attributes of cluster identifiers, in turn encouraging application (profile) recycling. Furthermore, manufacturers may offer extensions to public profiles. In other words, a manufacturer can provide specific extensions to public profile **A**, for example, which can then be added to their own implementation of that profile. Naturally, devices that do not support these manufacturer extensions (public profile **A**) would only advertise services specific to their core features.

8.7.2 Endpoints

A maximum of `0xFE` individual objects can be stored within the application framework, each with its own unique *endpoint* identifier – each object is labeled uniquely from `0x01` to `0xFE`. However, endpoints `0xF1` to `0xFE` can only be used with the approval received from the ZigBee Alliance. Furthermore, note that two endpoints are reserved, namely endpoint `0x00`, for the interface with ZDO, and `0xFF`, which is reserved whilst broadcasting data to all application objects within the application framework.

8.7.3 Device Descriptors

An application profile will contain a collection of device descriptors that, when combined, form the intended application. For example, a thermostat on one node or device will communicate with a heating system on another node that, in turn, forms the heating application profile. The unique identifier retained by the device descriptor is used as part of the discovery procedure. As we touched upon earlier, a profile identifier has a device descriptor with a 16-bit value, which permits a possible 65 536 device descriptors from one profile identifier. Our ZigBee devices, which we discussed earlier in Section 8.5.1, "ZigBee Devices," each contain a data structure held within the device descriptors themselves. In total there are five descriptors, namely *node*, *node power*, *simple*, *complex*, and *user*, which are summarized in Table 8.3. Table 8.3 also presents the actual transmission order: the node (the first item in the table) is transmitted first and the last field, at the bottom of the table (user), is transmitted last. The node, node power,

Table 8.4. The fields that make up the node descriptor

Name	Description	Size
Logical type	This 3-bit field specifies the device type of the ZigBee node, as listed in Table 8.5.	3
Complex descriptor available	If this bit is set to 1, then a complex descriptor is present on the device (if it is set to 0, then no descriptor is present).	1
User descriptor available	If this bit is set to 1, then a user descriptor is present on the device (if it is set to 0, then no descriptor is present).	1
Reserved	Currently unused.	3
APS flags	Normally this field specifies the capabilities of the APS layer, but it's not used and should be set to 0.	3
Frequency band	This field specifies what frequency bands are supported by the IEEE 802.15.4 radio, and these are listed in Table 8.6.	5
MAC capability flags	This field indicates the capabilities, as mandated by IEEE 802.15.4-MAC sub-layer, and we list these flags in Table 8.7.	8
Manufacturer code	This 16-bit manufacturer code field is allocated by the ZigBee Alliance.	16
Maximum buffer size	This 8-bit field, with a range of `0x00` to `0x7F`, specifies the maximum size in octets of the *Network Sub-layer Data Unit* (NSDU) for the node.	8
Maximum incoming transfer size	This 16-bit field, with a range of `0x0000` to `0x7FFF`, specifies the maximum size in octets of the *Application Sub-layer Data Unit* (NSDU) for the node.	16
Server mask	The system server capabilities are defined and used to discover services of other nodes within the network. The 16-bit field has settings which are listed in Table 8.8.	16
Maximum outgoing transfer size	This 16-bit field, with a range of `0x0000` to `0x7FFF`, specifies the maximum size in octets of the *Application Sub-layer Data Unit* (NSDU) for the node.	16
Descriptor capability field	This 8-bit field is used to discover capabilities of features of other devices in the network.	8

Table 8.5. The logical type field values

Value	Description
`0x00`	Coordinator.
`0x01`	Router.
`0x02`	End device.

and simple descriptors are mandatory features within the descriptor, whilst the complex and user descriptors are optional.

8.7.3.1 Node Descriptor

The node descriptor is mandatory and contains information relating to the capability of the node; only one descriptor is contained within a node. The fields that form this descriptor are listed in Table 8.4, along with their descriptions and sizes.

Table 8.6. The frequency band field values

Bit	Description
0	868–868.6 MHz
1	Reserved.
2	902–928 MHz
3	2400–2483.5 MHz
4	Reserved.

Table 8.7. The MAC capability flag field

Bit	Description
0	Alternative PAN coordinator is set to 1 if the node is capable of becoming a PAN coordinator.
1	The device type flag is set to 1 if the node is FFD; otherwise it's set to 0 if the node is a RFD.
2	The power source field is set to 1 if the node power source is mains powered.
3	The receiver when idle field is set to 1 if the device is not capable of disabling its receiver during idle periods.
4	Reserved.
5	Reserved.
6	The security capable field is set to 1 if the node can send and receive data securely.
7	The allocate address field is set to either 0 or 1.

Table 8.8. The server mask bit assignments

Bit	Description
0	Primary trust center.
1	Backup trust center.
2	Primary binding table cache.
3	Backup binding table cache.
4	Primary discovery cache.
5	Backup discovery cache.
6	Network manager.
7–15	Reserved.

8.7.3.2 Node Power Descriptors

The node power descriptor is mandatory and provides dynamic information relating to the power status of the device; only one descriptor is contained within a node. The fields that form this descriptor are listed in Table 8.9, along with their descriptions and sizes.

8.7.3.3 Simple Descriptors

The simple descriptor is mandatory and provides information relating to each endpoint contained in the device; only one descriptor is contained within a node. The fields that form this descriptor are listed in Table 8.13, along with their descriptions and sizes.

Table 8.9. The fields that make up the node power descriptor

Name	Description	Size
Current power mode	The current power mode 4-bit field values specify the current sleep/power-saving mode of the device and are listed in Table 8.10.	4
Available power sources	The available power sources 4-bit field values specify the power source available on this node and are listed in Table 8.11.	4
Current power source	The current power source 4-bit field values specify the power source being used on this node and are listed in Table 8.11.	4
Current power source level	The current power source level 4-bit field values specify the level of charge on this node and are listed in Table 8.12.	4

Table 8.10. The current power mode field values

Bit	Description
0	Receiver synchronized with receiver on when idle.
1	Receiver comes on periodically, as specified by the node power descriptor.
2	Receiver turns on when stimulated.
3–15	Reserved.

Table 8.11. The available and current power sources' field values

Bit	Description
0	Constant (mains) powered.
1	Rechargeable battery source.
2	Disposable battery source.
3	Reserved.

Table 8.12. The current power sources' level field values

Value	Description
`0x00`	Critical.
`0x04`	33%
`0x08`	66%
`0x0C`	100%

8.7.3.4 Complex Descriptors

The complex descriptor is optional and provides additional information for each of the device descriptors contained in the device; only one descriptor is contained within a node. The fields that form this descriptor are listed in Table 8.14, along with their descriptions and sizes. The format of the complex descriptor is presented in an XML form using compressed XML tags.

Table 8.13. The fields that make up the simple descriptor

Name	Description	Size
Endpoint	The 8-bit endpoint descriptor field specifies the endpoint within the node. Endpoints from `0x01` to `0xFE` can be used (see Section 8.7.2, "Endpoints"), but the use of endpoints `0xF1` to `0xFE` can only be approved by the ZigBee Alliance.	8
Application profile identifier	Profile identifiers are provided by the ZigBee Alliance; this 16-bit field specifies which profile is supported on this endpoint.	16
Application device identifier	Device description identifiers are provided by the ZigBee Alliance; this 16-bit field specifies which device description is supported on this endpoint.	16
Application device version	A 4-bit field which specifies the version of the device descriptor supported by the endpoint.	4
Reserved	Currently unused.	4
Application input cluster count	This 8-bit field specifies the number of input clusters supported by this endpoint.	8
Application input cluster list	This field specifies the number of list of input clusters supported by this endpoint.	$16 \times x$[a]
Application output cluster count	This 8-bit field specifies the number of output clusters supported by this endpoint.	8
Application output cluster list	This field specifies the number of list of output clusters supported by this endpoint.	$16 \times y$[b]

[a] x represents the value of the application *input* cluster count.
[b] y represents the value of the application *output* cluster count.

Table 8.14. The fields that make up the complex descriptor

Name	Description	Value
Reserved	Currently unused.	`0x00`
Language and character set	This 3-octet field is used to specify the language and character set used by the strings in the complex descriptor.	`0x01`
Manufacturer name	The name of the manufacturer in string format.	`0x02`
Model name	The name of the model in string format.	`0x03`
Serial number	The manufacturer's serial number in string format.	`0x04`
Device URL	A URL where more information about the device can be located.	`0x05`
Icon	The icon can be used if the device is being accessed on a computer, gateway, or tablet device.	`0x06`
Icon URL	A URL where more information about the icon can be located.	`0x07`
Reserved	Currently unused.	`0x08` – `0xFF`

8.7.3.5 User Descriptor

The user descriptor is optional and contains information that would enable a user to recognize the device. A user-friendly name (character string) can be used to identify the device, for example "Lounge TV" or "Living room light." The user descriptor contains one field and is described in Table 8.15.

Table 8.15. The field used in the user descriptor

Name	Description	Length
User description	ASCII character set with a maximum of 16 octets.	16

8.8 ZigBee Device Objects

The APL provides generic functionality for such activities as *binding*, *device*, and *service discovery*, which are all undertaken by the ZDO, as an application. ZDOs utilize APS primitives to implement coordinators, routers, and end devices, whereas common behavior and functionality are provided by the ZigBee *Device Profile* (ZDP), which we discuss in greater detail in Section 8.8.1, "Device Profile." The ZDO is conceptually located between the application framework and APS (see Figure 8.9). It also has the responsibility of instantiating the APS, NWK, and SSP, as well as deriving essential configuration data applications to form a better understanding of discovery, security, network, and binding management.

8.8.1 Device Profile

In essence, the ZigBee Device Profile behaves like any other ZigBee profile using clusters, as discussed earlier in Section 8.7.1, "Application Profiles." However, clusters used within the device profile differ from clusters in application profiles in that the clusters defined here are used and supported across all ZigBee devices to provide common functionality. The profile also utilizes one *device description*. We discuss the device profile topology, along with the transmission of ZDP commands and its format, in Section 8.8.1.1, "The Device Profile Topology." The primary functions supported by the device profile are *device* (Section 8.8.1.2, "Device Discovery") and *service discovery* (Section 8.8.1.3, "Service Discovery"), *end device binding* (Section 8.8.1.4, "End Device Binding"), *bind/unbind* (Section 8.8.1.5, "Bind/Unbind"), *binding table management* (Section 8.8.1.6, "Binding Table Management"), and *network management* (Section 8.8.1.7, "Network Management").

8.8.1.1 The Device Profile Topology

The topology supported within the device profile is a client/server relationship. So, when a device issues a request via the *device profile* message, such as a device or service discovery, or binding or network management, then it takes on the role of a *client*. Naturally, the device servicing such requests has the role of *server*, although any device can take on the role of client or server. All commands, which are transmitted via the APS data service, are formatted using the following frame structure (as illustrated in Figure 8.10). The transaction *sequence number* field is 8-bits in length and carries the identification number for the ZDP transaction. The sequence number affords the transaction a correlation between a command response frame and a request frame. What's more, if the controlling device has issued a single or several commands, the sequence number(s) allow the device to match incoming response(s) to the original

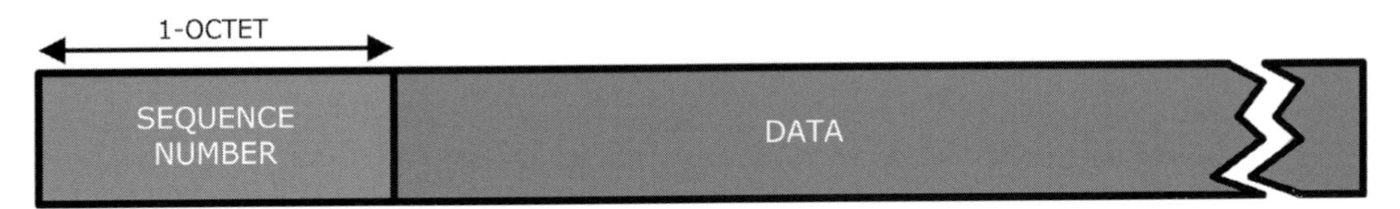

Figure 8.10. The ZDP frame format and structure.

command(s) request(s). The sequence number starts at `0x00`, but once `0xFF` is reached it is reset to `0x00`. The application object likewise maintains an 8-bit reference, which is copied into the transaction sequence number and incremented for each command transmitted. Lastly, the data field, which is of unspecified length since this is specific to the type of command being sent, will contain data for the actual ZDP transaction.

Table 8.16. How messages are used during device discovery

Message	Description
Broadcast	All devices should respond. The coordinator, router, and their associated devices will respond with their address, along with addresses of their associated devices. End devices, on the other hand, will simply respond with their address.
Unicast	Only the device specifically addressed should respond. So, a coordinator and router device will both respond with their addresses, along with addresses of their associated devices, which helps determine the topology in use. However, an end device will simply respond with its address.

8.8.1.2 Device Discovery

An individual device with "device address of interest" field enabled or a designated cached device may respond to a *discovery* request, although, in instances where both types of devices respond, the device with the "device address of interest" field enabled will take precedence. Device discovery ultimately ascertains the identity of other devices within the PAN, and in Table 8.16 we look at how device discovery messages are used during broadcast and unicast addressing.

8.8.1.3 Service Discovery

In a similar manner to device discovery, an individual device with "device address of interest" field enabled or a designated cached device may respond to a *service* request, although, in instances where both types of devices respond, the device with the "device address of interest" field enabled will take precedence. Service discovery ultimately ascertains the services offered by other devices within the PAN, and in Table 8.17 we look at how service discovery messages are used during broadcast and unicast addressing; *query types* are listed in Table 8.18.

8.8.1.4 End Device Binding

Within the ZigBee protocol, *binding* is the mechanism which permits the creation of a *uni-directional logical link* between a *source* endpoint or cluster identifier and a *destination* endpoint, which may be present on one or more devices. The end device

Table 8.17. How messages are used during service discovery

Message	Description
Broadcast	In an attempt to minimize the amount of data being returned during a broadcast, the individual or the discovery cache should respond with the relevant criteria.
Unicast	Only the device specifically addressed should respond. A coordinator and router device will cache service discovery criteria for their associated devices.

Table 8.18. Service discovery supported query types

Query type	Description
Active endpoint	The active endpoint, a device with an application, which is described by a simple descriptor, enables an enquiring device to ascertain active endpoints.
Match simple descriptor	This command can be broadcasted. It allows an enquiring device to supply a profile ID and, optionally, lists of input/output cluster IDs. Moreover, the command acquires the identity of an endpoint on the destination device which matches the criteria.
Simple descriptor	This unicast addressed command provides an enquiring device with information relating to the simple descriptor of the endpoint.
Node descriptor	This unicast addressed command provides an enquiring device with information relating to the node descriptor of the device.
Power descriptor	This unicast addressed command provides an enquiring device with information relating to the power descriptor of the device.
Complex descriptor	This unicast addressed command provides an enquiring device with information relating to the complex descriptor of the device.
User descriptor	This unicast addressed command provides an enquiring device with information relating to the user descriptor of the device.

binding feature offers a simpler method of identifying command/control pair devices when user interaction is required. In one such example, a user may use a two-button combination to complete an installation procedure; repeating the same procedure would unbind the devices.

8.8.1.5 Bind/Unbind

A *binding table* is created when cluster identifiers are designated as input/output in the simple descriptor. So, the ability to *bind* provides an entry into the table that corresponds to control messages and their destination, whereas the ability to *unbind* offers the opportunity to remove the entry from the binding table.

8.8.1.6 Binding Table Management

We provide in Table 8.19 the list of features available for the management of binding tables.

Table 8.19. Features available for binding table management

Feature	Description
Registration	A source device can notify its primary binding table cache to retain its own binding table.
Replacement	To replace one device with another (all instances).
Backup (single entry)	Offers backup of newly created single entries.
Removal (single entry)	Offers the ability to remove a single entry from the binding table once an unbind request has been received.
Backup (entire table)	Offers an ability to request a backup of an entire binding table using the backup binding table cache.
Restore (entire table)	Offers an ability to request a restoration of an entire binding table using the backup binding table cache.
Backup (primary binding table cache)	Offers an ability to request a backup of its entire source devices address table.
Restore (primary binding table cache)	Offers an ability to request a restoration of its entire source devices address table.

8.8.1.7 Network Management

A number of features are provided for effective network management, one of which is the ability to retrieve management information (including discovery results and cache content, binding and routing table contents, link quality, and energy detection results) from devices. A number of features are also available to define management control criteria, which include network "leave" and "join," allow joining, network fault, and updates notifications.

8.8.2 Device Objects

As we mentioned earlier in Section 8.8, "ZigBee Device Objects," ZDOs are applications that are conceptually located within the APL and sit above the APS, as illustrated in Figure 8.9. The ZDO has several responsibilities, which include initializing the APS, NWK, SSP, and any other layer, with the exception of end applications that occupy endpoints `0x01` to `0xFE`. Furthermore, we provide in Table 8.20 a summarized list of the six available device objects, along with their responsibilities that are undertaken by the ZDO. You may also recall from our earlier discussion that endpoint `0x00` is reserved. In fact, ZDO utilizes endpoint `0x00` using the APSDE-SAP, just like all the other profiles, for both the transmission and receipt of source and destination request frames and response frames, which, incidentally, are not fragmented. The ZDO interfaces with the APS using the APSDE-SAP as well as the APSME-SAP, and we discuss this now in Section 8.9, "Application Support Sub-layer."

8.9 Application Support Sub-layer

Following our discussion surrounding the manufacturer-specific application objects and the ZDO, we now look more closely at the interface and the services provided by the

Table 8.20. The six ZDO device objects

Object	Description
Device and service discovery	This object offers device and service discovery for all device types within the PAN, and is mandatory.
Security manager	If security is enabled, then the security manager will take responsibility for establish, transport, request, and switch keys and, update, remove, and authenticate devices. The security object is optional.
Network manager	The network manager has the responsibility of defining the logical device types for the coordinator and router end devices, and is mandatory.
Binding manager	This processes requests for adding/removing into/from the APS binding table and supports bind/unbind commands. The binding object is optional.
Node manager	Only applicable to coordinator and router devices, the node manager provides network discovery, leave the network, retrieve the routing or binding table, and permits joining or leaving particular routers or the Trust Centre. The node object is optional.
Group manager	Offers the ability to include or remove application objects with the local device into/from groups, and is optional.

APS. The APS provides the interface for, and sits between, the application layer and the NWK layer (see Figure 8.9), and provides a number of services for use by both application objects and the ZDO; it also provides the interface between a *Next Higher Layer Entity* (NHLE) and the NWK layer. In Figure 8.11, we provide a reference model which depicts the conceptual building blocks and interfaces of the APS.

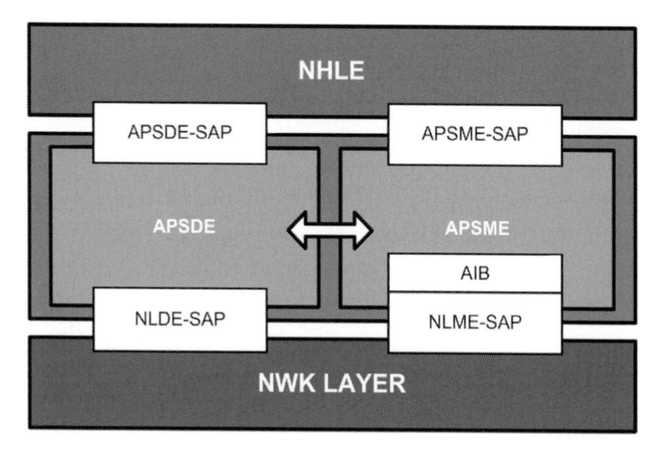

Figure 8.11. A reference model illustrating the conceptual building blocks and interfaces surrounding the APS.

Earlier, in Section 8.6, "The Application Layer," we touched upon the data entity (APSDE) and management entity (APSME) of the APS. The APSDE provides the data transmission service through the relevant SAP, namely the APSDE-SAP, while the APSME offers management services also through its associated SAP, that is, the APSME-SAP. What's more, the APSME maintains a database of managed objects,

Table 8.21. Services provided by the APSDE

Service	Description
PDU generation	An *APS PDU* (APDU) is generated, along with the relevant protocol overhead.
Binding	When two devices are bound, the APSDE can exchange messages between them.
Group address filtering	Filtering of group-addressed messages, which are based on endpoint group membership.
Reliable transport	Offering increased reliability of transactions.
Duplicate rejection	Limiting message transmission to being received only once.
Fragmentation	When messages exceed the NWK layer frame, then segmentation and reassembly is provided by this service.

which is referred to as the *APS Information Base* (AIB); Figure 8.11 shows the AIB in context. The APSDE offers a data service to the NWK layer, as well as to application objects and the ZDO, which supports the transportation of *Protocol Data Units* (PDUs) over the same network for a number of ZigBee devices; Table 8.21 lists the services provided by the APSDE. On the other hand, in Table 8.22, we list the services provided by the APSME, which, in turn, enables an application to interact with the ZigBee stack whilst having the capability of matching two devices based on their needs and services. This particular service is called the *binding service*, and the APSME offers the ability to construct and maintain a table to retain such information.

Table 8.22. Services provided by the APSME

Service	Description
Binding management	As already mentioned, the APSME has the ability to match two devices based on their needs and services.
AIB management	Defines and retrieves attributes within the AIB.
Security	Using secure keys, authenticated relationships can be created with other devices.
Group management	The declaration of a single address, which is shared by multiple devices, wherein such devices can be included or removed from a group.

8.9.1 APS Service Primitives

As we have mentioned already, the APS provides two services, both of which are offered through their respective service access points (APSDE-SAP and APSME-SAP) and are the supported interface between the NHLE and NWK layer; we will discuss the network interface later in Section 8.10, "The Network Layer." In the sections to follow, we discuss in more detail the APSDE-SAP primitives that are supported by the data entity, as well as the primitives supported by the APSME-SAP.

The APSDE-SAP offers the ability to transport APDUs between application peers, and we list these specific primitives in Table 8.23 and provide a message sequence chart in Figure 8.12 to illustrate how they might be used. In this particular

Table 8.23. APSDE-SAP primitives

Primitive	Behavior
`APSDE-DATA.request`	This primitive provides a transfer of an NHLE PDU from the local NHLE to one or more peers when requested.
`APSDE-DATA.confirm`	This primitive responds to and provides the results of the `APSDE-DATA.request` primitive.
`APSDE-DATA.indication`	This primitive acknowledges the transfer of a PDU from the APS to the local entity.

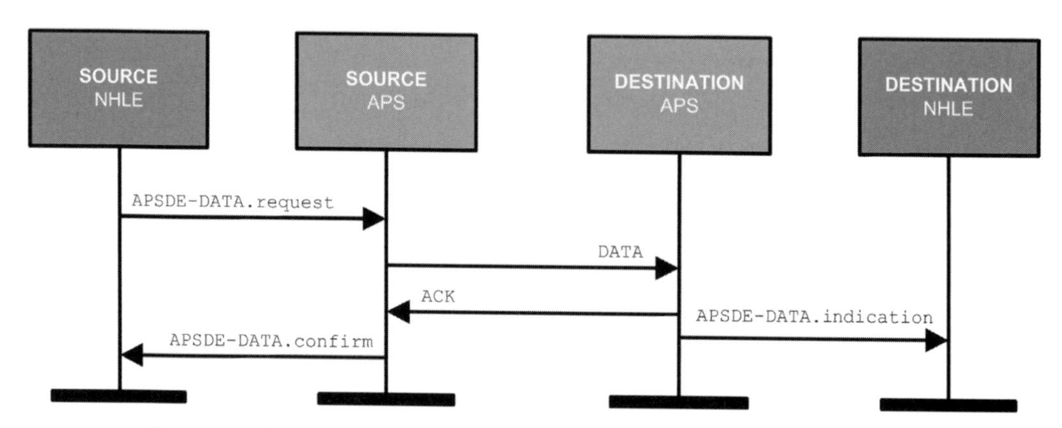

Figure 8.12. A message sequence chart showing how the APSDE-SAP primitives may be used.

illustration, an *Acknowledgement* (ACK) is provided, although this is optional. However, if an acknowledgement is not requested, then the source device assumes that the data frame transmission was successful.

The APSME-SAP provides a number of primitives to enable transfer of management commands between the NHLE and APSME. In particular, there are (un)bind and AIB and group management primitives. The (un)bind primitives, namely `APSME-BIND` and `APSME-UNBIND`, determine how the NHLE can include/exclude a binding record into/from its binding table, respectively, as we summarize in Table 8.24.

Table 8.24. APSME-SAP (un)bind primitives

Primitive	Behavior
`APSME-BIND.request`	Provides the NHLE with the ability to bind two devices or to bind a device to a group.
`APSME-BIND.confirm`	This primitive responds to and provides the results of the `APSME-BIND.request` primitive.
`APSME-UNBIND.request`	Provides the NHLE with the ability to unbind two devices or to unbind a device to a group.
`APSME-UNBIND.confirm`	This primitive responds to and provides the results of the `APSME-UNBIND.request` primitive.

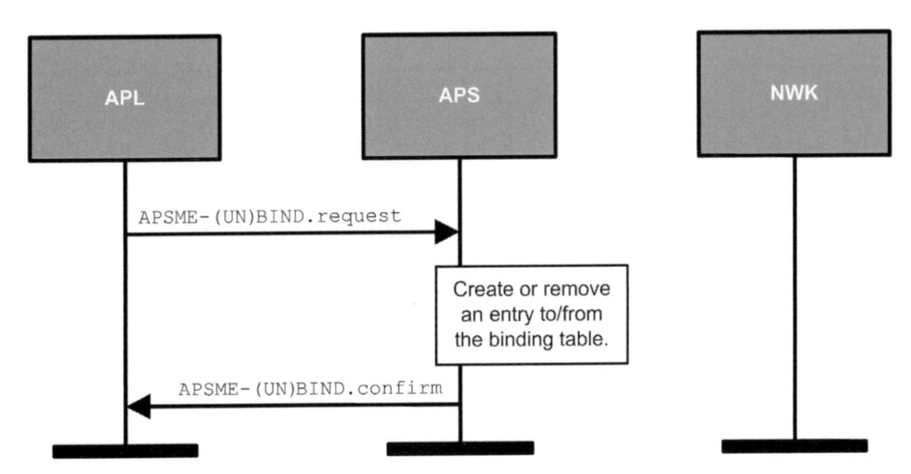

Figure 8.13. A message sequence chart depicting the possible activities of both the bind and unbind primitives.

So, the `APSME-BIND.request` and `APSME-UNBIND.request` are responsible either for creating a binding link or for removing it from a device; in Figure 8.13, we provide a message sequence chart which alludes to the typical activity for all four primitives, as outlined in Table 8.24.

Table 8.25. APSME-SAP AIB management primitives

Primitive	Behavior
`APSME-GET.request`	This primitive is issued to the APSME to allow the NHLE to read an attribute from the AIB.
`APSME-GET.confirm`	This primitive responds to and provides the results of the `APSME-GET`.request primitive.
`APSME-SET.request`	This primitive is issued to the APSME to allow the NHLE to write an attribute to the AIB.
`APSME-SET.confirm`	This primitive responds to and provides the results of the `APSME-SET.request primitive.`

The primitives listed in Table 8.25 provide the NHLE with the ability to read and write attributes from the AIB – these primitives form part of the APSME-SAP services. Likewise, the primitives illustrated in Table 8.26 offer the NHLE the ability to manage, in terms of adding or removing entries to/from group memberships for endpoints on the current device that support a binding table.

8.9.2 APS Frame Formats

We have discussed how APDUs are transmitted and received across the APS using the various services and associated primitives to action these exchanges. So, in this section, we take a closer look at how APS frames or payloads are formatted. Typically, an APS frame comprises an APS header and payload, a sequence of fields that are provided in a

Table 8.26. APSME-SAP group management primitives

Primitive	Behavior
`APSME-ADD-GROUP.request`	Provides the NHLE with the ability to add an endpoint to a particular group.
`APSME-ADD-GROUP.confirm`	This primitive responds to and provides the results of the `APSME-ADD-GROUP.request` primitive.
`APSME-REMOVE-GROUP.request`	Provides the NHLE with the ability to remove an endpoint from a particular group.
`APSME-REMOVE-GROUP.confirm`	This primitive responds to and provides the results of the `APSME-REMOVE-GROUP.request` primitive.
`APSME-REMOVE-ALL-GROUPS.request`	Provides the NHLE with the ability to remove memberships in all groups from an endpoint.
`APSME-REMOVE-ALL-GROUPS.confirm`	This primitive responds to and provides the results of the `APSME-REMOVE-ALL-GROUPS.request` primitive.

specific transmission order. It is worth noting that all reserved fields are set to 0, but if these fields are unequal to zero, then the frame is discarded.

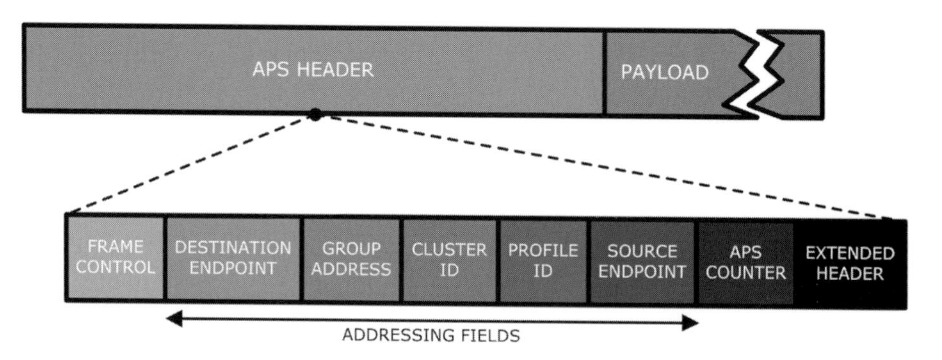

Figure 8.14. The general APDU frame format showing the makeup of the APS header.

8.9.2.1 General APDU Format

In Figure 8.14, we illustrate the generic APDU. As you can see, the structure comprises the APS header and payload where this appears in a specific sequence; however, the addressing fields may not necessarily be present in all frames. Of course, the APS (frame) payload is of variable length and will ultimately contain information relevant to the type of frame being transmitted. In Table 8.27, we look at the specific fields that make up the APS header, along with size of the field in bits.

8.9.2.2 Data, APS Command, and Acknowledgement Frames

In Section 8.9.2.1, "General APDU Format," we looked at the generic frame format for APS traffic; in this section, we review the specific types for *data*, *APS command*, and *acknowledgement frames*. The data frame abides to the generic APDU frame format, as previously illustrated in Figure 8.14, whilst adhering to the same transmission order.

Table 8.27. Fields that make up the APS header

Field	Description	Size
Frame control	The frame control field is 8-bits in length and is further broken down in Figure 8.15 and Table 8.28.	8
Destination endpoint	If the delivery mode is set to 0 (unicast, see Table 8.30) then the destination endpoint offers the endpoint that will receive the frame. What's more, a destination endpoint of `0x01` to `0xFE` addresses the frame to a specific application on that endpoint, whereas, if the destination endpoint is `0x00`, then the frame is destined for the ZDO.	8
Group address	If the delivery mode is set to `0x03` (broadcast, see Table 8.30) then this 16-bit field will contain a group address field to address all endpoints within a specific group.	16
Cluster ID	This 16-bit field is only present for data and acknowledgement frames, and provides the cluster ID of the frame source.	16
Profile ID	This 16-bit field is only present for data and acknowledgement frames, and provides the profile ID of the frame destination.	16
Source endpoint	The source endpoint 8-bit field specifies the originator of the frame. The value `0x00` would indicate that the originator is ZDO, whereas a value between `0x01` and `0xFE` would indicate that the frame originated from an application endpoint.	8
APS counter	This 8-bit value is incremented for every new transmission, as discussed in Section 8.8.1.1, "The Device Profile Topology."	8
Extended header	The extended header frame comprises three further sub-fields, as illustrated in Figure 8.16 and discussed in Table 8.31.	24
Frame payload	The frame payload contains the individual frame type, which is of variable length.	

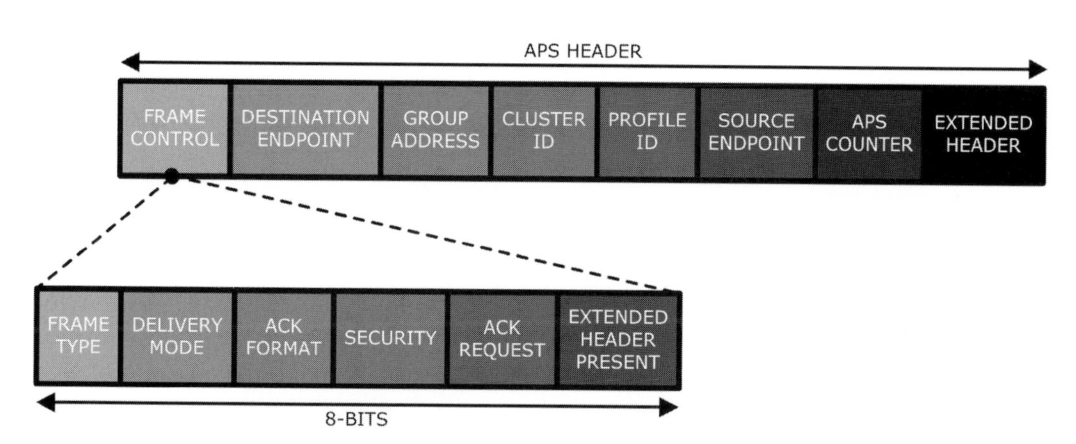

Figure 8.15. The makeup of the frame control field, which is explained in Table 8.28.

All fields of the APU payload will be included, and the frame type field will be set to `0x00` (data), as indicated in Table 8.29.

The APS command frame format, as shown in Figure 8.18, differs from the general APDU structure in that the payload includes just the *frame control*, *APS counter*, *APS*

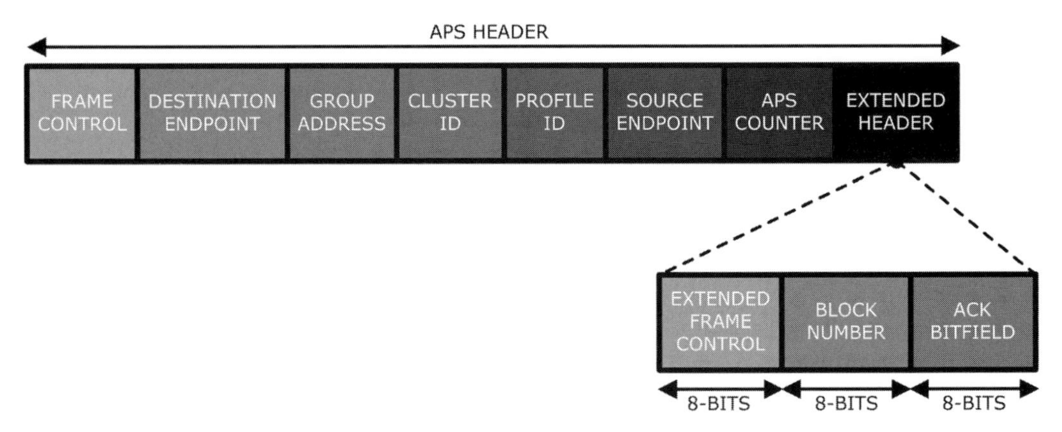

Figure 8.16. The extended header, along with its three sub-fields.

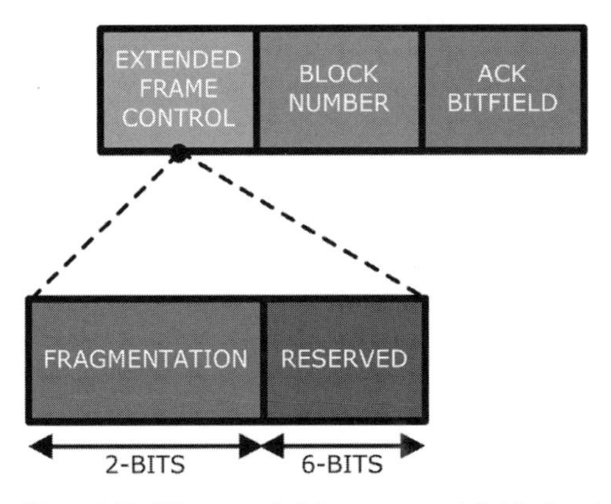

Figure 8.17. The extended frame control field, showing the format of the fragmentation sub-field.

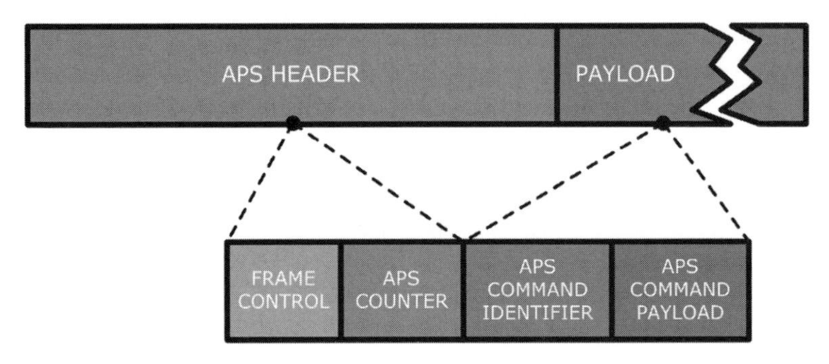

Figure 8.18. The APS command frame format.

command identifier, and the *APS command payload* fields. The frame type field will be set to `0x01` (command), as indicated in Table 8.29. Finally, the acknowledgement frame format, which we illustrate in Figure 8.19, obeys the same transmission order, as used for general APDUs.

Table 8.28. Fields that make up the APS frame control field

Field	Description	Size
Frame type	The frame type field is 2-bits in length and is set according to Table 8.29.	2
Delivery mode	The delivery mode field is 2-bits in length and is set according to Table 8.30.	2
ACK format	This single-bit field is set to 0 for data frame acknowledgement and 1 for APS command frame acknowledgement.	1
Security	We discuss security later, in Section 8.11, "ZigBee Security."	1
ACK request	The ACK request 1-bit field is set to 0 for no acknowledgement or if the message has been broadcast. However, if set to 1, then an acknowledgement is sent back to the source device.	1
Extended header present	The extended header present field is set to 1 if the extended header is present (see Figure 8.15, Figure 8.16, and Table 8.31).	1

Table 8.29. Frame type value

Value	Description
0x00	Data.
0x01	Command.
0x02	Acknowledgement.
0x03	Reserved.

Table 8.30. Delivery mode value

Value	Description
0x00	Unicast: delivered to a specific endpoint.
0x01	Reserved.
0x02	Broadcast: delivered to all devices that are selected for the broadcast address.
0x03	Group addressing: delivered to endpoints that are part of a specific group membership.

Table 8.31. Fields that make up the extended header field

Field	Description	Size
Extended frame control	This 8-bit field is formatted as illustrated in Figure 8.17, and its associated values are listed in Table 8.32. Only 2-bits are used to define the fragmentation status; the remaining 6-bits are reserved.	8
Block number	The block number is an 8-bit field (only 2-bits are used), used to manage fragmentation, as shown in Table 8.32.	8
ACK bitfield	An 8-bit field used to acknowledge which blocks of a fragmented APDU have been received.	8

Table 8.32. Fragmentation value

Value	Description
0x00	Transmission is not fragmented.
0x01	First frame of fragmented transmission.
0x02	Part of frame of fragmented transmission.
0x03	Reserved.

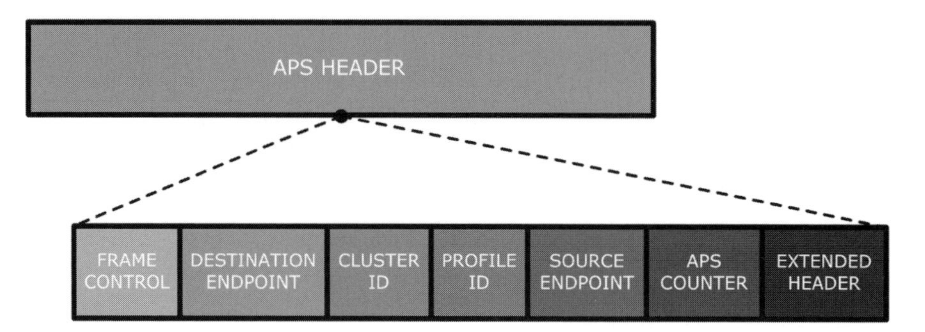

Figure 8.19. The acknowledgement frame format.

In this instance, in the frame control field, the sub-field frame type field will be set to `0x02` (acknowledgement), as indicated in Table 8.29. You may recall earlier, from Section 8.9.2.1, "General APDU Format," that we discussed the ACK format field (see Table 8.28), where this single-bit field is set to 0 for a data frame acknowledgement and 1 for an APS command frame acknowledgement. So, in the instance where this field is set for a data frame acknowledgement, the destination endpoint, the cluster and profile IDs, the source endpoint, and the extended header fields will all be provided and should reflect the intended use. The extended header field will match the value that was defined in the sub-field of the frame control field. Conversely, when the ACK format field is set for APS command acknowledgement, the destination endpoint, the cluster and profile IDs, the source endpoint, and the extended header fields are not included. However, when an APS data frame is being acknowledged, the source endpoint field must match the destination endpoint field of the acknowledged frame. Likewise, the cluster and profile IDs and the APS counter should also reflect the values in the acknowledged frame.

8.10 The Network Layer

The NWK layer offers functionality to support the MAC sub-layer, which also extends to a serviceable interface for the application layer. We discuss the MAC layer further in Section 8.12, "MAC and PHY Layers." In a similar manner to the APS layer, service entities are conceptually provided; again, these entities relate to data and management services. So, the *NWK Layer Data Entity* (NLDE) provides a data transmission service through an associated SAP, namely the NLDE-SAP, and the *NWK Layer Management*

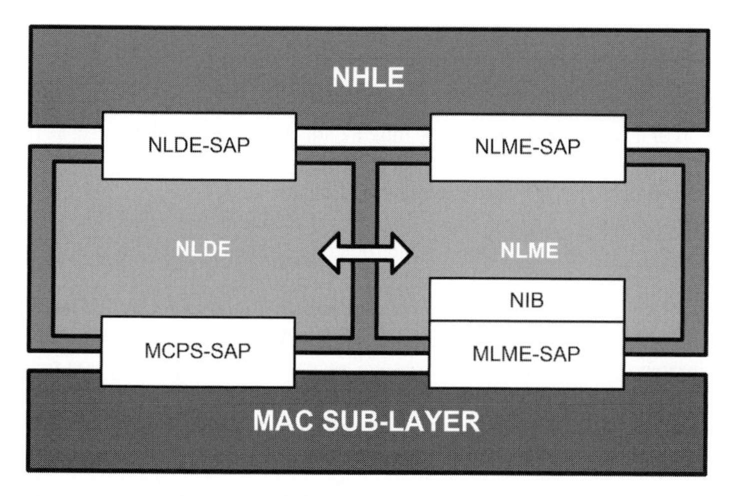

Figure 8.20. A reference model illustrating the conceptual building blocks and interfaces surrounding the NWK.

Entity (NLME) offers management services through its associated SAP, the NLME-SAP. What's more, the NLME relies upon the NLDE to undertake some management tasks, whilst maintaining a database of managed objects, referred to as the *Network Information Base* (NIB). In Figure 8.20, we illustrate the conceptual entities of the NWK layer, along with the NIB in context.

Table 8.33. Services provided by the NLDE

Service	Description
PDU generation	A *Network PDU* (NPDU) is generated from an APS PDU, along with the relevant protocol overhead.
Topology-specific routing	Describes the ability to transmit an NPDU to a device that's either the final destination or to a device that is en route to the final destination device.
Security	A service which offers both authentication and confidentiality of data transmission across the network.

As we already mentioned, the NLDE offers a data service that supports the transportation of APDUs over the same network for a number of ZigBee devices; Table 8.33 lists the services provided by the NLDE. In Table 8.34, we list the services provided by the NLME, which, in turn, enables an application to interact with the ZigBee stack.

8.10.1 NWK Service Primitives

The NWK provides two services, both of which are offered through their respective service access points (NLDE-SAP and NLME-SAP) and are the supported interface between the application and MAC sub-layer via the MCPS-SAP and MLME-SAP interfaces (see Figure 8.20); we will discuss the MAC sub-layer interface later, in Section 8.12, "MAC and PHY Layers." In the sections to follow, we discuss in more detail the

Table 8.34. Services provided by the NLME

Service	Description
Configuration of a new device	The ability to configure adequately the stack for use, as required.
Initiating a network	The establishment of a new network.
Joining, rejoining, and leaving a network	This service offers the ability to join, rejoin, and leave a network; a coordinator and router device can specifically request a device to leave a network.
Addressing	The ability for coordinators and routers to assign an address for a new device on joining a network.
Neighbor discovery	The ability to acquire, record, and report information relating to a one-hop neighbor device.
Route discovery	The ability to retrieve and record a path through a network, affording effective message delivery.
Reception control	The control of a receiver in terms of duration to enable MAC sub-layer synchronization or direct reception.
Routing	This service offers the ability to utilize several routing mechanisms to route data effectively within the network.

Table 8.35. NLDE-SAP primitives

Primitive	Behavior
`NLDE-DATA.request`	This primitive provides a transfer of a *Network Sub-layer Data Unit* (NSDU) from the local APS sub-layer entity to one or more peers when requested.
`NLDE-DATA.confirm`	This primitive responds to and provides the results of the NLDE-DATA.request primitive.
`NLDE-DATA.indication`	This primitive acknowledges the transfer of a NSDU from the NWK to the local APS sub-layer entity.

NLDE-SAP primitives that enable the transportation of APDUs between peer application objects (see Table 8.35), as well as the primitives supported by the NLME-SAP to permit the transportation of management commands between the NHLE and the NLME (see Table 8.36). In Figure 8.21, we demonstrate the primitives in use that are responsible for establishing a new network.

In a similar manner to the AIB management primitives, the NWK-specific primitives listed in Table 8.37 provide the NHLE with the ability to read and write attributes from the NIB – these primitives form part of the NLME-SAP services.

8.10.2 NWK Frame Formats

NPDUs are transmitted and received across the NWK using various services and associated primitives to action these exchanges. In this section, we take a closer look at how NWK frames or payloads are formatted. Typically, a NWK frame comprises a NWK header and payload, a sequence of fields that is provided in a specific transmission order.

Table 8.36. NLME-SAP management primitives

Primitive	Behavior
NLME-NETWORK-DISCOVERY.request	Allows the NHLE to make a request to the NWK to discover networks operating within the *Personal Operating Space* (POS).
NLME-NETWORK-DISCOVERY.confirm	This primitive responds to and provides the results of the NLME-NETWORK-DISCOVERY.request primitive.
NLME-NETWORK-FORMATION.request	Allows the NHLE to make a request to initiate a new network, whilst assigning itself as the coordinator, along with making amendments to the overall configuration.
NLME-NETWORK-FORMATION.confirm	This primitive responds to and provides the results of the NLME-NETWORK-FORMATION.request primitive.
NLME-PERMIT-JOINING.request	Allows the NHLE (coordinator or router) to open the network up for a fixed period to accept devices into its network.
NLME-PERMIT-JOINING.confirm	This primitive responds to and provides the results of the NLME-PERMIT-JOINING.request primitive.
NLME-START-ROUTER.request	Allows the NHLE of a ZigBee router to commence activities, such as routing of data frames and route discovery, and to accept requests to join the network from other devices.
NLME-START-ROUTER.confirm	This primitive responds to and provides the results of the NLME-START-ROUTER.request primitive.
NLME-ED-SCAN.request	Permits the NHLE to request an energy scan to ascertain where there are operating channels in the local area.
NLME-ED-SCAN.confirm	This primitive responds to and provides the results of the NLME-ED-SCAN.request primitive.
NLME-JOIN.request	Allows the NHLE to request to join or rejoin a network, or to alter the operating channel.
NLME-JOIN.confirm	This primitive responds to and provides the results of the NLME-JOIN.request primitive.
NLME-JOIN.indication	Allows the NHLE (coordinator or router) to be notified when a device has joined its network.
NLME-DIRECT-JOIN.request	This primitive optionally allows the NHLE (coordinator or router) to request another device to join its network.
NLME-DIRECT-JOIN.confirm	This primitive responds to and provides the results of the NLME-DIRECT-JOIN.request primitive.
NLME-LEAVE.request	Allows the NHLE to request that it wishes, or wishes another device, to leave the network.
NLME-LEAVE.confirm	This primitive responds to and provides the results of the NLME-LEAVE.request primitive.
NLME-LEAVE.indication	Allows the NHLE to be notified when a device has left the network.
NLME-RESET.request	Allows the NHLE to request that the NWK perform a reset operation.
NLME-RESET.confirm	This primitive responds to and provides the results of the NLME-RESET.request primitive.
NLME-SYNC.request	Allows the NHLE to retrieve information from its coordinator or router.
NLME-SYNC.confirm	This primitive responds to and provides the results of the NLME-SYNC.request primitive.

(*continued*)

Table 8.36 (*continued*)

Primitive	Behavior
NLME-SYNC-LOSS.indication	Allows the NHLE to be notified when a device has lost synchronization at the MAC sub-layer.
NLME-NWK-STATUS.indication	Allows the NHLE to be notified of network failures and so on.
NLME-ROUTE-DISCOVERY.request	Allows the NHLE to commence route discovery.
NLME-ROUTE-DISCOVERY.confirm	This primitive responds to and provides the results of the NLME-ROUTE-DISCOVERY.request primitive.

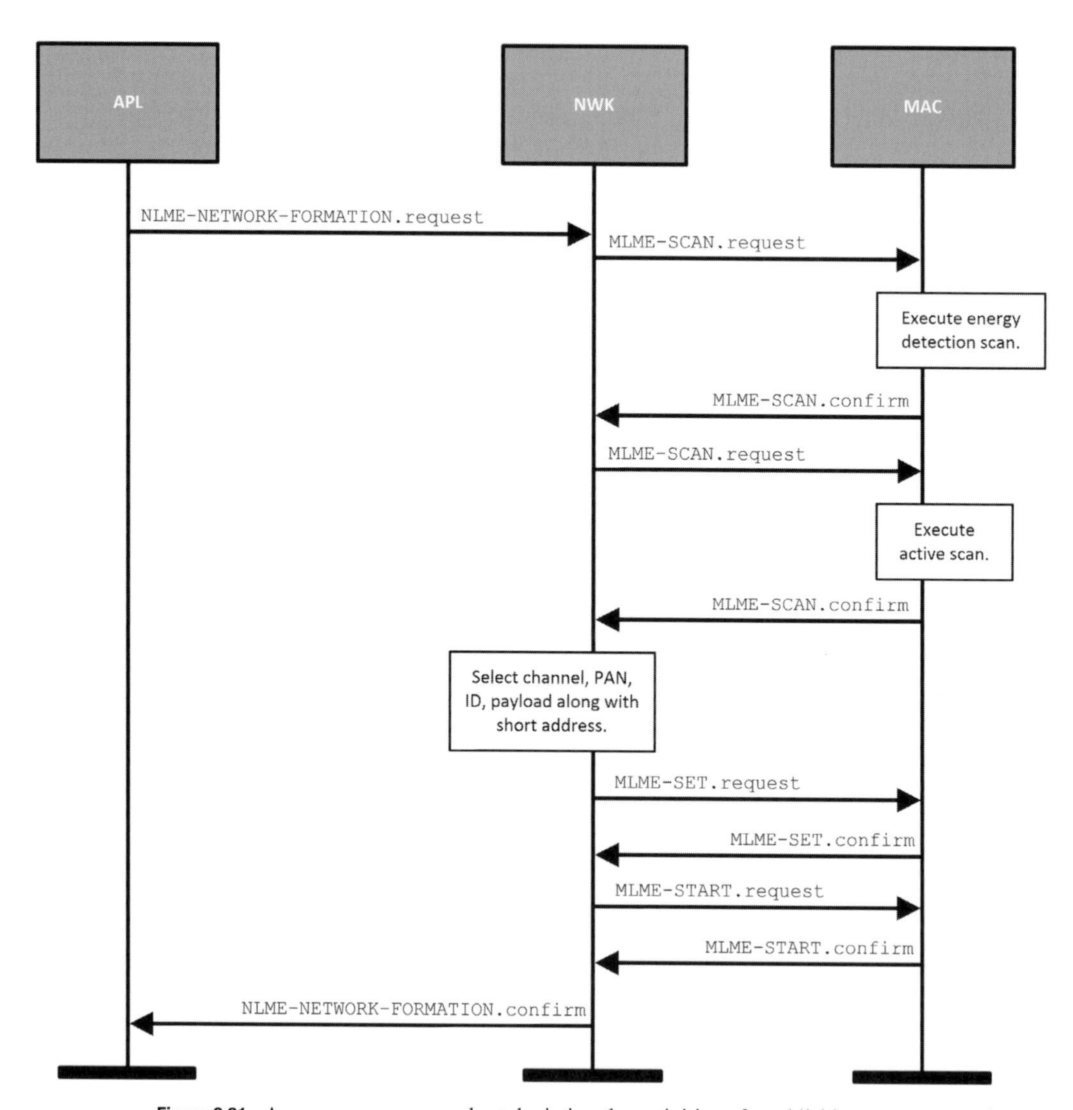

Figure 8.21. A message sequence chart depicting the activities of establishing a new network.

Table 8.37. NLME-SAP NIB management primitives

Primitive	Behavior
`NLME-GET.request`	This primitive is issued to the NLME to allow the NHLE to read an attribute from the AIB.
`NLME-GET.confirm`	This primitive responds to and provides the results of the `NLME-GET.request` primitive.
`NLME-SET.request`	This primitive is issued to the NLME to allow the NHLE to write an attribute to the AIB.
`NLME-SET.confirm`	This primitive responds to and provides the results of the `NLME-SET.request` primitive.

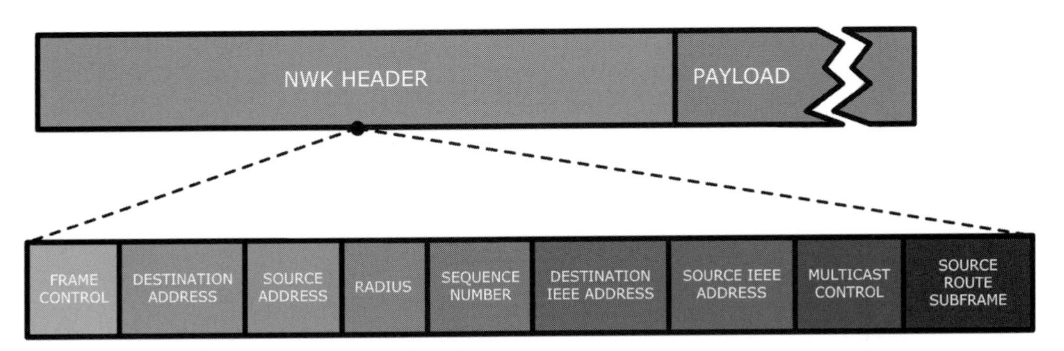

Figure 8.22. The general NPDU frame format showing the makeup of the NWK header.

It is worth noting that all reserved fields are set to 0, but if these fields are unequal to zero, then the frame is discarded.

8.10.2.1 General NPDU Format

In Figure 8.22, we illustrate the generic NPDU. As you can see, the structure comprises the NWK header and payload, where it appears in a specific sequence; however, the addressing fields may not necessarily be present in all frames. Of course, the NWK (frame) payload is of variable length and will ultimately contain information relevant to the type of frame being transmitted. In Table 8.38 we look at the specific fields that make up the NWK header, along with size of the field in bits.

8.10.2.2 Data and NWK Command Frames

We looked at the generic frame format for NWK in Section 8.10.2.1, "General NPDU Format," and in this section we shall review the specific types for *data* and *NWK command frames*. In Figure 8.25, we illustrate the data frame format, where, in the NWK header, we can see that there is a frame control field. The frame type will be set to `0x00` (data), as shown in Table 8.40. The routing fields are of variable length and hold a combination of address and broadcast fields. Finally, the data (NWK) payload, again of variable length, will contain the frame that the NHLE has requested to transmit.

Table 8.38. The fields that make up the NWK header

Field	Description	Size
Frame control	The frame control field is 16-bits in length and is further broken down in Figure 8.23 and Table 8.39.	16
Destination address	This mandatory field is 16-bits in length and contains the destination or broadcast address when the multicast flag field is set to 0. However, if the multicast flag field is set to 1, then the destination address field will contain a 16-bit value for the group ID for the destination multicast group.	16
Source address	This mandatory field is 16-bits in length and contains the originating source address.	16
Radius	The radius field is mandatory and is 8-bits in length.	
Sequence number	The sequence number field is mandatory for every frame and is 8-bits in length. This field is incremented by 1 for each frame transmitted.	8
Destination IEEE address	A 64-bit address, in turn corresponding to the 16-bit network destination address field.	64
Source IEEE address	A 64-bit address, in turn corresponding to the 16-bit network source address field.	64
Multicast control	This 8-bit field is present if the multicast flag value is set. The field is illustrated in Figure 8.24, and the multicast mode values are listed in Table 8.43. The non-member radius sub-field provides the range of member modes that is relayed by the devices that are non-members of the destination group, whilst the max non-member radius sub-field offers a maximum value for the non-member radius sub-field.	8
Source route subframe	This variable field is present when the source route sub-field is set.	
Frame payload	The frame payload contains the individual frame type, which is of variable length.	

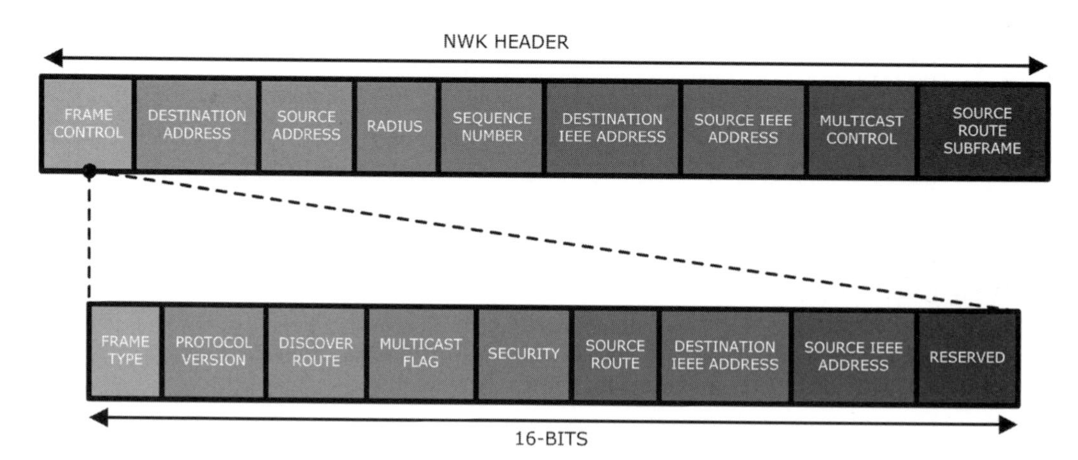

Figure 8.23. The makeup of the NWK frame control field, which is explained in Table 8.39.

Table 8.39. The fields that make up the NWK frame control field

Field	Description	Size
Frame type	This 2-bit field will be set to a value which is provided in Table 8.40.	2
Protocol version	Typically, the majority of ZigBee products will use the value `0x02` in the NWK header; however, a list of values is provided in Table 8.41.	4
Discover route	The value set here determines route discovery procedures for the frame being transmitted (Table 8.42).	2
Multicast flag	This 1-bit field is set to 1 if the frame is multicast, whereas a 0 signifies the frame is either unicast or broadcast.	1
Security	This 1-bit frame is set to 1 if NWK security operations are required.	1
Source route	If the source route subframe is present in the NWK header, then this 1-bit field is set to 1.	1
Destination IEEE address	If the destination IEEE address has been included in the NWK header, then this 1-bit field is set to 1.	1
Source IEEE address	If the source IEEE address has been included in the NWK header, then this 1-bit field is set to 1.	1

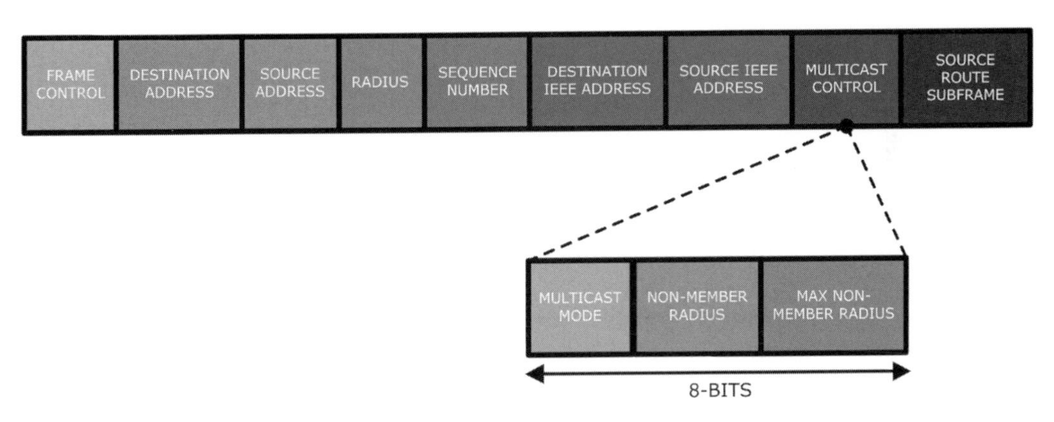

Figure 8.24. The format of the multicast control field.

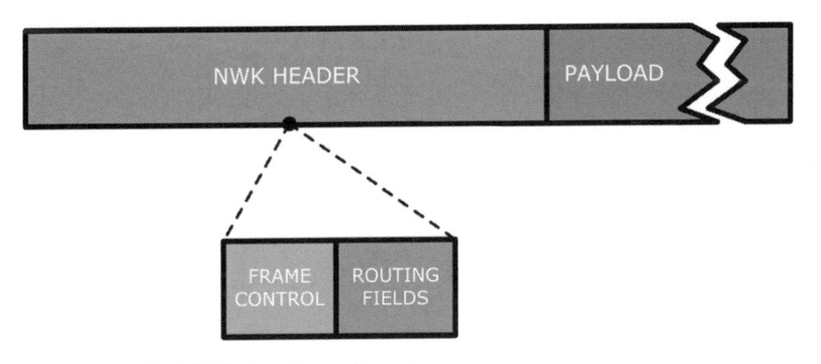

Figure 8.25. The NWK data frame format.

In the NWK command frame, as shown in Figure 8.26, the NWK header again comprises the frame control and routing fields. The frame type, within the frame control field, has the value `0x01` (NWK command), as shown in Table 8.40. The NWK command identifier contains the NWK command in current use and is shown in Table 8.44; the NWK command payload contains information relating to the NWK command.

Table 8.40. Frame type value

Value	Description
0x00	Data.
0x01	NWK command.
0x02	Reserved.
0x03	Reserved.

Table 8.41. Protocol version value

Specification	Version	Description
Current	0x02	Backward compatibility is supported with ZigBee-2006; not required with ZigBee-2004.
ZigBee-2006	0x02	Backward compatibility support with ZigBee-2004 is not required.
ZigBee-2004	0x01	The original specification.

Table 8.42. Discover route value

Value	Description
0x00	Suppress route discovery.
0x01	Enable route discovery.
0x02	Reserved.
0x03	Reserved.

Table 8.43. Multicast mode value

Value	Description
0x00	Non-member node.
0x01	Member node.
0x02	Reserved.
0x03	Reserved.

8.11 ZigBee Security

The ZigBee network is implicitly secure. Nevertheless, ZigBee, like any other wireless system, may be prone to attack, and, despite ZigBee's infrequent data transmissions, some larger networks may be more susceptible. The basic premise of the security architecture within ZigBee focuses on the security and confidentiality of *keys*. Essentially, keys are

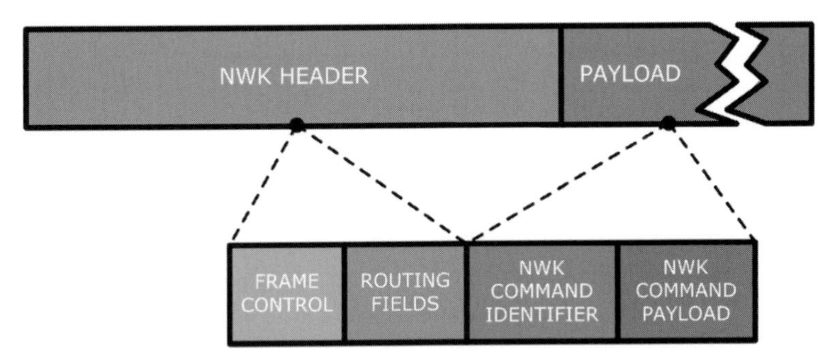

Figure 8.26. The NWK command frame format.

Table 8.44. Command identifier value

Identifier	Command
0x01	Route request.
0x02	Route reply.
0x03	Network status.
0x04	Leave.
0x05	Route record.
0x06	Rejoin request.
0x07	Rejoin response.
0x08	Link status.
0x09	Network report.
0x0A	Network update.
0x0B–0xFF	Reserved.

never (or only ever inadvertently) transmitted across a channel unless it's secured – this forms the founding basis of security within the ZigBee architecture. However, there is one exception to this rule: during the initial configuration and set-up of a new device, the security key is momentarily open to interrogation by an alien device. However, more secure commissioning methods are available that never expose the security keys. Within the ZigBee protocol stack, the APS and NWK layers are solely responsible for securing their frames, whereas the ZDO manages the security policies and configuration of a device.

Table 8.45. A summary of the 802.15's seven working groups; in particular, 802.15.4 and its numerous subset standards provide ZigBee with its air-interface

Standard (Task Group)	Scope of activity
802.15.(1)	Bluetooth
802.15.(2)	Coexistence.
802.15.(3)	High-rate MAC and PHY WPAN standard.
802.15.(4)	Low-rate MAC and PHY WPAN standard.
802.15.(5)	Mesh networking.
802.15.(6)	*Body Area Networking* (BAN).
802.15.(7)	*Visible Light Communication* (VLC).

8.12 MAC and PHY Layers

In Chapter 2, "What is a Personal Area Network?," we discussed the IEEE 802.15 working group, which comprises seven task groups, as listed in Table 8.45. The working group's objective[6] is to establish industry standards and recommend industry practices, whilst addressing issues surrounding interoperability and coexistence for the wireless personal area networking space market sector. The IEEE typically develops standards for the PHY and MAC layers, but these are not always adopted by the industry. However, the ZigBee Alliance adopted the 802.15.4 standard, and the PHY and MAC layers now underpin ZigBee technology.

ZigBee technology relies upon the IEEE 802.15.4 specification to provide its air-interface – something we have already touched upon earlier in this chapter. The specification specifies both the PHY and MAC layers. The IEEE 802.15.4 specification also offers two PHY layers using the *Industrial, Scientific, and Medical* (ISM) frequency bands, namely 868/915 MHz and 2.4 GHz. The lowest (868 MHz) frequency band is used throughout Europe and supports one channel with a maximum data rate of 20 kbit/s; the 915 MHz spectrum is used within the United States and Australia, and supports up to ten channels with a maximum data rate of 40 kbit/s; finally, the 2.4 GHz band is used predominately worldwide and can support up to 16 channels with a maximum data rate of up to 250 kbit/s. *Direct Sequence Spread Spectrum* (DSSS) is used across all frequencies; however, the modulation scheme employed for both 868 and 915 MHz frequency bands is *Binary Phase Shift Keying* (BPSK), whilst the 2.4 GHz band uses *Orthogonal-Quadrature Phase Shift Keying* (O-QPSK).

8.12.1 Coexistence

In Section 8.5, "Networking Topology," we touched upon beacon and non-beacon network environments, along with their associated significance. In a non-beacon environment, devices can transmit ad hoc and, as such, this may cause interference with other devices within proximity. For devices operating in a non-beacon mode, a *Carrier Sense Multiple Access with Collision Avoidance* (CSMA-CA) technique is used to minimize interference; however, and more significantly, an acknowledgement is provided for successfully transmitted packets. In a beacon-enabled network, beacon packet transmission is governed by predetermined intervals, in which allocated time slots ensure contention-free access and guarantee delivery of payloads, in turn ensuring overall *Quality of Service* (QoS) – other data packets rely on CSMA-CA.

8.13 ZigBee RF4CE

ZigBee's RF4CE specification seeks to liberate the classic *Infra-red* (IR) remote control that was introduced as far back as the 1970s. Nowadays, some innovators are looking at

[6] IEEE 802.15 Working Group for Wireless Personal AreaNetworks.

Table 8.46. ZigBee RF4CE device types

Device	Description
Target	A target device can initiate or join a ZigBee RF4CE network.
Controller	A controller device can join a network once it has been initiated by the target node.

voice commands or (hand) gestures to control a new generation of consumer electronic products. Nonetheless, ZigBee, building on its success with home and commercial automation, smart energy, and health care applications, aims to extend its presence within the home and industry with a specification that will populate and liberate a new generation of consumer electronic products. The specification relies upon the MAC/PHY IEEE 802.15.4 standard utilizing the ISM 2.4 GHz frequency band, which affords the ZigBee RF4CE specification its robust coexistence techniques. The 2.4 GHz band can be overcrowded, and therefore a small subset of channels (15, 20, and 25) is used to combat interference. A target device will select the best available channel during initialization and set-up by performing an energy detection scan. What's more, if a channel deteriorates or becomes compromised during operation, then the target device can choose to select another channel in an attempt to improve quality of experience. With this in mind, and accompanied by a simplified networking layer, manufacturers can build a range of interoperable products specifically targeted toward remote control applications.

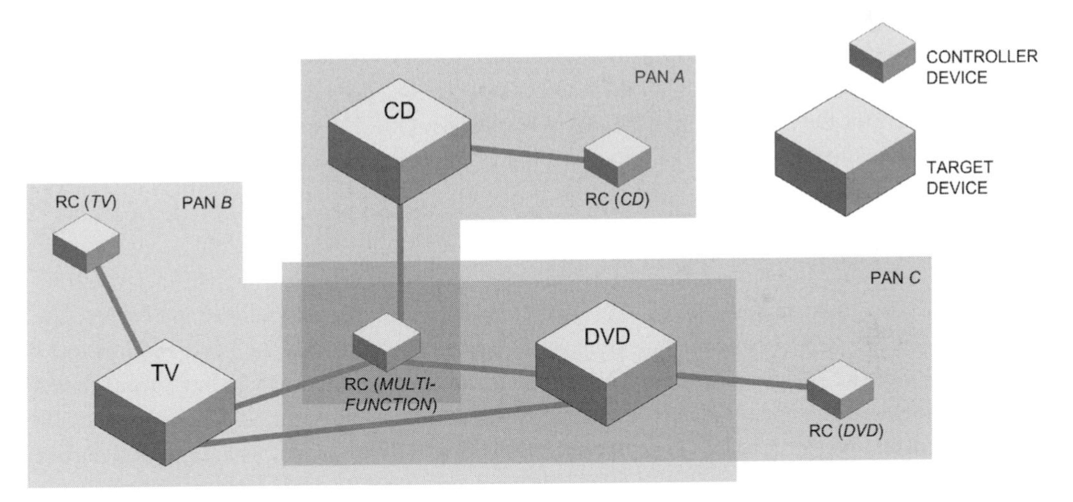

Figure 8.27. The ZigBee RF4CE networking topology depicting RC controller devices operating in multiple PANs.

8.13.1 Networking Topology

The ZigBee RF4CE topology supports two types of devices within its *Remote Control* (RC) PAN, namely *target* and *controller* nodes; see Table 8.46. A target device is capable of initiating a network and configuring the stack, whereas a controller node simply configures the stack and operates as normal. A target or controller device can join a

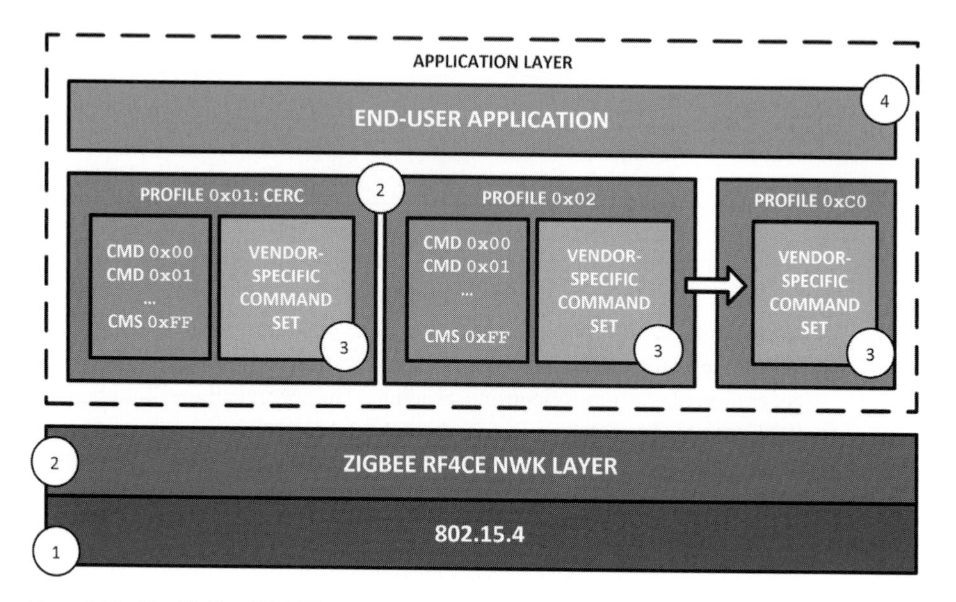

Figure 8.28. The ZigBee RF4CE software protocol stack.

network once initiated, and in Figure 8.27 we illustrate multiple RC PANs in operation, as the RC can communicate interchangeably between PANs. Figure 8.27 shows three target devices, namely the TV, DVD, and CD, each with their own RC or controller device; in turn, this forms its own, dedicated RC PAN. However, the multi-function RC is capable of controlling all three target devices. So, PAN **A** comprises the CD and its RC plus the multi-function RC; PAN **B** comprises the TV and its RC, DVD player, and the multi-function RC; and, finally, PAN **C** comprises the DVD and its RC along with the multi-function RC.

8.13.2 The ZigBee RF4CE Application Layer

In a similar manner to the ZigBee architecture, we take a top-down approach to the ZigBee RF4CE stack architecture. In Figure 8.28, we provide the key building blocks of the RF4CE architecture – you will note that these building blocks have been labeled numerically to ease identification and classify responsibility. So, looking at Figure 8.28 again, blocks labeled (**1**) represent the MAC and PHY layers, which provide ZigBee RF4CE with its air-interface, as we have already mentioned; (**2**) represents the building blocks that are RF4CE responsible – in essence, this is the RF4CE protocol standard; (**3**) are vendor-specific, which form standard profiles, but also allow vendors to extend or build proprietary profiles. The *ZigBee Remote Control* (ZRC) standard provides common commands and procedures to empower CE products; and finally, (**4**) represents building blocks that are application-specific, namely the end applications. The ZigBee RF4CE architecture is composed in a similar fashion to the core stack of the ZigBee protocol, in that each layer forming the software building blocks is layered to offer services to the next higher or lower layer, as we can see in Figure 8.28. The application layer within the ZigBee RF4CE stack offers a profile and application component.

8.13.3 The ZigBee RF4CE Network Layer

We can make similar comparisons between the ZigBee RF4CE NWK layer operation and the core NWK layer, which we discussed back in Section 8.10, "The Network Layer." The ZigBee RF4CE provides two services, namely data (via NLDE) and management, both of which interface to/from the NLME entity. Again, these services are accessed through their respective SAPs, NLDE-SAP and NLME-SAP. Naturally, the NLDE-SAP service permits the reception and transmission of NPDUs across the MAC data service.

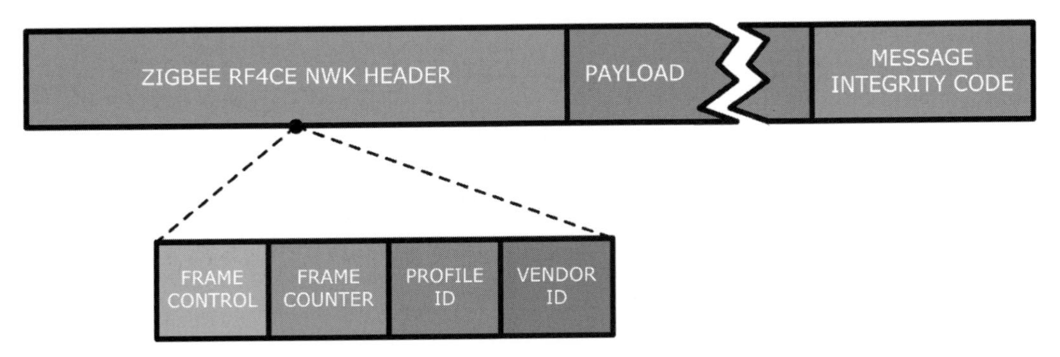

Figure 8.29. The ZigBee RF4CE NWK NPDU frame format.

8.13.3.1 ZigBee RF4CE NWK Frame Format

ZigBee RF4CE NPDUs are transmitted and received across the NWK using the various services and associated primitives to action these exchanges. In this section, we take a closer look at how ZigBee RF4CE NWK frames or payloads are formatted. Typically, a ZigBee RF4CE NWK frame comprises a ZigBee RF4CE NWK header, payload, and a message integrity code, a sequence of fields that is provided in a specific transmission order. In Figure 8.29, we illustrate the generic ZigBee RF4CE NPDU. As you can see, the structure comprises the NWK header, payload, and message integrity code where it appears in a specific sequence. In Table 8.47 we look at the specific fields that make up the ZigBee RF4CE NWK header, along with size of the field in bits.

8.13.3.2 Data (Standard and Vendor-specific) and ZigBee RF4CE NWK Command Frames

In the previous section, we looked at the frame format for the ZigBee RF4CE NWK NPDU, and in this section we shall review the specific types for *data*, standard, and vendor-specific, as well as *ZigBee RF4CE NWK command frames*. The data for both standard and vendor-specific frame abides by the generic NPDU frame format, as we previously illustrated in Figure 8.29. Furthermore, in the frame type field, the value is set either to `0x01` (standard) or `0x03` (vendor-specific), as shown in Table 8.49.

The ZigBee RF4CE NWK command frame format, as shown in Figure 8.31, differs from the general NPDU structure in that the payload includes just the *frame control*, *frame counter*, *command identifier*, and the *command payload* fields. The frame type

Table 8.47. The fields that make up the APS header

Field	Description	Size
Frame control	The frame control field is 8-bits in length and is further broken down in Figure 8.30 and Table 8.48.	8
Frame counter	The frame counter field is 24-bits in length and maintains the sequence identifier of the frame being transmitted.	24
Profile ID	The profile ID field is 8-bits in length and specifies the ID of the command being transported.	8
Vendor ID	The vendor ID is 16-bits in length and specifies the vendor identifier of the command being transported and is only applicable to vendor-specific frames.	16
Frame payload	The frame payload is of variable length and will contain information relating to the individual frame.	
Message integrity code	The message integrity code is present if the security enabled field (see Table 8.48) is set to 1.	24

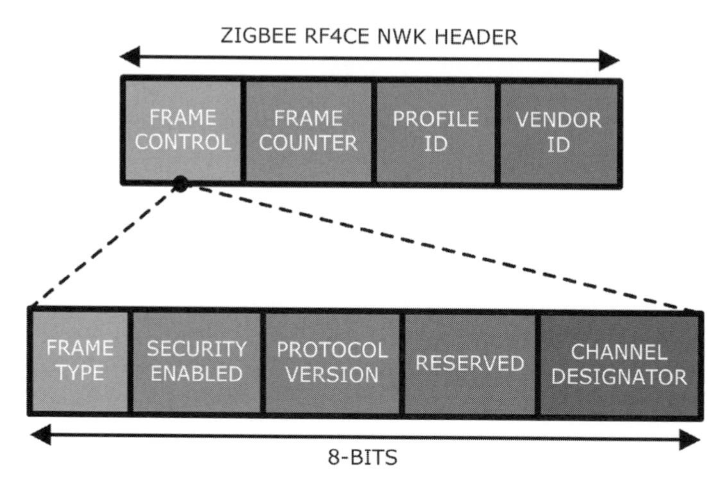

Figure 8.30. The makeup of the ZigBee RF4CE NWK frame control field, which is explained in Table 8.48.

Table 8.48. The fields that make up the APS frame control field

Field	Description	Size
Frame type	The frame type field is 2-bits in length and is set according to Table 8.49.	2
Security enabled	This 1-bit field is set to 1 if it is to be protected by the ZigBee RF4CE NWK layer.	1
Protocol version	The protocol version field is 2-bits in length.	2
Reserved	Reserved.	1
Channel designator	The channel designator field is 2-bits in length and is set according to Table 8.50. It specifies the channel of the transmitting device.	2

Table 8.49. Frame type value

Value	Description
0x00	Reserved.
0x01	Standard data frame.
0x02	NWK command frame.
0x03	Vendor-specific data frame.

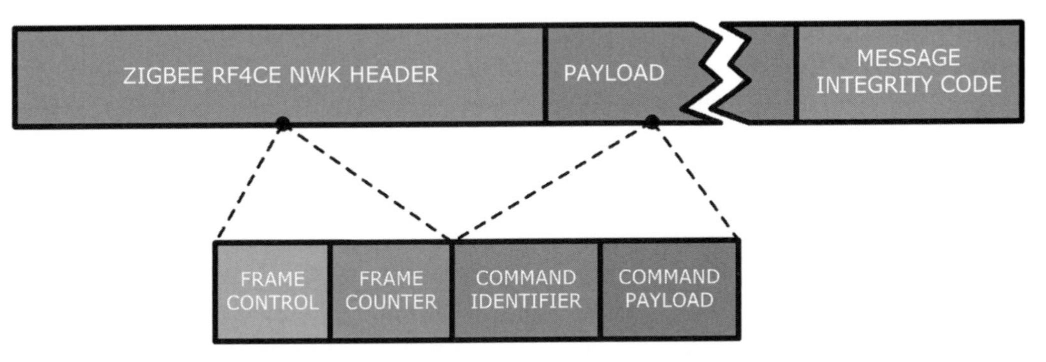

Figure 8.31. The ZigBee RF4CE NWK command frame format.

Table 8.50. Channel designator value

Value	Description
0x00	Unspecified.
0x01	Channel 15.
0x02	Channel 20.
0x03	Channel 25.

Table 8.51. Command identifier value

Identifier	Command
0x01	Discovery request.
0x02	Discovery response.
0x03	Pair request.
0x04	Pair response.
0x05	Unpair request.
0x06	Key seed.
0x07	Ping request.
0x08	Ping response.
0x09 – 0xFF	Reserved.

field is set to `0x02` (command), as indicated in Table 8.49, whilst the channel designator field shall be set to `0x00` (unspecified). The command identifier field specifies the ZigBee RF4CE NWK command being used and is set to a value as indicated in Table 8.51, while the command payload specifies the command itself.

9 Green, Smart, and Wireless

EnOcean

The low power, wireless sensor technology portfolio would not be complete without the inclusion of EnOcean's award-winning patented batteryless, wireless sensor radio technology. In this chapter, we discuss EnOcean's inception and moderately short history, and we provide a review of the Alliance and promoter members' benefits. EnOcean uses a patented technique to self-power its wireless sensors, and we'll delve deeper into the company's innovative technology and how the technology is being used within the commercial sector later on in the chapter. Likewise, we'll explore EnOcean's product portfolio, the market scope, and numerous user scenarios that place the technology into the real world, ultimately to gain a better understanding of its application-base and purpose. Similarly, we'll look at how EnOcean compares with other low power wireless sensor technologies and review some key differentiators that set EnOcean apart from its competitors.

9.1 Overview

Briefly, EnOcean's innovative technology prides itself on a technique known as *energy harvesting* – the ability of a sensor to derive energy from its environment to power its sensor technology, as opposed to using batteries or any other fixed power source. The technology, or indeed the notion, isn't new; you may recall from Chapter 5, "Introducing Low Power and Wireless Sensor Technologies," that we discussed numerous techniques used to source energy from natural resources. *Renewable energy* describes the conversion of natural resources, such as the wind, sunlight, rain, and so on, into a sustainable energy form which can be used to self-power a device. The energy harvesting technique used by EnOcean detects minor variations in the environment, such as change in temperature, ambient light, and the turning on and off of a switch, whereby such differences are sufficient to be converted to enable an EnOcean sensor to accumulate and store energy.

The EnOcean standard affords its range of EnOcean-enabled products a high reliability of wireless transmission when installed in large buildings, where all sensors can operate simultaneously. This is due to the periodic (approximately once a minute) transmission of *telegrams*, which reduces the likelihood of a telegram collision, in turn increasing the reliability of the EnOcean ecosystem. With such a low energy footprint, the range offered by EnOcean products can extend to 30 meters in a building environment (up to 300 meters in free field), although repeaters can be used to increase this range. The

Figure 9.1. The EnOcean trademark and logo. (Courtesy of EnOcean GmbH.)

technology utilizes the regulated frequency bands 315 MHz (for North America) and 868 MHz (for Europe).

Let's begin by turning our attention to the formation of the company, along with its motivation and agenda.

9.1.1 Venture-funding

EnOcean GmbH (see Figure 9.1 for the company's trademark and logo) is a venture-funded company, EnOcean GmbH (enocean.com), with its head office based in Germany (near Munich) and with offices in the United States. The company was established in 2001 as a spin-off from Siemens AG. EnOcean GmbH maintains that it is the originator of the patented self-powered technology which combines a number of miniaturized energy convertors to harness fluctuations in the ambient environment in order to power the various wireless sensors or nodes that we have already touched upon. The technology (to date) has been successfully deployed in over 200 000 buildings and has been selected by over 100 manufacturers worldwide.

Figure 9.2. The EnOcean Alliance trademark and logo. (Courtesy of the EnOcean Alliance.)

9.1.2 The EnOcean Alliance

The EnOcean Alliance (enocean-alliance.org), a non-profit consortium of companies, is responsible for the development and promotion of the self-powered technology worldwide (see Figure 9.2 for the EnOcean Alliance trademark and logo). It is also responsible for the standardization of the technology, in turn ensuring the formalization of successful interoperable wireless products for residential, commercial, and industrial buildings, whilst encouraging third-party developers to commercialize the technology further within the global marketplace. The EnOcean Alliance, which was formed in April 2008, was created by a number of companies across Europe and North America.

In a green and eco-conscious world, the EnOcean Alliance continues to promote the enablement of sustainable buildings through its interoperable wireless standard, along with its mission to define a generation of wireless monitoring and controlling products. The Alliance, comprising over 250 committed companies, purports to have the largest proven installed base of wireless building automation networks in the world.

Table 9.1. The EnOcean Alliance offers potential participants three levels of membership class, to include Promoters, Participants, and Associates

Membership	Scope
Promoter	This highest level of membership allows participants to define and steer the Alliance, along with other Promoter members, who form the board of directors, as well as enjoying all the benefits of Participant membership level.
Participant	This level offers its members the opportunity to create and use interoperable profiles, primarily in home and building monitoring and control, as well as in metering and industrial applications. Members can propose new technical features or profiles to the standard via the *Technical Working Group* (TWG) or user education programs via the *Marketing Working Group* (MWG), and may use all Alliance collateral and branding. Typically, this is the standard membership level for OEMs.
Associate	This basic membership level offers access to approved standards and other documentation once published, along with access to rudimentary information. Other benefits include the use of the Alliance brand and other facets, which are available by special invitation only. Normally, Associate membership is aimed at building professionals, academics, and distribution channels.

9.1.2.1 Membership and the EnOcean Alliance

In short, the Alliance offers potential participants three classes of membership (as shown in Table 9.1); this enables companies to decide on how they wish to participate within the Alliance. Like most standards-based committees, the offer of membership is structured, so that a participant can decide on how much involvement they wish to have – from *Associate*, a basic level where the member can access rudimentary documentation when it's made publicly available, through to *Promoter*, where board level membership is given and members can participate in the future direction of the technology.

9.1.3 EnOcean's Timeline

The infographic shown in Figure 9.3 provides a snapshot of EnOcean's technology timeline, from the company's inception to where the technology is today. EnOcean has matured considerably over the last decade or so. It was originally conceived at Siemens Research in Munich, Germany, during the 1990s, at which time several patents surrounding energy harvesting emerged. EnOcean GmbH created and released various v0.x and v1.x revisions between the period 2003 and 2007. However, it wasn't until after the EnOcean Alliance was established (April 2008) that its first official *EnOcean Equipment Profiles* (EEP) v2.0 specification appeared (January 2009). EnOcean products entered the marketplace circa 2002, and installation of EnOcean-enabled products into

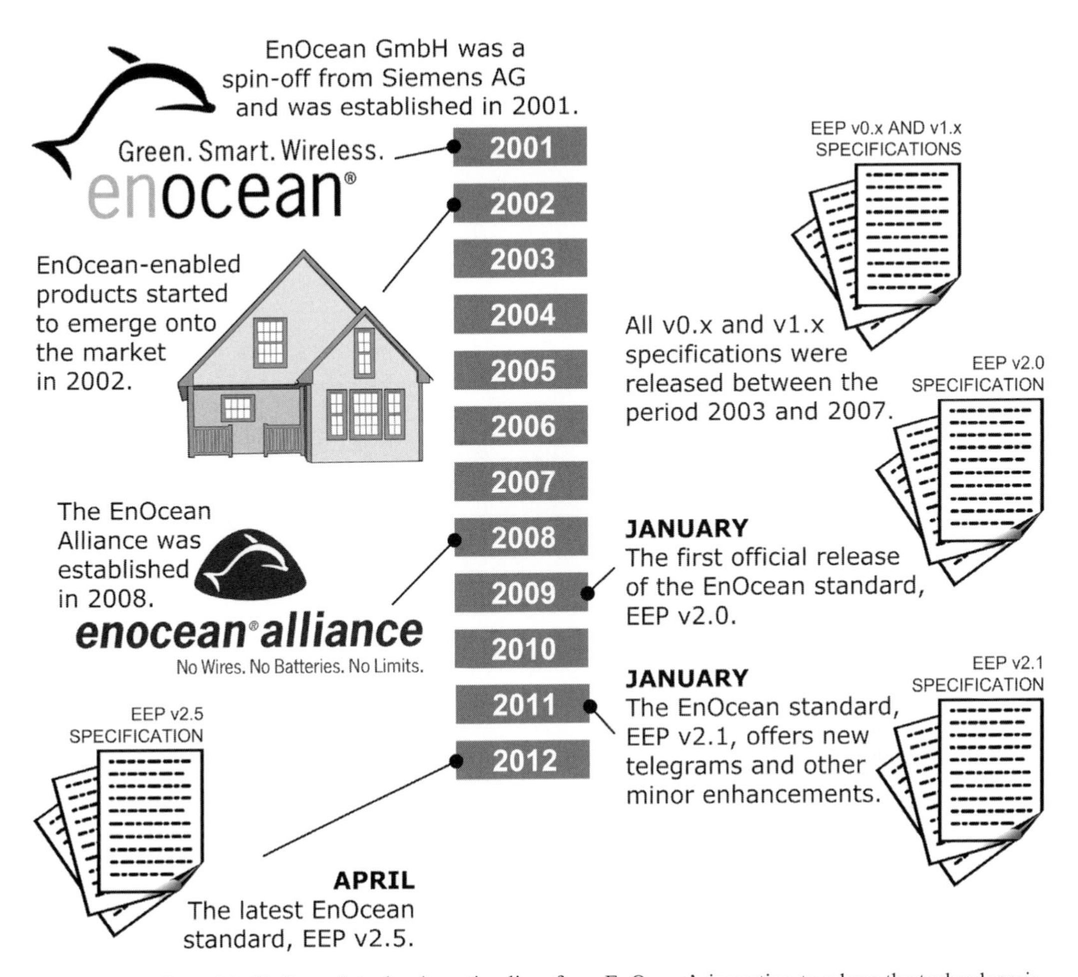

Figure 9.3. EnOcean's technology timeline, from EnOcean's inception to where the technology is today with its EEP v2.5 specification.

buildings occurred during early 2003. The current version of the EnOcean specification, EEP v2.5, offers new telegrams and other minor enhancements. We'll pick up on telegram use within the EEP specification later in this chapter. In the meantime, we shall review EnOcean's market and application portfolio.

9.2 EnOcean's Market

As we mentioned in Chapter 8, "Control Your World with ZigBee," the number of wireless technologies which feature in Part III, "The Classic Personal Area Network," typically serve high-end, high data rate products normally suited to audio-, voice-, and data-heavy-centric applications. Conversely, EnOcean technology empowers building and industry installations, ensuring environmental responsibility; furthermore, advocates of the technology have, at their fingertips, self-powered

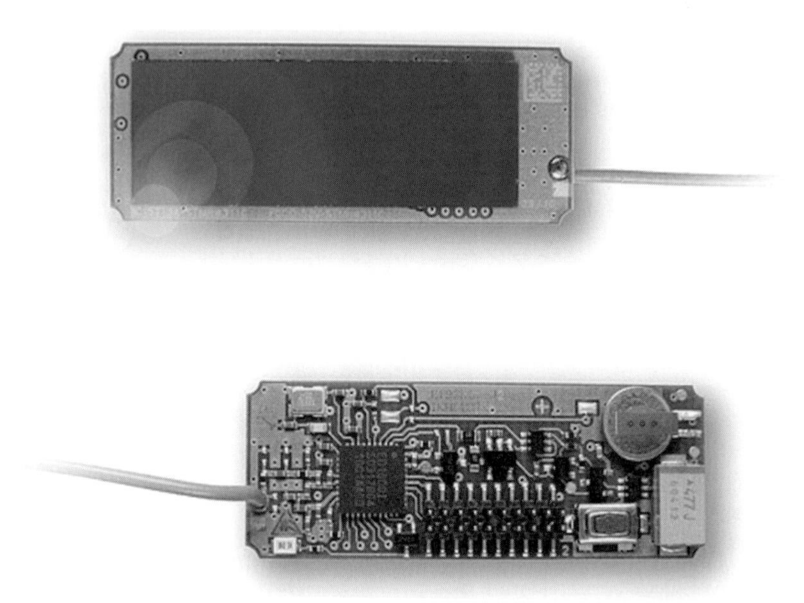

Figure 9.4. The STM 310, a thermo convertor wireless sensor utilizing either 315 MHz or 868 MHz frequencies. (Courtesy of EnOcean GmbH.)

wireless sensor technology, which minimizes and even possibly negates the need to maintain the EnOcean ecosystem. The technology also plays an essential role in supporting the *Internet of Things* (IoT). Above all, the EnOcean ecosystem affords its users an economic use of lighting, heating, and climate control, as well as offering general building automation. There are many overall benefits; for example, there is no need to retro-fit cabling; there is a reduced need for equipment to be on standby; and the system can provide single-room heating solutions, along with camera surveillance, all of which can be managed via a smartphone or over the Internet. In addition, the EnOcean ecosystem provides reduced, or indeed may eliminate, associated cabling costs. What's more, most EnOcean installations can adapt a Feng Shui approach!

Like ZigBee technology, EnOcean requires small or low data rates when communicating within its topology. EnOcean's unique energy harvesting technique eliminates the need for batteries. These self-powered wireless sensors capture the minutest changes in the environment, such as the pressing of a light switch, the vibrations from motor vehicles and even from people in motion, and the variance in ambient temperature and luminance. Singularly, or when combined, these small environmental changes are adequate to generate sufficient energy to transmit data. The EnOcean *Dolphin* system architecture (more about this later) empowers developers of the EnOcean ecosystem to create effective wireless solutions for the building and industry sectors. EnOcean has a wealth of products that have emerged, all utilizing the EnOcean standard; in Figure 9.4, we illustrate EnOcean's self-powered wireless sensors, the STM 310, a *thermo* energy harvesting sensor, and in Figure 9.5 we illustrate the ECO 200, a *motion* energy harvesting sensor.

Figure 9.5. The ECO 200, a motion convertor wireless sensor utilizing either 315 MHz or 868 MHz frequencies. (Courtesy of EnOcean GmbH.)

9.3 EnOcean's Application Portfolio

If we refer back to Figure 9.3, we can see that, like most technologies discussed in Part II, EnOcean has a lifespan of just over a decade. Nevertheless, the technology has adequately matured to capture a competitive market. You may recall from Chapter 7, "Bluetooth low energy: The *Smart* Choice," and Chapter 8, that Bluetooth and ZigBee both corner a similar market to EnOcean. The consortium of companies that comprise the Alliance is proactively establishing EnOcean as a standard for sustainable buildings and the advancement of self-powered interoperable wireless building and industrial control systems. EnOcean has already been installed in over 100 000 buildings, and the company boasts that it has "the largest installed base of proven interoperable wireless building automation networks in the world."[1] In this section, we'll explore further how the technology is being used to gain a better understanding of the number of applications supported by such an ecosystem.

The EnOcean wireless sensor technology is uninhibited by location and, as such, can be installed in residential homes, industrial buildings, hospitals, historic buildings, airports, commercial aircraft, and even yachts! In Figure 9.6, we depict a typical EnOcean installation; in this particular example, the EnOcean ecosystem manages and supports lighting and *Heating, Ventilation and Air Conditioning* (HVAC) systems through a series of wireless sensors, which can be managed through a touch panel (**8**) located on the desk, via the Internet, or remotely, via a cellular (smart)phone (**9**). Elsewhere in this installation, we have a series of batteryless wireless switches and lights for blinds (**1**); occupancy (**2**) and outdoor light (**3**) sensors; room temperature (**4**) and humidity/CO_2

[1] Schneider, A., "EnOcean Company Presentation," 2010.

Figure 9.6. A typical EnOcean ecosystem powering lighting and HVAC systems.

(**5**) sensors; window handle (**6**) and contact (**7**) sensors; and finally a line-powered gateway device (**10**). The occupancy sensors automatically switch off lights in rooms that are unoccupied, as well as manage the ambient temperature. The window handles and contact sensors cut off the heating or the air conditioning when a window is opened.

The various specifications that are discussed in the sections to follow demonstrate EnOcean's acumen and foresight within this competitive market. As such, the EnOcean Alliance has developed a standard, along with a number of supportive specifications, which collectively address a need to satisfy consumers, businesses, industries, and governments. Overall, the greener view from the Alliance continues to help improve efficiency for homes and offices.

9.4 EnOcean and its Competitors

As we have already touched upon in Chapter 5, the ability to offer energy efficient technologies to empower intelligent *green* buildings has its cost benefits, and consequently EnOcean faces competition from other technologies, including low power Wi-Fi, Z-Wave, and ZigBee – all of which target a similar, if not the same, application-base. With *Bluetooth low energy* (BLE), ZigBee, Z-Wave, and low power Wi-Fi all vying for the same application space, EnOcean has set itself apart from the competition with its energy harvesting patented technology. The technology's unique ability to capture changes in its ambient environment to self-power its wireless sensors to transmit and receive data arguably sets the technology apart from its competitors. In essence, EnOcean has removed the need for batteries, and has reduced, or eliminated, the need to maintain the EnOcean ecosystem. Since EnOcean's inception, it has enjoyed success of its

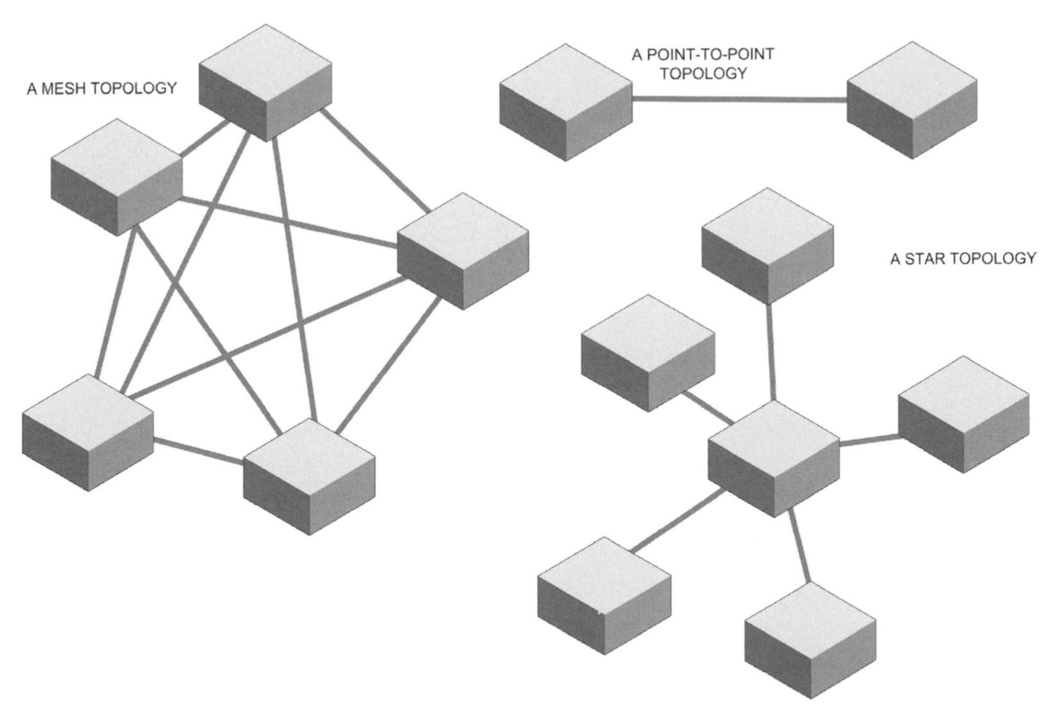

Figure 9.7. EnOcean supports mesh, point-to-point, and star networking topologies.

installed application-base, and the Alliance continues to establish the technology as a de facto low energy standard for sustainable buildings. As we have already intimated elsewhere in Part II, the Bluetooth *Special Interest Group* (SIG) has essentially re-invented Bluetooth wireless technology with its introduction of the new Specification of the Bluetooth System: Core, v4.0, wherein the SIG hopes to expand into other market sectors, to include home and building automation, health, sport, and fitness.

9.5 Networking Topology

EnOcean supports *mesh*, *point-to-point*, and *star* networking topologies, some examples of which are illustrated in Figure 9.7. Within an EnOcean network, the ecosystem may comprise *energy harvesting switches* and *sensors*; *actuators* and *controllers*; *a gateway* and *building management system*; and miscellaneous *wireless modules*, *sensors*, and *repeaters*, as previously discussed and illustrated in Section 9.3, "EnOcean's Application Portfolio." In Figure 9.8, we illustrate a topology in which a series of batteryless wireless sensors communicate with their respective line-powered controllers, which, in turn, manage the light, shading, and HVAC systems.

In Figure 9.9, we depict a similar topology, with a series of batteryless wireless sensors all communicating with an EnOcean gateway device. The gateway is connected through a backend bus network that, in turn, manages the light, shading, and HVAC systems.

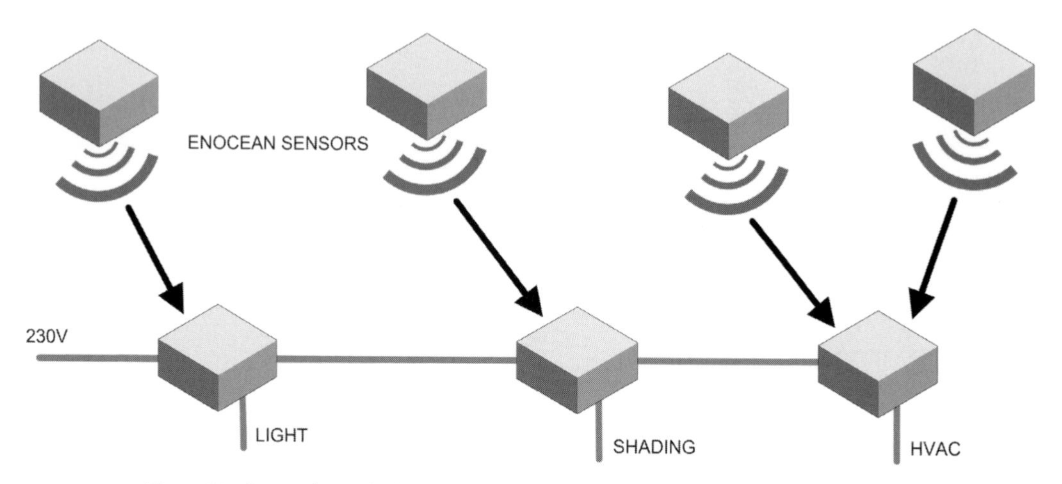

Figure 9.8. Batteryless wireless sensors communicate with their respective line-powered controllers.

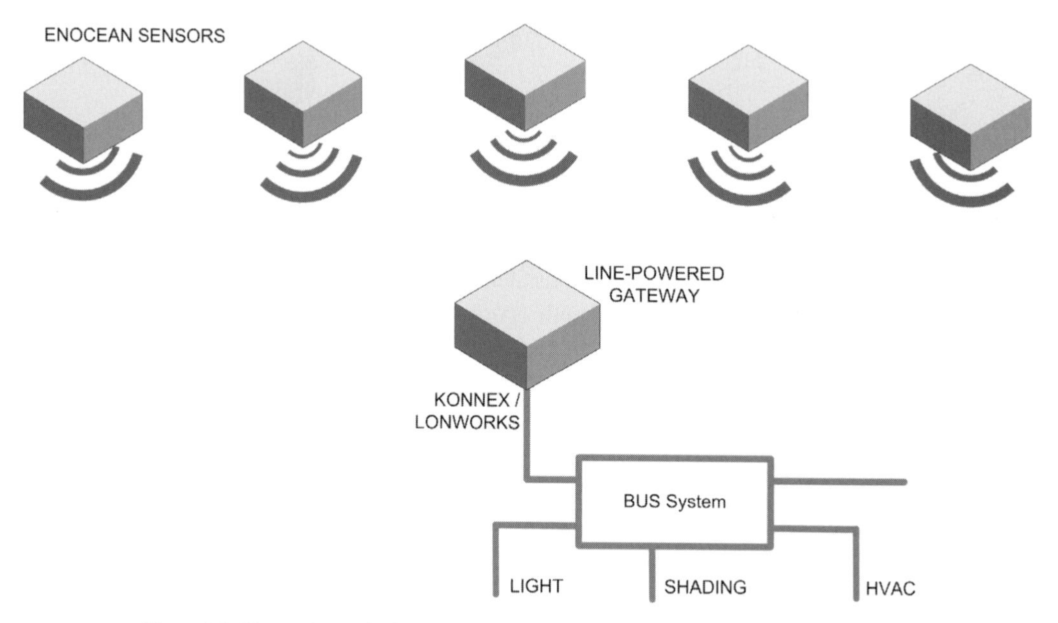

Figure 9.9. Batteryless wireless sensors communicate with a gateway device that, in turn, manages the light, shading, and HVAC systems.

The gateway supports *LonWorks*[2] (LON) and *Konnex*[3] (KNX)/*European Installation Bus* (EIB) protocols. Alternatively, the gateway supports both BACnet[4] and TCP/IP protocols. In short, these protocols form automation standards that provide automation

[2] LonWork is a networking protocol that was developed and created by the Echelon Corporation.
[3] Konnex is a network communications protocol standard and is managed by the KNX Association.
[4] BACnet is a communications protocol used for intelligent buildings.

within buildings to include lighting and HVAC, essentially realizing the concept of *intelligent building* or building automation. In fact, you may recall from Chapter 5, "Introducing Low Power and Wireless Sensor Technologies," that we introduced the concept of the *smart home* or *maison intelligente*.

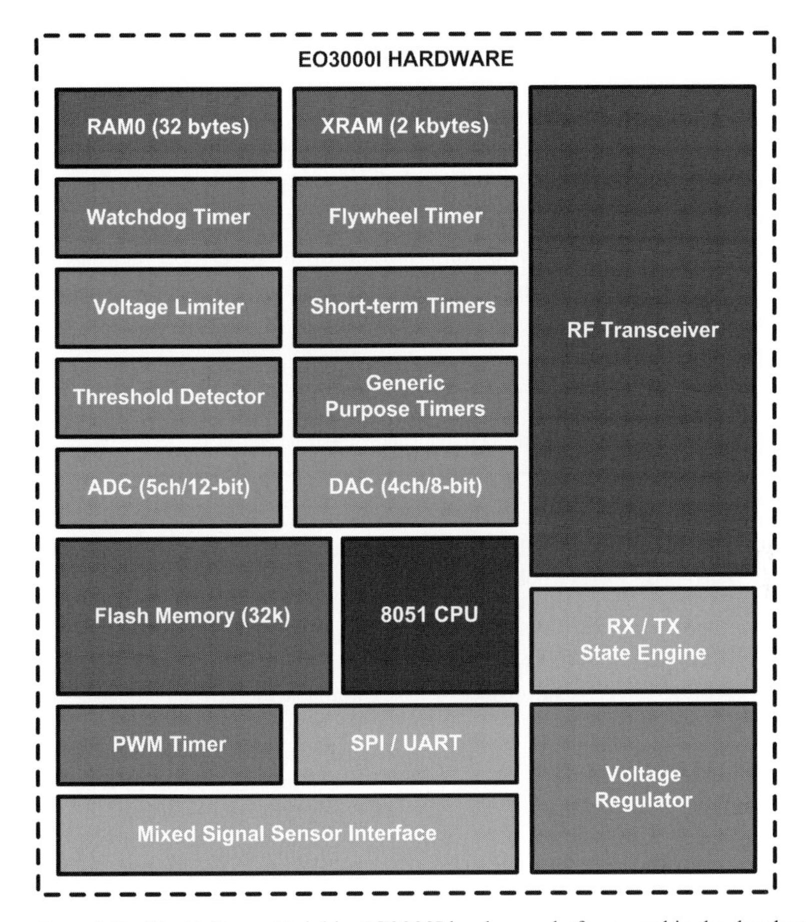

Figure 9.10. The EnOcean Dolphin EO3000I hardware platform used in the development and design of numerous EnOcean-enabled products.

9.6 The Dolphin Platform

In the sections to follow, we take a bottom-up approach when discussing the EnOcean architecture, and, in this section, we'll take a closer look at the hardware and development environment provided by EnOcean, the *EnOcean Radio Protocol* (ERP), and the various layers that make up the stack architecture.

The EnOcean *Dolphin* platform is a complete *System-on-Chip* (SoC) transceiver solution that provides bi-directional *Ultra-low Power* (ULP) applications; it comprises an RF transceiver, with a data rate of up to 125 kbit/s and an 8051 microcontroller core, along with associated peripherals and numerous unique ULP power management blocks. In Figure 9.10, we illustrate the building blocks of the EO3000I hardware, and in Table 9.2 we provide a summary of its functionality. The Dolphin platform empowers its users to design and develop a breadth of switches, sensors, receivers, and transceivers for the home, commercial, and industry automation industries, all self-powered using EnOcean's energy harvesting technology, as discussed earlier in Section 9.3, "EnOcean's Application Portfolio." The platform can also be used in various line-powered EnOcean receivers with switched output or gateways.

Furthermore, EnOcean offers a comprehensive development environment, along with an *Application Programming Interface* (API) – named the *DolphinAPI*. The API permits users to develop their own specific firmware, wherein management of the system is extended to the control and configuration of the processor, facilitates in the reception and transmission of radio telegrams based on the ERP (see Section 9.6.1, "The EnOcean Radio Protocol"), and manages generic I/O and the control of various power modes.

Table 9.2. A summary of EnOcean's EO3000I hardware building blocks

Block	Name	Description
ULP	Voltage limiter.	Limits the supply voltage of the Dolphin system.
	Threshold detector.	An ULP on / off threshold detector.
Timers	Watchdog.	An ULP timer, which is based on an internal oscillator.
RF	Transceiver and state engine.	Both the 315 MHz and 868 MHz flavors are available to EnOcean products. The configurable RF transceiver, along with the integrated state engine, permit the transmission and reception of telegrams based on the ERP.
CPU and peripherals		An 8051 processor accompanied by a transceiver and state engine, CPU and system timers, memory, and serial interface.
Mixed signal sensor interface		A mixed signal sensor interface capable of supporting up to ten configurable I/O lines.
Operating modes	"CPU mode," "OFF mode," "Deep sleep mode," "Flywheel sleep mode," "Short-term sleep mode," and "Standby mode."	

9.6.1 The EnOcean Radio Protocol

In Figure 9.11, we map the *Open Systems Interconnection* (OSI) model against the EnOcean stack, along with its associated high-level modules, to provide a better understanding of the significant components that make up the overall EnOcean stack architecture, which is explored in greater detail later in Section 9.7, "The Dolphin Architecture."

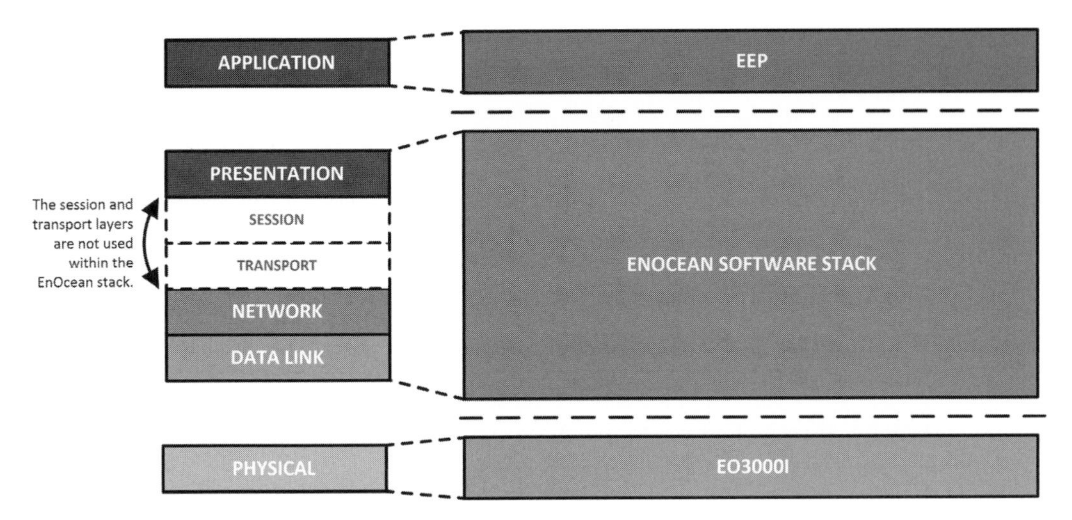

Figure 9.11. The OSI model mapped against EnOcean's architectural building blocks – the session and transport layers are not used within the EnOcean ecosystem, but are shown here for clarity.

9.6.1.1 Sub-telegrams

The ERP is a packet-based protocol and takes a hierarchical form of three possible different data unit types, namely a *frame*, a *sub-telegram*, or a *telegram*. The frame type is the lowest type of data unit within the ERP, as it represents data at the physical layer. It will include information relating to the control and synchronization of the receiver from where the unit is transmitted as a bit serial sequence. A sub-telegram, as shown in Figure 9.12, is managed at the data link layer. The EnOcean system software automatically manages the encoding/decoding process, such as the preamble, *Start of Frame* (SoF), inverse bits, and *End of Frame* (EoF), ultimately to ensure encoding quality. Encoding occurs prior to transmission to the physical layer; likewise, during the decoding process, the padding is removed and discarded. The ERP is predominately a uni-directional protocol, but bi-directional capability is also available, especially with SmartACK, which we discuss later. The ERP offers no handshaking, although to ensure reliability within the ERP, three identical sub-telegrams are typically transmitted over a certain period. In Table 9.3, we expand on and explain further the fields that form the sub-telegram data unit, and we include their size in bits. We discuss the third data unit, telegrams, later, in Section 9.8, "The EnOcean Equipment Profiles," but before we discuss the sub-telegram timings, we should note for future reference that *Radio* telegrams are grouped *ORG*anizationally (RORG).

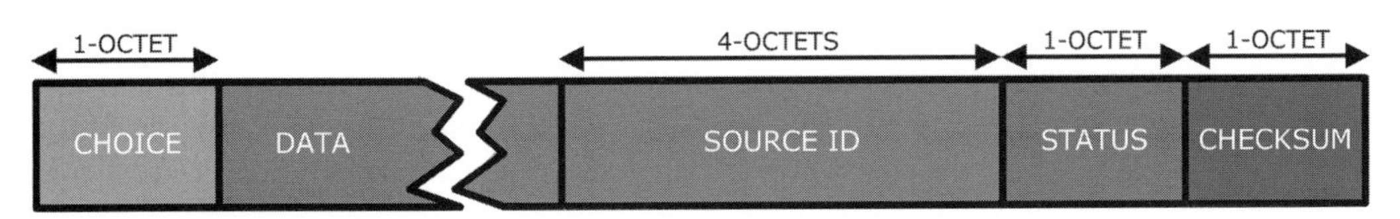

Figure 9.12. The structure of the sub-telegram data unit.

Table 9.3. The universal fields that make up the sub-telegram structure

Field	Description	Size
Choice	Defines the type of sub-telegram, has a field size of 8-bits, and matches the RORG value at the EEP.	8
Data	The actual payload being transmitted by the sub-telegram, and is of variable length.	
Source ID	Identifies the source transmitter, and is 32-bits in length.	32
Status	This status value, which is 8-bits in length, determines whether the sub-telegram was transmitted from a repeater, and identifies the type of integrity control mechanism used.	8
Checksum	Offers a data integrity check on the data being transmitted, and is 8-bits in length.	8

Table 9.4. The maturity window for both TX and RX periods

Maturity (ms)	Description
40	maximum TX maturity time
100	RX maturity time

9.6.1.2 Sub-telegram Timings

The timing of sub-telegrams within the ERP is significant insofar as it is important to avoid unnecessary telegram collision over the air-interface; as such, a sub-telegram is transmitted over a different time range, which is determined by *transmit* (TX) and *receiver* (RX) *maturity* periods. A maturity period defines the window in which all sub-telegrams have to be completed and received, and their associated timings are provided in Table 9.4. So, a telegram uses three sub-telegrams, and each sub-telegram, that is, start to finish, must not exceed the TX maturity period.

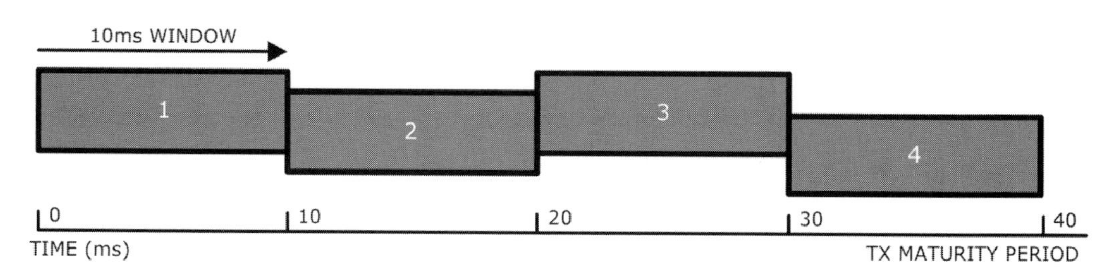

Figure 9.13. The TX maturity period, which is divided into four 10 ms windows.

Furthermore, a repeater uses a different timing method compared with that of the originating transmitter and, as such, each sub-telegram, again from start to finish, must not exceed the RX maturity period. What's more, to provide an effective scheduling mechanism, the TX maturity period is divided into four windows, where each window has a 10 ms time slot, as depicted in Figure 9.13. The TX period offers the opportunity for a maximum of three sub-telegrams to be transmitted, where the exact timing of the transmission is determined by a sequence relating to the current window in use.

Table 9.5. The allocation of time slots to the appropriate telegram

Status	First sub-telegram	Second sub-telegram	Third sub-telegram
Original	0	1...9	20...39
Level 1 repeated	10...19	20...29	
Level 2 repeated	0...9	20...29	

Additionally, to assist in collision avoidance when using repeaters, the timing of the original versus the repeated telegrams differs. We demonstrate the allocation of time slots for the various telegrams in Table 9.5.

All sub-telegrams are transmitted in sequence. In other words, a second or third sub-telegram may only commence transmission once the first sub-telegram has finished. However, if the wireless channel is occupied, perhaps by a neighboring transmission or interference, then the *Listen Before Talk* (LBT) mechanism can delay the transmission until the end of the TX maturity period.

9.6.1.3 Listen Before Talk

The LBT mechanism is a collision avoidance technique used generally within wireless communications in which a transmitting device ascertains if there is an ongoing transmission prior to the device sending its frame. If an ongoing conversation is detected, then the EnOcean device suspends transmission for a random period, after which the device will try again. Naturally, if the device determines that the frame can be transmitted, then the sub-telegram is sent. The LBT is optional, although is highly recommended.

Table 9.6. Status field values for repeater levels

Value	Description
`0x00`	Original sender.
`0x01`	Sub-telegram was repeated once.
`0x02`	Sub-telegram was repeated twice.
`0x0F`	Sub-telegram is not repeated.

9.6.1.4 Repeater Behavior

A repeater device is used in an environment where the distance between a sender and receiver device is too great to achieve a satisfactory wireless connection. In a relay race fashion, the repeater device receives the frame from the source or sender device and relays that sub-telegram to the destination or receiver device. It is possible to have up to two repeaters in an EnOcean ecosystem. Repeated sub-telegrams within the ecosystem are managed by modifying the status field (see Table 9.6) of the ERP, especially when one or more repeaters are present. You may recall that we illustrated the sub-telegram in Figure 9.12 and described its purpose in Table 9.3. What's more, two repeater levels are used to distinguish and restrict the amount of repeated sub-telegrams in an environment – we characterize the two repeater levels in Table 9.7. We also discussed sub-telegram

Table 9.7. The two repeater levels that aid in limiting repeated sub-telegrams

Level	Description
1	Repeat only received original sub-telegrams.
2	Repeat only received original sub-telegrams or once-repeated sub-telegrams.

timings, which help with collision avoidance, earlier, in Section 9.6.1.2, "Sub-telegram Timings."

9.6.1.5 The ISO/IEC 14543-3-10 Standard

The *International Organization for Standardization* (ISO)/*International Electrotechnical Commission* (IEC) has used the ERP as a basis for a new international standard, namely the ISO/IEC 14543-3-10 Information technology – *Home Electronic Systems* (HES), Part 3–10 "*Wireless Short-Packet* (WSP) protocol optimized for Energy Harvesting – Architecture and lower layer protocols."[5]

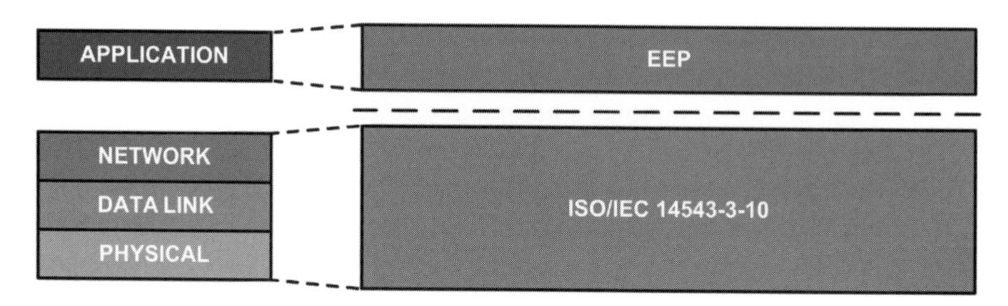

Figure 9.14. The ISO/IEC 14543-3-10 provides the physical, data link, and the network layers.

The standard covers OSI layers one to three (physical, data link, and networking), as illustrated in Figure 9.14. It is the first and only wireless standard focused on optimized ultra-low energy consumption that is also suitable for use with energy harvesting. The new standard, along with the EnOcean EEP, together provide an holistic, interoperable, open technology useable worldwide, just like other standards, for example technologies such as Bluetooth, Zigbee, Wi-Fi, and so on. The standard was ratified and published by the ISO/IEC in April 2012.

9.6.2 The EnOcean Serial Protocol

The *EnOcean Serial Protocol* (ESP) permits serial communication between a host (for example, an external microcontroller or PC) and an EnOcean module – the topology of which is illustrated in Figure 9.15. The interface shared between a host and the EnOcean device is based on an RS232 serial interface; in particular, a 2-Wire *Universal Asynchronous Receiver/Transmitter* (UART) connection, comprising receive, transmit, and ground, along with software handshake and full-duplex.

[5] The ISO/IEC 14543–3-10 is an open standard and is generically available for download at iso.org.

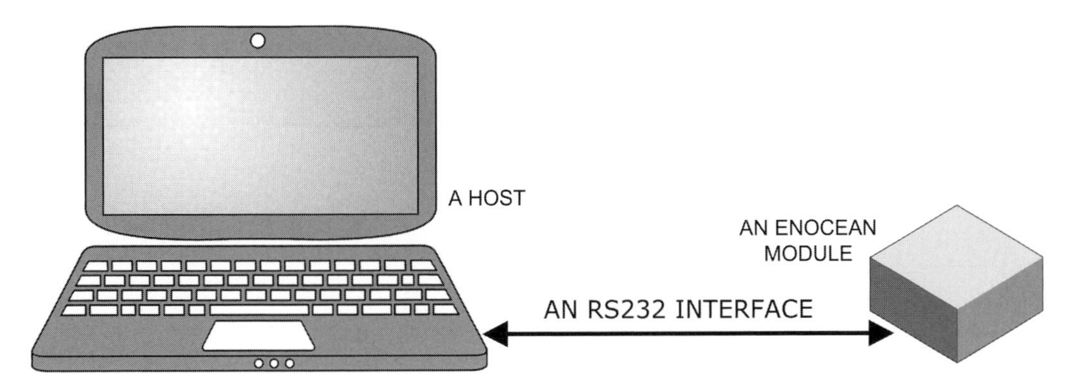

Figure 9.15. The host (PC) and an EnOcean wireless module communicating bi-directionally over an RS232-like interface.

The ESP utilizes a point-to-point topology with a data structure, as depicted in Figure 9.16. The ESP data structure encapsulates user data, command, response, and event messages. In Figure 9.16, you will see that the packet comprises a header, length, packet type, data, and an optional data payload, in addition to a sync-byte (start of frame), CRC8 header, and data fields, which are explained further in Table 9.8.

Table 9.8. The fields that make up the ESP frame structure

Field	Description	Size
Sync-byte	An 8-bit field used in synchronization of the serial connection, always set to `0x55`. We discuss synchronization in Section 9.6.2.1, "Synchronization."	8
Data length	This 16-bit field specifies the number of bytes that need to be addressed in the data field.	16
Optional length	This 8-bit field specifies the number of bytes that need to be addressed in the optional data field.	8
Packet type	This 8-bit field specifies the packet type contained within the data and optional data fields. The possible packet types are listed in Table 9.9.	8
CRC8H	The CRC8H (header) 8-bit field contains the checksum for fields, data, optional lengths, and type.	8
Data	This field will contain the actual payload, and is of variable length.	
Optional data	This optional field will contain the actual payload, and is of variable length.	
CRC8D	The CRC8D (data) 8-bit field contains the checksum for fields, data, and optional data.	8

9.6.2.1 Synchronization

Once a sync-byte value (`0x55`) is identified, the data and optional length fields, along with the packet type field (four bytes in total), are validated against the CRC8H (header)

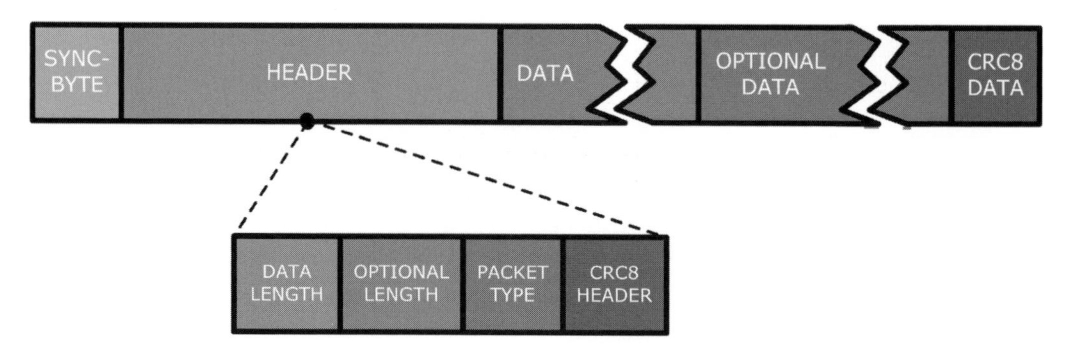

Figure 9.16. The ESP packet structure.

Table 9.9. The packet type field value

Value	Description
`0x00`	Reserved.
`0x01`	Radio telegram.
`0x02`	Response to any packet.
`0x03`	Radio sub-telegram.
`0x04`	Event message.
`0x05`	Common command.
`0x06`	SmartACK command.
`0x07`	Remote management command.
`0x08--0x7F`	Reserved (EnOcean use).
`0x08--0xFF`	Manufacturer-specific command and messages.

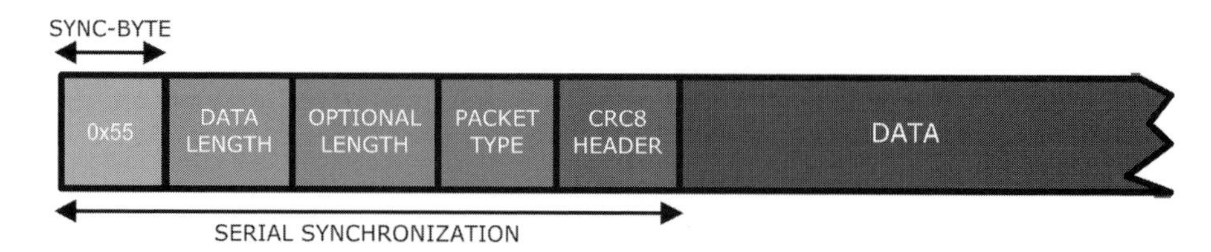

Figure 9.17. The sync-byte is used to denote a new payload, as long as the CRC8H field corresponds to the following four bytes.

field, as we illustrate in Figure 9.17. If there's a match, then the new payload is processed and passed along. However, if there's no match, then the sync-byte value does not correspond to a new payload, and the whole process is repeated.

9.7 The Dolphin Architecture

The Dolphin system architecture is modular in nature; we illustrate a high-level perspective in Figure 9.18. The functionality offered within the architecture is grouped into a number of software modules, where each module supports its own functions

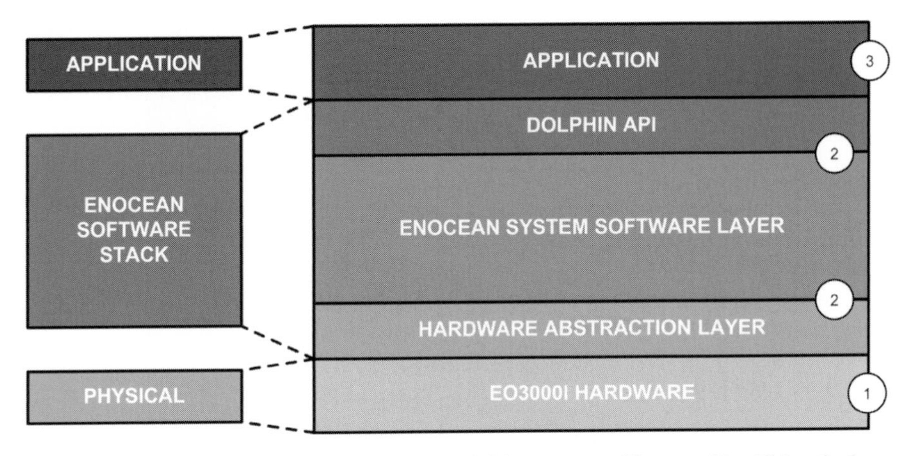

Figure 9.18. The high-level perspective of the Dolphin system architecture identifying the layers that make up the EnOcean software stack.

and parameters – something we touch upon in the following sections. The majority of these software modules are independent and support an interface that can be directly accessed by the application, where each accessible software module has two types of interfaces, namely *initialization* and *functional*. The functional interface provides a consistent method by which to determine whether the function was executed successfully (or not, along with a reason), in addition to providing parameters of the function, which, in turn, is used as an input/output mechanism. However, some core software modules within the Dolphin architecture are not necessarily accessible from the application layer, for example modules such as the schedule and interrupt handlers (see Figure 9.19).

Looking back at Figure 9.18, where we provide the key building blocks of the EnOcean software stack, you will note that these building blocks have been labeled numerically to ease identification and classify responsibility. The block labeled (**1**) represents the *Media Access Control* (MAC) and *Physical* (PHY) layers, which we discussed earlier in Section 9.6, "The Dolphin Platform," which is EnOcean responsible; (**2**) represents the building blocks that comprise the crux of the EnOcean architecture, to include the DolphinAPI (see Section 9.7.1, "The DolphinAPI"), the *EnOcean System Software Layer* (ESSL), and the *Hardware Abstraction Layer* (HAL), which are all EnOcean responsible; and finally, (**3**) represents the building block that enables third-party manufacturers to develop their own end-applications utilizing the latest EEP specification.

9.7.1 The DolphinAPI

The DolphinAPI provides the environment in which a developer has an opportunity to develop two applications types, namely *line-powered* and *self-powered ULP* applications. A developer should decide which one to use, as the concepts cannot be combined. A line-powered application is normally intended for modules that require a permanent power supply, such as controllers, repeaters, and gateways – typically, the radio receiver needs to remain on and doesn't enter a sleep state, with the exception of standby. On the

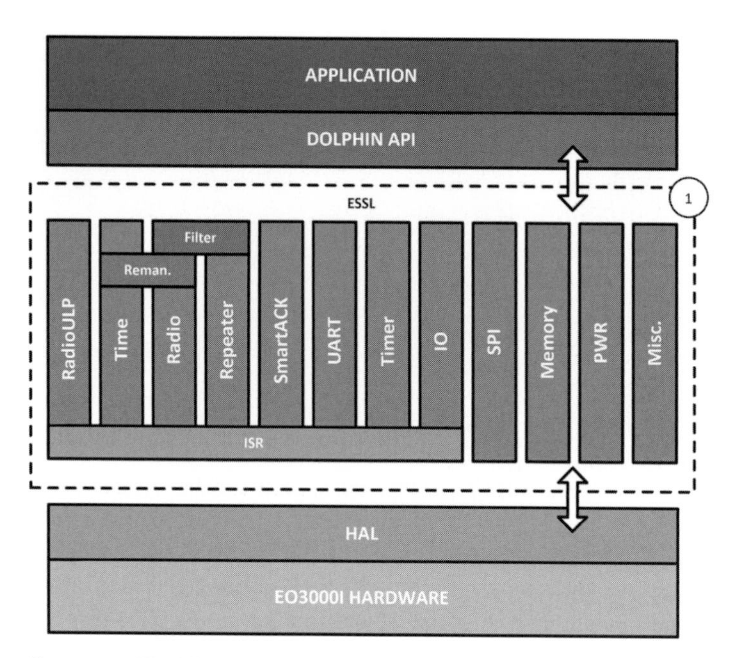

Figure 9.19. The ESSL within the Dolphin architecture comprises a number of interactive software modules.

other hand, a self-powered ULP application is intended for modules that are powered solely by energy harvesting techniques. In this instance, the module is predominately switched off or in a deep sleep state, awaking only periodically. Typical self-powered applications include sensors and actors.

9.7.1.1 Application Types

The two applications (line-powered and self-powered) differ in that the API provides different methods in functionality for energy optimization. For example, ordinarily, a line-powered device, whilst energy efficient, can perhaps become somewhat carefree with its energy consumption, by which we mean that a telegram can be transmitted using the *scheduler* (see Section 9.7.2.1, "Scheduler") irrespective of the time lapsed. Conversely, a self-powered ULP device has to adopt a more conservative approach to its energy consumption, as telegram transmission has to be executed quickly and efficiently. So, whilst both applications offer identical functions, line-powered applications tend to utilize the system scheduler to execute functionality effectively.

9.7.2 The EnOcean System Software Layer

In Figure 9.19, we depict a number of boxes that are shown horizontally, which represent numerous software modules that form the ESSL and are denoted by the numeric reference (**1**). So, moving from left to right, we have the following modules: RadioULP;

several timer modules; a *Serial Peripheral Interface* (SPI); a *Universal Asynchronous Receiver/Transmitter* (UART); SmartACK (see Section 9.7.3, "Smart Acknowledge"); generic I/O; *Power Management* (PWR); and miscellaneous software module resources. These modules interact with each other, but, of course, some modules are not directly accessible from the application layer, as we have already touched upon. The ESSL layer is responsible for housing the scheduler, power management, the serial and radio protocol stacks, and other peripheral components. Most notably, the Dolphin architecture does not implement a user/kernel topology and, as such, the DolphinAPI and other library objects are all linked to the application during compilation time. What's more, memory management is not required within the Dolphin architecture, as this is also allocated during compilation.

Table 9.10. The functionality supported by the internal scheduler

Scheduler functionality
Transmit and receive telegrams.
Apply RX/TX maturity time management.
Offer levels 1 and 2 repeating.
SmartACK postmaster functionality (see Section 9.7.3, "Smart Acknowledge").
LBT (see Section 9.6.1.3, "Listen Before Talk").
Software timing.
Remote management.

9.7.2.1 Scheduler

In Table 9.10, we summarize the functionality offered by the scheduler. The scheduler has the ability to execute tasks concurrently within the EnOcean system. The DolphinAPI provides a unique preemptive priority scheduler at the system level and supports both single application and multiple system tasks.

In essence, system tasks are executed concurrently, but they can also be executed preemptively compared with the application task. Incidentally, there is only one application task. Typically, system tasks execute to completion and are unable to interrupt ongoing system tasks. However, the UART system task possesses higher priority and has the ability to preempt other system tasks.

Synchronous System Tasks

The scheduler is executed on a *round robin* basis, whereby the application is interrupted and a task is provided with a 1 ms opportunity – synchronous system tasks are executed sequentially and are triggered by the timer hardware interrupt. Typically, the average duration for a synchronous system task to complete is circa 50 to 150 μs. In some instances, a synchronous system task may take longer than 1 ms, but this is typically subject to active features, radio buffers, traffic load, and the length of a sub-telegram.

9.7.3 Smart Acknowledge

Smart ACKnowledge (SmartACK) is a bi-directional communications protocol. Communication occurs between two *actors*: an energy autarkic sensor and a line-powered controller. SmartACK is used to optimize data transmission and reception within the EnOcean ecosystem. Normally, a sensor in receive mode causes undue energy consumption and, as such, message synchronization is used to alleviate receiver on-time, where the message flow is performed in a predefined interval; for example, the *actual reclaim period*. Once a sensor transmits its telegram, it expects an immediate response and, as a consequence, the receiver on-time is available for a very short period. In circumstances where the sensor hasn't had the opportunity to receive a response, it can *reclaim* it later from a *mailbox*, which is located on the controller device. Likewise, on occasions, repeaters may cause an unknown delay and, in turn, disrupt synchronization. To overcome this potential issue, mailboxes are established with line-powered devices, which have direct contact with the autarkic device. These line-powered devices are termed *postmasters* and may also behave as a controller or a repeater.

In the SmartACK protocol there are three actors, including a sensor, repeater, and controller; these are typically *actual* devices within an EnOcean ecosystem. Both repeater and controller devices can adopt the role of postmaster, housing a sensor's mailbox, and, whilst the functionality may be similar across these devices, controllers are capable of *learning*. A repeater may also provide other functionality, such as a light actuator. A SmartACK sensor is a self-sufficient device and is also capable of learning; as a consequence, both the controller and the sensor are aware of each other's existence.

Table 9.11. The two types of mailbox that can be in use

Type	Description
Temporary	A temporary mailbox is used during learning.
Normal	A normal mailbox is used in standard operation.

9.7.3.1 Mailboxes

A mailbox has the capacity to retain one telegram at a time, whereas a postmaster will retain one of two types of mailbox, namely a *temporary* or a *normal* mailbox (see Table 9.11). The postmaster also has the responsibility of notifying a sensor if its mailbox is empty or if it doesn't exist. A sensor, such as a controller device, can hold multiple (as many associations as it may have with other sensors within the EnOcean ecosystem) mailboxes.

9.7.3.2 Reclaiming

As you may recall, the reclaim process offers the opportunity to reduce the need for radio-on time when receiving incoming telegrams. As such, time synchronization is established between the sensor and the postmaster where the "radio on" is only enabled for the actual transfer. In circumstances where a reclaim has failed, the sensor can

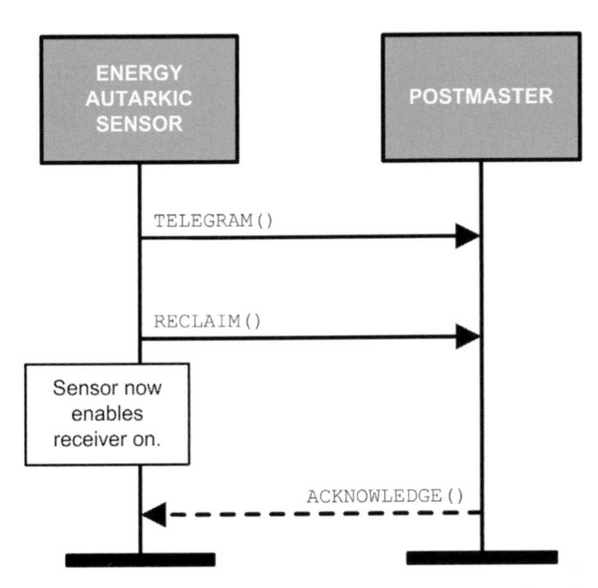

Figure 9.20. The sensor posts a reclaim message to retrieve its telegram, at which point the radio is switched on for an answer (acknowledge) within the actual reclaim period.

retry up to three times. The period between a retry is mandated by the length of a sub-telegram, as previously discussed in Section 9.6.1.2, "Sub-telegram Timings." In Figure 9.20, we provide a message sequence chart depicting when the sensor transmits a reclaim telegram and switches the radio (receiver) on. As you can see, as soon as the sensor sends a request to reclaim its telegram, the radio is turned on; likewise, following the actual reclaim period, the sensor switches off the radio. The "acknowledge" in this sequence essentially forms the conceptualization *SmartACK*.

Table 9.12. The different types of telegram use when operating in normal or learning operation

Telegram	Normal operation	Learning behavior
Initial	Data.	Learn request.
Reply	Data reply.	Learn reply.
Reclaim	Data reclaim.	Learn reclaim.
Acknowledge	Data acknowledge.	Learn acknowledge.

9.7.3.3 Normal Operation and Learning Behavior

We need to make a distinction between *normal operation* and *learning behavior*; in Table 9.12, we list the SmartACK telegram differences. Normal operation refers to the generic data transfer that is typically performed between two devices, and in Figure 9.21 we illustrate the message sequence for a simple operation where the sensor and postmaster are in direct radio contact. Conversely, in Figure 9.22 we illustrate a sequence where the sensor is not in direct radio contact; you can clearly see from the illustration

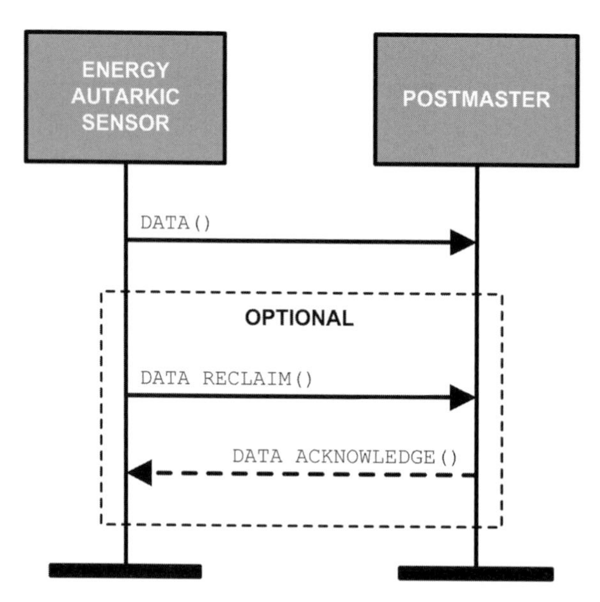

Figure 9.21. The normal operation message sequence in which the sensor is in direct radio contact. The acknowledge reclaim is optional.

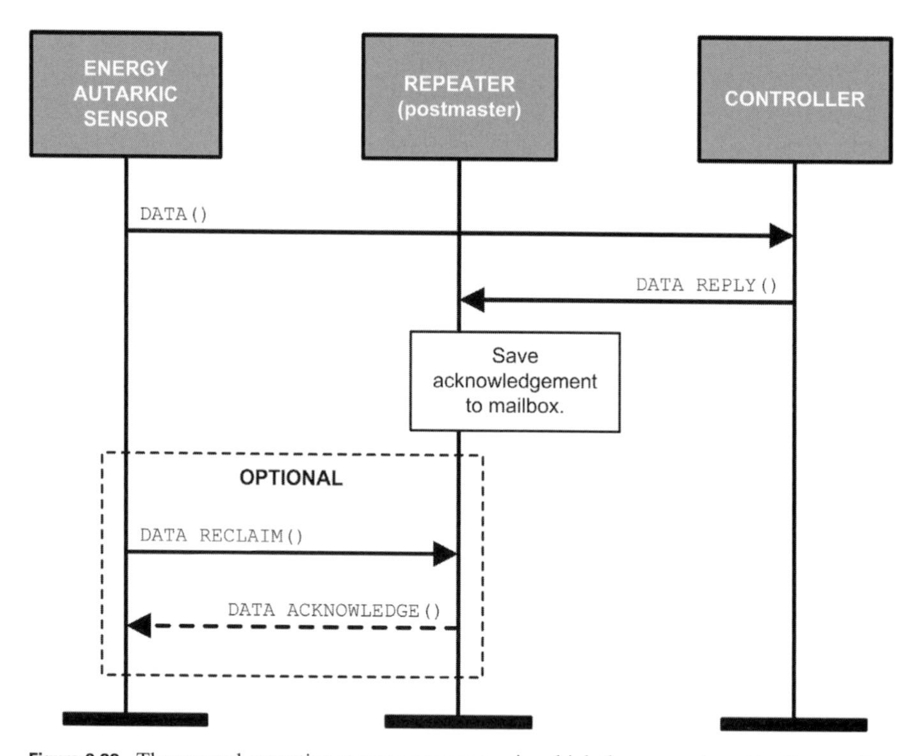

Figure 9.22. The normal operation message sequence in which the sensor is not in direct radio contact and a repeater is used. The acknowledge reclaim is optional.

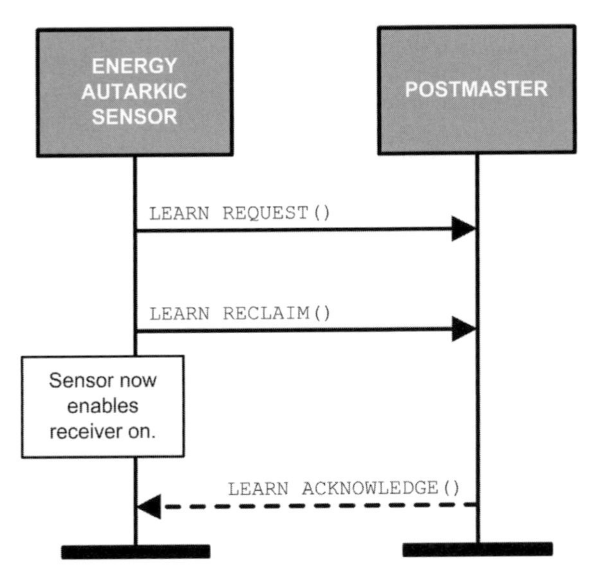

Figure 9.23. The learning behavior message sequence in which the sensor is in direct radio contact.

that a repeater is used – we'll come back to how this distinction is made in a moment. What's more, the acknowledgement provided in both instances is optional.

So, learning refers to a sensor's ability to understand more about neighboring sensors by exchanging information; in turn, this results in *learning in* or *learning out* – more about this later. Both a controller and an autarkic sensor enter learning mode, where the latter device posts a *learn request* telegram (see the message sequence chart in Figure 9.23 and Figure 9.24, which shows a more complex scenario in which a repeater is involved), since the sensor is not in direct radio contact. Let's revisit how a device understands whether it is in direct radio contact. When a sensor issues a learn request telegram, a *Radio Signal Strength Index* (RSSI) value is used to denote whether the signal strength is acceptable. If the value provided is significant, then the device is considered to be within direct contact.

As we have already mentioned, SmartACK controllers or repeaters are both capable of retaining a sensor's mailbox. Once a sensor ascertains the nearest device, a postmaster is chosen such that the controller (or repeater) device will then decide if the learning is "in" or "out." Naturally, for both normal operation and learning behavior processes, the optimal scenario is to determine a simple mode, that is, the sensor in either scenario is in direct radio contact.

Learn In and Learn Out

In short, "learn in" denotes a situation such that a controller needs to sustain a relationship with its sensor, whereas "learn out" refers to a context in which a controller wishes to cease the relationship. A relationship may be determined if the controller is already aware

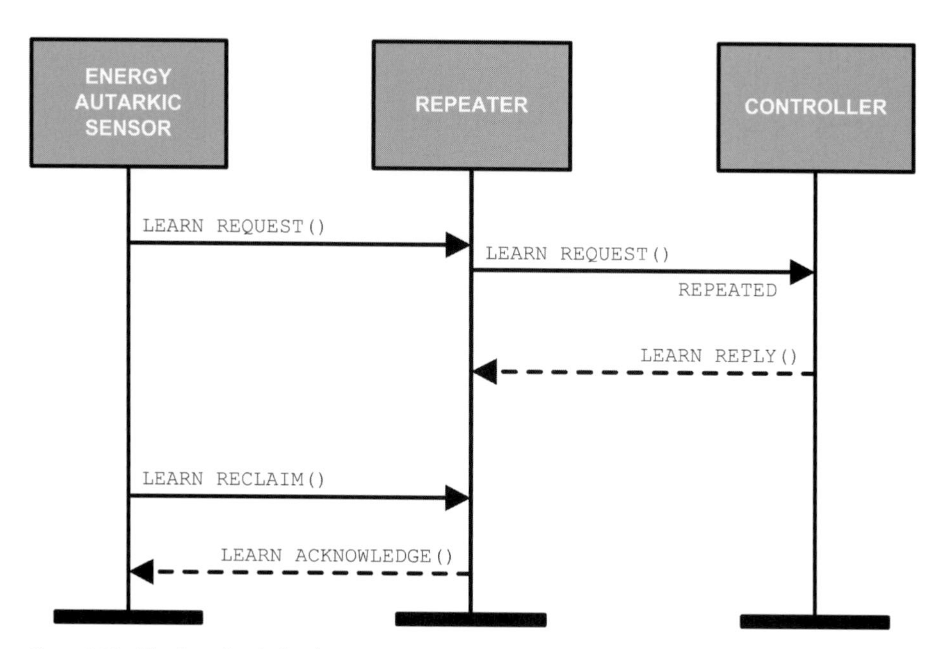

Figure 9.24. The learning behavior message sequence in which the sensor is not in direct radio contact and a repeater is used.

of an existing relationship or if the controller accepts EEP telegrams from neighboring sensors.

9.7.4 The Hardware Abstraction Layer

The HAL forms the lowest layer within the EnOcean stack (see Figure 9.18 and Figure 9.19), and it offers hardware independence and abstraction for the underlying Dolphin platform.

9.8 The EnOcean Equipment Profiles

The EEP essentially forms the application layer, as we discussed earlier, in Section 9.7, "The Dolphin Architecture." As already discussed, the ERP defines the radio telegram and is optimized to transmit data using relatively little power. The transmission window for a radio telegram is under 1 ms. With infrequent data transmission, the likelihood of data collision is considerably reduced, ensuring a reliable ecosystem. In this section, we learn more about the EEP telegram structures used in the user data or payload that, in turn, offer interoperability across manufacturers. The EnOcean Alliance has defined the key characteristics of a device by creating three profile elements, which are common to all profiles – we introduce these elements in Table 9.13 and illustrate them in Figure 9.25.

Table 9.13. The three common profile elements used in the payload

	Description
1	ERP radio telegram type, "RORG."
2	Basic functionality of the data content, "FUNC."
3	Type of device, "TYPE."

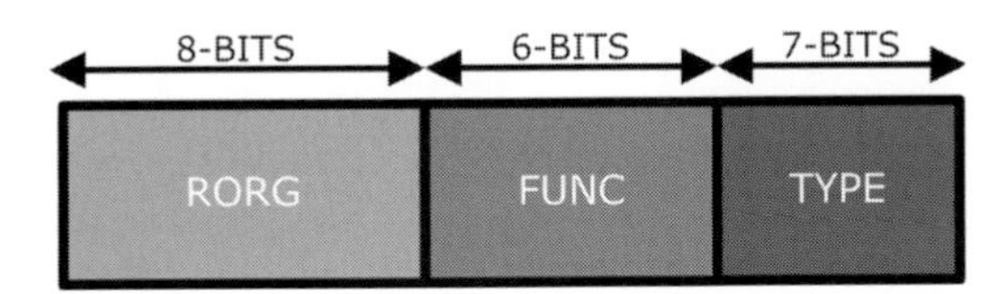

Figure 9.25. The three profile elements that make up all EnOcean devices.

Table 9.14. The telegram types used within the EEP

RORG	Telegram	Definition
`0xF6`	RPS	Repeated switch communication.
`0xD5`	1BS	1-byte communication.
`0xA5`	4BS	4-byte communication.
`0xD2`	VLD	Variable length data.
`0xD1`	MSC	Manufacturer-specific communication.
`0xA6`	ADT	Addressing destination telegram.

You may recall from Section 9.6.1.1, "Sub-telegrams," that we touched upon the RORG value corresponding to "Choice" at the radio telegram level, and in Table 9.14 we list the telegram types used at the EEP layer.

9.8.1.1 RPS and 1BS Telegram Types

The RPS and 1BS telegram types offer a one-byte data payload; as such, the frame structures are comparable, as you can see in Figure 9.26. However, the two telegrams differ, as the 1BS telegram uses an LRN-bit in its data field.

To demonstrate an RPS data telegram in use, we have provided, in Figure 9.27, an example EnOcean device, which is based on the actual PTM200 transmitter module. You should note that the image refers to "state" and "channel," as these distinctions influence the functionality and values set in the status field. What's more, there are two different message types, namely the *N*-message and the *U*-message, which are also uniquely identified in the status field, as shown in Figure 9.26. As such, not only is the data type provided within the data telegram, but also the T21 and NU bits of the status field are used to convey the state and channel distinction; and finally, the status field also includes a *Repeater Count* (RC). In Table 9.15 and Table 9.16, we list the range of application sets used for the RPS and 1BS telegram types, respectively.

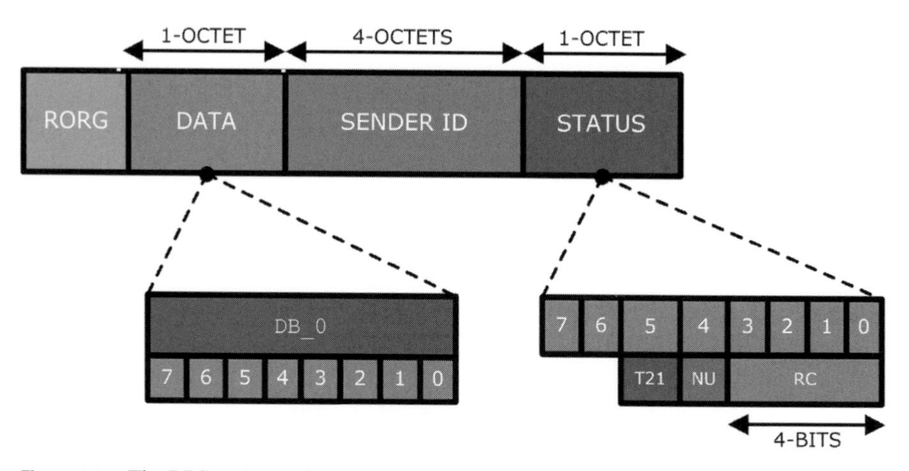

Figure 9.26. The RPS and 1BS frame structure are comparable.

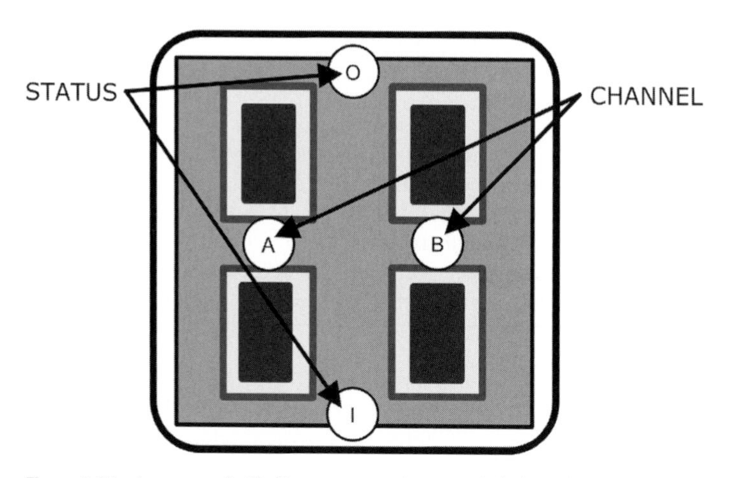

Figure 9.27. An example EnOcean transmitter module based on the PTM200.

Table 9.15. Application types that utilize the RPS telegram

RORG	FUNC	Description	TYPE	Description
0xF6	0x02	Rocker switch, two rockers.	0x01	Light and blind control, style 1.
	0x02	Rocker switch, two rockers.	0x02	Light and blind control, style 2.
	0x03	Rocker switch, four rockers.	0x01	Light and blind control, style 1.
	0x03	Rocker switch, four rockers.	0x02	Light and blind control, style 2.
	0x04	Position switch, home and office application.	0x01	Key card activation switch.
	0x10	Mechanical handle.	0x00	Window handle.

Table 9.16. Application types that utilize the 1BS telegram

RORG	FUNC	Description	TYPE	Description
0xD5	0x00	Contact and switches.	0x0	Single input contact.

Table 9.17. Application types that utilize the 4BS telegram

RORG	FUNC	Description	TYPE	Description
0xA5	0x02	Temperature sensors.	0x01–0x1B	Variable temperature ranges.
	0x04	Temperature and humidity sensors.	0x01	Range 0 °C to 40 °C and 0 to 100%.
	0x06	Light sensor.	0x01	Range 300 lux to 60.000 lux.
	0x06	Light sensor.	0x02	Range 0 lux to 1.020 lux.
	0x07	Occupancy sensor.	0x01	Occupancy.
	0x08	Light, temperature, and occupancy sensors (ceiling).	0x01	Range 0 lux to 510 lux; 0 °C to 51 °C and occupancy.
	0x08	Light, temperature, and occupancy sensors (wall).	0x02	Range 0 lux to 1020 lux; 0 °C to 51 °C and occupancy.
	0x08	Light, temperature, and occupancy sensors (outdoor).	0x03	Range 0 lux to 1530 lux; −30 °C to 50 °C and occupancy.
	0x09	Gas sensor.	0x01	Co sensor.
	0x09	Gas sensor.	0x04	Co_2 sensor.
	0x10	Room operating panel.	0x01–0x1E	Various: voltage monitor; temperature sensor; set point; fan speed control; humidity; occupancy control.
	0x11	Controller status.	0x01	Lighting controller.
	0x11	Controller status.	0x02	Temperature controller output.
	0x12	*Automated Meter Reading* (AMR).	0x00–0x03	Counter: electricity, gas, and water.
	0x13	Environmental applications.	0x01–0x06	Weather station; sun intensity, northern hemisphere; data exchange; time and day exchange; direction exchange; geographic position exchange.
	0x20	HVAC components.	0x01–0x03	Battery powered, basic, and line-powered actuators.
	0x20	HVAC components.	0x10	Generic HVAC interface.
	0x20	HVAC components.	0x11	Generic HVAC interface (error control).
	0x20	HVAC components.	0x12	Temperature controller input.
	0x37	Energy management.	0x01	Demand response.
	0x38	Central command.	0x08	Phc gateway.
	0x3F	Universal.	0x00	Radio link test.

9.8.1.2 4BS Telegram Type

The 4BS telegram type provides a four-byte data payload. We illustrate the 4BS frame structure in Figure 9.28. In this example, we show a 10-bit temperature profile demonstrating how a range from 0 to 1023 can be addressed. In Table 9.17, we list the range of application sets used for the 4BS telegram type.

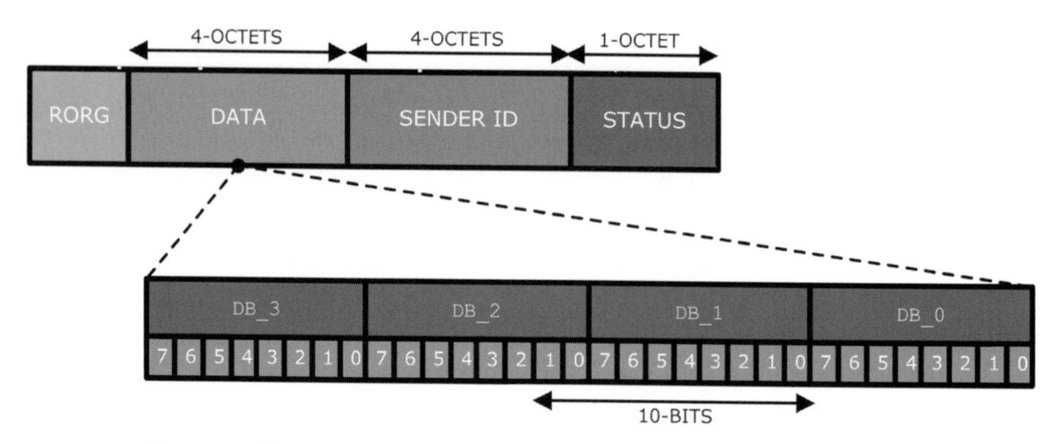

Figure 9.28. The 4BS frame structure.

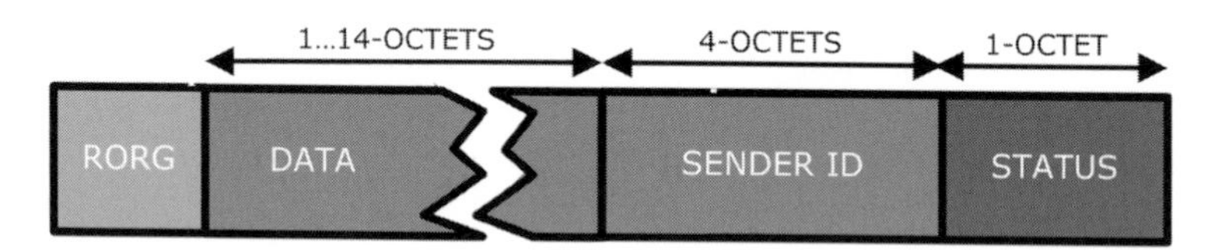

Figure 9.29. The VLD frame structure.

9.8.1.3 VLD Telegram

The VLD telegram type offers a variable data payload of 1–14-bytes. We illustrate the VLD frame structure in Figure 9.29. In Table 9.18, we list the range of application sets used for the VLD telegram type. The application type shown utilizes the SmartACK protocol, which we discussed earlier in Section 9.7.3, "Smart Acknowledge."

Table 9.18. Application types that utilize the VLD telegram

RORG	FUNC	Description	TYPE	Description
0xD2	0x00	*Room Control Panel* (RCP).	0x01	RCP with temperature measurement and display.

10 The Power of Less

ANT

Last, but not least, we complete Part II, "The Wireless Sensor Network," with ANT Wireless (thisisant.com). This ultra-low power, coin-cell battery operated wireless technology operates in the unlicensed *Industrial, Scientific, and Medical* (ISM) 2.4 GHz frequency band and is an adaptive isochronous ad hoc wireless protocol providing an extensible, self-adaptive, and practical mesh network that, in turn, offers flexibility and scalability. As with other technologies, we review ANT's history as well as the ANT+ Alliance benefits. Similarly, we'll also review ANT's product portfolio and its market scope, and we'll place the technology into a real-world context to form a better understanding of its role within its market sector. What's more, we'll compare ANT Wireless with other competing technologies within the same market in order to understand what separates ANT from its competitors.

Figure 10.1. ANT+ trademark and logo. (Courtesy of Dynastream Innovations Inc.)

10.1 Overview

ANT (see Figure 10.1 for the technology's logo) is a proprietary wireless technology supported and developed by Dynastream Innovation Inc. (dynastream.com). Dynastream, established in 1998, have their head office located in Alberta, Canada, and became a subsidiary of Garmin International Inc. (garmin.com) in December 2006. Dynastream named their wireless technology "ANT" after the tiny insect known for its productivity and strength; they believe that ANT Wireless is analogous to the insect's activity from both a visible and an unseen perspective, operating as it does at these two levels. ANT Wireless technology emerged as far back as 2000 as a consequence of Dynastream's initial development of the Triax Elite foot pod, which aimed to solve the problem of how to create an ultra-low power, low cost wireless link. Working with Nike, Dynastream identified a need for a more generic ultra-low power wireless networking solution, and in 2004 Dynastream aligned with Nordic Semiconductor (nordicsemi.com), and later Texas Instruments (ti.com), to create fully integrated ANT radio silicon to advance their theoretical concept into a practical and real-world wireless solution. ANT has evolved very differently than other technologies such as Bluetooth and Zigbee, which were initially created as generic platforms by standards bodies. ANT was initially created to solve commercial problems, and then generalized and made available to a broader audience. The technology has evolved through several generations and has become the dominant wireless technology used within the sports and fitness markets. ANT also offers a set of features that enable it to penetrate other markets, including wellness, home health, and industrial sectors. *ANT+* builds upon the core ANT protocol and offers an interoperability function specifically targeted toward manufacturers of a range of data-centric sensor products. Later in this chapter, we'll lift the lid on ANT Wireless and explore in greater detail many of the technology's features, but, in the meantime, let's review the ANT+ Alliance structure and membership benefits.

Figure 10.2. The ANT+ Alliance member trademark and logo. (Courtesy of Dynastream Innovation Inc.)

10.1.1 The ANT+ Alliance

The ANT+ Alliance is open to all individuals and companies wishing to utilize the technology. The Alliance is responsible for the development and promotion of the proprietary wireless technology worldwide, and in Figure 10.2 we illustrate the ANT+ Alliance member trademark and logo. It is also responsible for the standardization of the de facto technology for sport, fitness, and health, in turn enabling manufacturers utilizing the technology to create successful interoperable wireless products. The

ANT+ Alliance, which was created in April 2005, was founded by Dynastream and comprises a diverse number of manufacturers spanning the globe; the current membership exceeds 400, with over 25 million devices shipped.

Table 10.1. The ANT+ Alliance offers potential participants two levels of membership classes to include Full and Adopter

Membership	Scope
Full	Full membership entry permits access to the adopter benefits as well as full technical support, invitations to attend Alliance symposiums, *business-to-business* (B2B) opportunities, participation in TWGs, cooperative marketing, promotional opportunities with the Alliance, and so on.
Adopter	Adopter entry permits free access to ANT+ device profiles and associated reference code and software simulations, in addition to ANT-FS design and reference code. The provision of these benefits is provided through the ANT+ adopter's agreement.

10.1.1.1 Membership and the ANT+ Alliance

The ANT+ Alliance offers potential participants two classes of membership (as shown in Table 10.1), which enable companies to decide on how they wish to participate within the Alliance. *Adopters* can access device profile documentation and design and reference code; *Full* members receive direct technical support, are permitted to participate in the ANT+ *Technical Working Groups* (TWGs), and cooperate in the future development and direction of ANT+ device profiles.

10.1.1.2 ANT+ Certification

As we have already mentioned elsewhere in Part II, the majority, if not all of, the manufacturers of technologies featured in this book require their products to be *certified*. The ANT+ Certification program offered by the Alliance encourages interoperability of ANT+-enabled products and ensures the highest inherent quality. Products that demonstrate conformance to the device profiles as well as successful interoperability are allowed to use the ANT+ logo on their product and/or software, and in any associated marketing material.

The ANT+ Alliance provides three certification categories or types, which vary in testing granularity and scope based on new or modified implementations, or on changes that are unrelated to the ANT+ profiles, such as cosmetic or user interface modifications. The Alliance offers four steps to help simplify and achieve certification: *preparation*, *self-certification*, *submission*, and finally *testing*.

10.1.2 ANT's Timeline

The infographic shown in Figure 10.3 provides a snapshot of ANT's technology timeline, from its inception to where the technology is today. ANT has matured over the last decade or so, from where it was first conceived back in 2000 as a simple broadcasting protocol

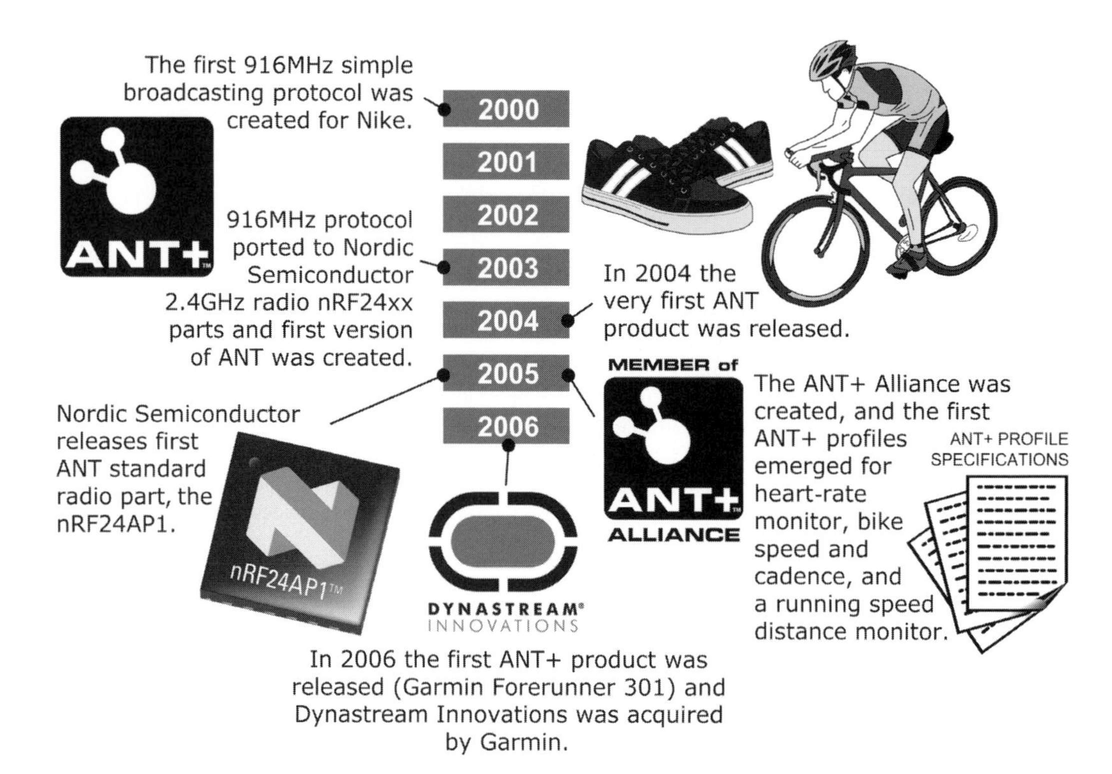

Figure 10.3. ANT's technology timeline, from its inception to where the technology is today with its numerous ANT+ profiles.

spawned from a Nike project. Later, in 2002, the initial protocol evolved to support file transfer, which is now known today as ANT *File Share* (ANT-FS). But it wasn't until 2003, when the protocol was ported to the initial generation of Nordic 2.4 GHz chipsets, namely the nRF24*xx* family, that the first official version of ANT was created. In Figure 10.4, we illustrate the new generation chipset from Nordic Semiconductor (the nRF24AP2 eight-channel chipset used in ANT+ product development).

Figure 10.4. The Nordic nRF24AP2 eight-channel chipset used in ANT+ product development. (Courtesy of Nordic Semiconductor.)

In 2005, Nordic Semiconductor released the first standard radio part, the nRF24AP1, and also in 2005 the ANT+ Alliance was formed, where it soon offered ANT+ *Profiles* for *Heart-rate Monitors* (HRMs), bike speed and cadence sensors, and *Speed and Distance Monitors* (SDMs). The first ANT+ product was released in 2006, and Dynastream Innovations was acquired by Garmin in the same year.

Figure 10.5. Garmin's Forerunner 910XT. (Courtesy of Garmin International Inc.)

10.2 ANT's Market

As we have touched upon elsewhere in Part II, "The Wireless Sensor Network," the wireless technologies that feature in Part III, "The Classic Personal Area Network," typically serve high-end, high data rate products. ANT wireless technology has become the de facto standard used within the sports and fitness markets, uniquely empowering products with ultra-low power sensor technology. As we mentioned in Section 10.1, "Overview," ANT is a proprietary technology and has been specifically engineered to offer simplicity and efficiency, in turn ensuring ultra-low power consumption that extends battery life. What's more, the ANT ecosystem empowers its users with a range of sports, fitness, and consumer health products such as heart-rate belts, watches, bicycle speed, cadence, and power, running speed and distance, weight scales, blood pressure monitors, and activity monitors, which all uniquely form a *Body Area Network* (BAN) that permits its users to monitor their performance directly through access to a *Personal Area Network* (PAN) or via the *Wider Area Network* (WAN) or Internet. In Figure 10.5, we illustrate Garmin's Forerunner 910XT device, which offers an all-in-one GPS product "that provides detailed swim metrics and tracks distance, speed/pace, elevation and heart rate for running and cycling."[1]

[1] Garmin International Inc. (garmin.com).

ANT offers a compact protocol stack that, in turn, minimizes microcontroller resources and overall system costs. Additionally, ANT provides reliable communication, offers flexible and adaptive network operations, and affords a "cross-talk" immunity or collision avoidance/interference solution.

As we have already touched upon, ANT technology has singularly cornered the sports and fitness markets. Like most of the technologies featured in Part II, ANT provides low-latency, optimized data transmission techniques and support for broadcast, burst, and acknowledged transactions, all managed at up to 60 kbit/s (the over the air-interface can manage a data rate throughput of 1 Mbit/s primarily for low duty cycle operation). The production-proven protocol used in over 25 million devices also offers bi-directional communication over adaptive isochronous channels, which support multiple frequencies, high node density broadcasting, and flexible networking capabilities for a number of networking topologies. Nowadays, ultra-low power technology faces competition from Bluetooth low energy (more about this later), but it has an established position in the market and has adapted to new competition within the *Wireless Sensor Networking* (WSN) market in providing a wealth of proven applications that enable it to provide a robust application-base.

10.3 ANT's Application Portfolio

Let's refer back to Figure 10.3. As we can see, ANT has had a lifespan of less than a decade – similar to some of the other technologies featured within Part II. Nonetheless, in this relatively short period, ANT Wireless has adequately matured to capture a unique market space. The consortium of companies that comprise the ANT+ Alliance has holistically established the technology as a de facto standard for the sports and fitness markets, as we have already touched upon. ANT technology has a proven consumer-base, inasmuch as it has been integrated into over 25 million consumer electronic products, by companies such as Garmin, Suunto (suunto.com), and Timex (timex.com). In this section, we'll explore further how the technology is being used to form a better understanding of the number of applications and products supported by the ANT eco-system.

The ANT wireless sensor networking technology typically forms a BAN, as the majority of devices will occupy your person, or perhaps be integral to your body or part of your clothing (*wearable* technology). You may recall from Chapter 2, "What is a Personal Area Network?," that we discussed the divergence of the personal area networking space, which has become somewhat diluted with numerous topologies that uniquely characterize scope and function. You may also recall that we discussed and characterized the diverse number of area networking topologies that have become popular in mainstream computing, and in Figure 10.6 we portray the segmented makeup of the most popular area networks over distance. In Chapter 2, we provided several "colloquial" topologies, which are used to portray a specialized, ad hoc, or unique form of area network. We introduced such terms as *Vehicle Area Network* (VAN) and *Wireless Video Area Network* (WVAN) as examples. The BAN, *Home Area Network* (HAN), and VAN are increasingly

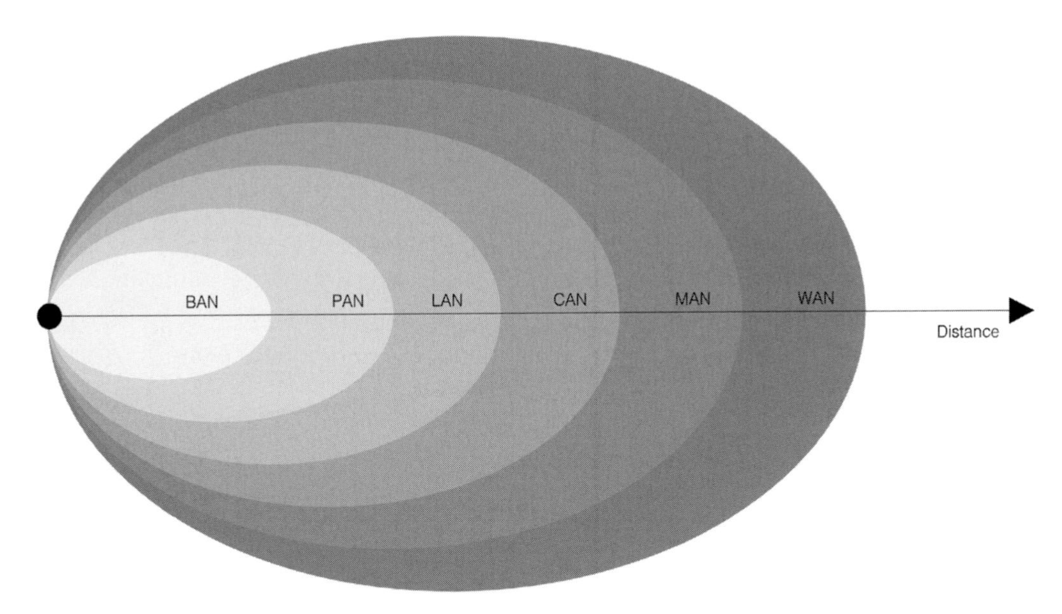

Figure 10.6. The most popular area networks commonly used within the computing industry today. An area network is qualified by distance or proximity, and alludes to a particular type of infrastructure.

becoming part of mainstream conventionalism. Naturally, these conventions are used to characterize the function, scope, and purpose of the devices that form such a topology. What's more, in Chapter 2, "What is a Personal Area Network?," we argued that there still very much exists a PC-centric perspective, and that, despite manufacturers attempts to steer away from PC-dependent technologies and applications, the PC still remains integral to most topologies. We also suggested that the BAN, HAN, and VAN are merely *extended* PAN topologies.

In Figure 10.7, we illustrate an example where a runner uses several wireless products to monitor speed (**1**) and endurance (**2**), which also offer her the ability to listen to music, (**3**) and (**4**), all of which uniquely form her BAN. She may wish to synchronize her data with her PC (**6**) to view statistics about her day and overall performance, in addition to updating her music playlist (**5**).

In essence, her BAN is an extended PAN.

In Figure 10.8, we provide a more exhaustive ANT ecosystem demonstrating the flexibility and diversity of the ANT product range. Let's begin with our runner and move around the ecosystem in an anti-clockwise direction. In the center of our ANT ecosystem, our PAN is made up of a notebook and desktop computer (**21**), as well as a smart or cellular phone and tablet device (**22**). The PAN has access to the Internet (**17**), where data can be shared with other devices or users; if you like, the PAN can behave as an Internet gateway for the numerous BANs, as shown. As we have already mentioned, the runner's BAN is an extended PAN, where she can listen to music (**1**), capture data with her heart-rate monitor (**2**), and capture information relating to her speed and distance (**3**) – all these products occupy her person, and she can synchronize

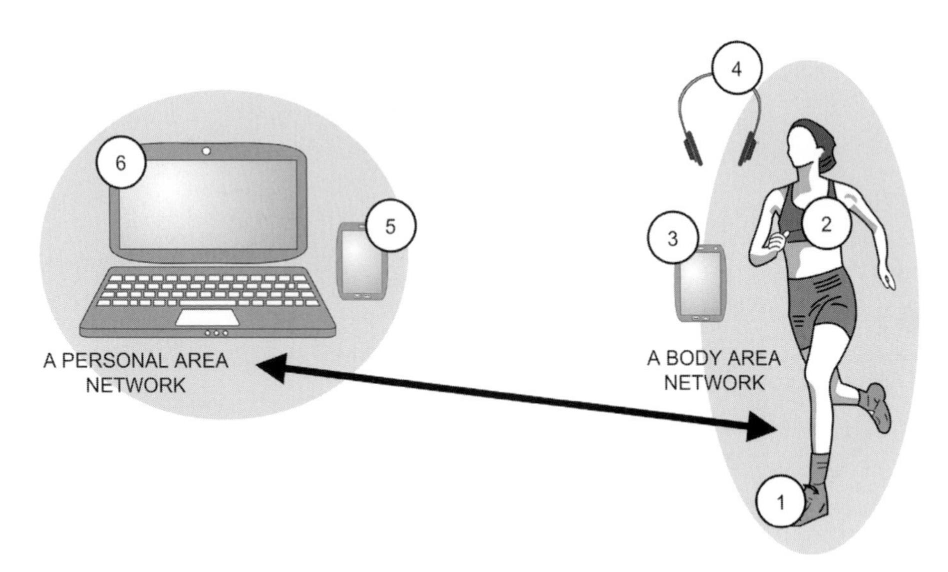

Figure 10.7. The runner has several devices about her person that interconnect to form a BAN. She synchronizes captured data with her PAN.

her gathered data later with her PAN, (**21**) or (**22**). Similarly, the runner on the treadmill (**6**) has the same ability either to view performance statistics directly on the treadmill itself (**5**) or to view information on his watch (**4**). The treadmill runner can synchronize his data later with his PAN, (**21**) or (**22**). The cyclist has about his person a number of products that likewise collect statistical data about his performance. He has a bicycle computer (**7**) to view statistics first hand; a heart-rate monitor (**8**); a speed and distance monitor (**9**); and a power sensor (**10**). In the next image, we see a woman standing on a weighing scale or a body composition product that is wirelessly connected to her PAN, along with a smart or cellular phone (**11**) that may also connect with her PAN. The man standing has a collection of products, including an activity monitor (**13**); a watch, cellular device or MP3 player (**14**); a heart-rate monitor (**15**); and a wireless stereo headset (**16**). In this instance, he has the ability to synchronize his personal data via his PAN or directly to the WAN (**17**) or Internet. Finally, in our ANT ecosystem, we can see ANT's own home area network, which has extended connectivity to the Internet (**17**). ANT's new generation of products include security (**18**) and other sensors (**19**), as well as a home console (**20**).

10.3.1 ANT+ Device Profiles

In the subsections to follow, we describe ANT+ device profiles to provide a better understanding of the applications supported by the ANT+ ecosystem. The profiles encourage interoperability with multiple manufacturers and use an associated icon to help end-users recognize supported capabilities within their chosen product and to identify which ANT+ devices will successfully interoperate. ANT+ devices that are compliant with such profiles may use the icon on the product and any associated documentation.

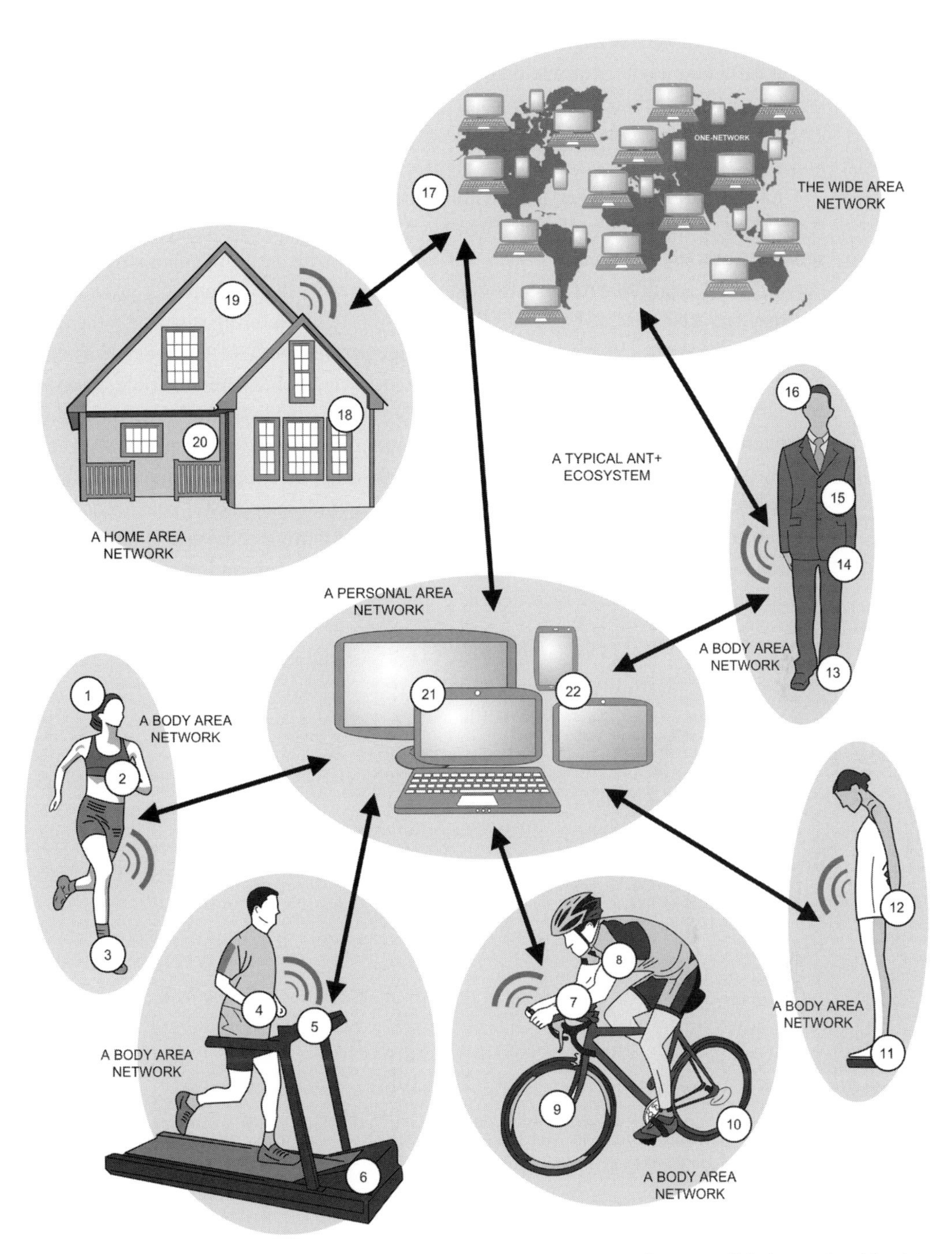

Figure 10.8. A typical ANT+ ecosystem powering the sports, fitness, well-being, and residential market sectors.

Furthermore, in the ANT+ device profiles there are numerous characteristics that define how a particular application should behave, along with specific attributes as to how and what it should communicate or transmit to its peer node. We discuss later, in Section 10.6.2, "ANT Channels," how communication is established between ANT-enabled nodes.

The most common sensor used within an ANT+ ecosystem often provides a data stream that is viewed in real-time. For example, a heart-rate belt would provide a constant real-time data stream. As such, sensors that provide real-time data streams have very different requirements when compared with sensors that periodically provide data transfer. The ANT+ sensor assumes the responsibility of capturing new data rather than prioritizing old data. What's more, this advantage permits the ANT+ network to support multiple nodes that do not need to re-transmit lost data, and this, in turn, reduces power consumption. Any lost data in the ANT+ ecosystem is received from interpolation of cumulating values, thereby assuring that the system receives important new data. The ANT adaptive channel system has been designed to support a high density of broadcasting nodes in the same proximity, which, in turn, provides successful coexistence with neighboring devices. This type of scenario, whilst permitting coexistence, for example, naturally eliminates the need for a central network manager, and empowers multiple concurrent displays or data recorders listening to the same sensor simultaneously.

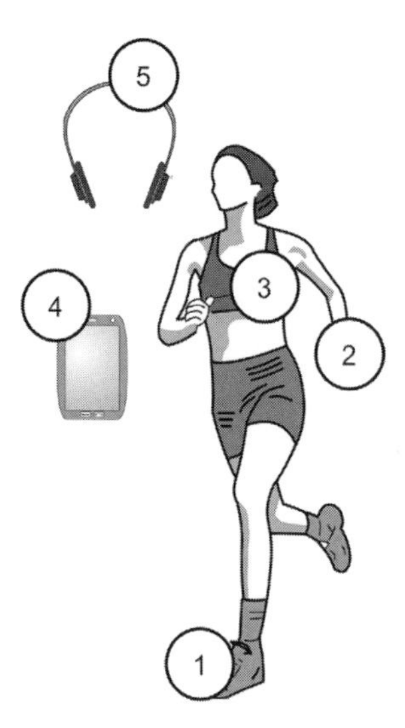

Figure 10.9. The runner may use the audio control feature within the watch to adjust volume, shuffle tracks, or to play, stop, and pause music.

10.3.1.1 Audio Control

In Figure 10.9, we show our runner utilizing a speed and distance sensor (**1**), a wireless watch (**2**), a heart-rate sensor (**3**), an MP3 player (**4**), and a wireless stereo headset (**5**). Both sensors wirelessly transmit data to the watch to permit the runner to view

data in real-time and to store metrics. The watch can also behave as a remote control, allowing the runner to manage and control the connected devices, and also offers audio controls to shuffle, advance, play, and stop/pause music. Likewise, the runner may also control volume and receive information relating to the track number and time, current status of the MP3 player, and so on. The audio control profile has an associated icon (as shown) – primarily the icon informs the end-user that it offers the ability to control audio devices remotely.

Figure 10.10. The bicycle power sensor is mounted on the bicycle to capture data relevant to power output, which is measured in watts.

10.3.1.2 Bicycle Power

In Figure 10.10, we show a cyclist using a bike power sensor (**1**) which has been mounted on his bicycle to capture data relevant to his power output. The force required to provide momentum to the bicycle is measured in watts. The bike power sensor may transmit information to a mounted display or wireless watch (**2**) – the cyclist is also using a heart-rate monitor (**3**). The purpose of the bicycle power profile is to establish a common ground from which multiple manufacturers of sensor and display units can offer successful interoperability for their products.

There are two types of bicycle power sensors, namely *power measurement* and *power information update*. In the former instance, the sensor type supports the measurement of power directly via the measurement of torque and rotational velocity at the crank and the measurement of torque and rotational velocity at the wheel. In the second sensor type, two methods are provided for information update: *event-synchronous*, offering an update based on detecting a motion event, and *time-synchronous*, offering updates at a fixed time interval, from which power can be determined. A bicycle power device displays an associated icon (as shown) – the icon informs the end-user that it offers the ability to capture and measure data, as well as store and display such relevant information.

10.3.1.3 Bicycle Speed and Cadence

Like the bicycle power sensor, bike speed sensors are also mounted on a bicycle. The sensor specifically measures the speed at which the bicycle is traveling. Typically, a magnet is connected to the wheel spokes, where it is detected by the bike speed sensor. A bike cadence sensor operates in a similar manner, but in this case a magnet is attached to the pedal shaft along with a cadence sensor to determine the speed at which the cyclist is pedaling. In both cases, the measured data is transmitted to a display unit, such as an onboard bicycle computer, a wireless watch, or a smart or cellular phone.

10.3.1.4 Blood Pressure

A *Blood Pressure* (BP) monitor is used to measure an end-user's blood pressure using a sphygmomanometer cuff that's normally fitted to the user's upper arm, at approximately the same height as the heart. The cuff is inflated to stop momentarily the flow of blood in the brachial artery, and the cuff is then deflated, allowing the flow of blood to return. The arterial pressure is recorded, potentially along with other data, including heart rate and any possible pulse irregularities. An ANT-enabled blood pressure device has the ability to save, display, and track such data and possibly to forward it to another device. Such devices may include a smart or cellular phone, or a PC, which may have access to the Internet. A blood pressure device can be used in a variety of situations, from a clinical environment through to a home context. The blood pressure profile uses an associated icon (as shown) to demonstrate that features, compliance, and interoperability are assured.

10.3.1.5 Fitness Equipment

The ANT+ *Fitness Equipment* (FE) profile prescribes how an FE should communicate with a wireless watch or other display device. The profile also discusses the *pairing zone*; an end-user must enter this zone to pair their display device with the FE. As such, if an end-user is wearing a watch, for example, pairing between the equipment (**1**) and watch (**2**) will occur once the user has entered this pairing zone, as denoted by the "Link Here" logo, as we illustrate in Figure 10.11 (left). Once pairing has been successfully established, the wireless range between

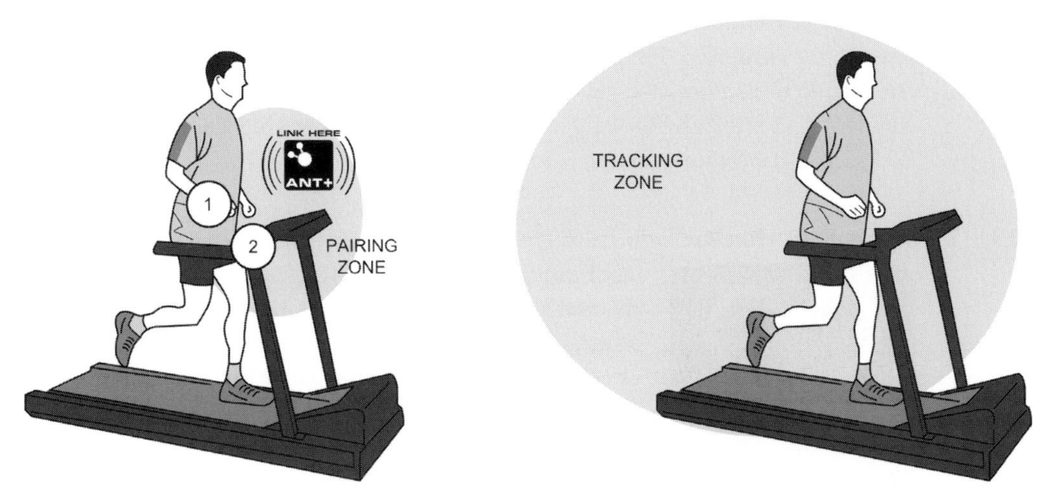

Figure 10.11. The end-user enters the pairing zone, as denoted by the "Link Here" logo, and, when successfully paired, enters the tracking zone.

the equipment and watch is extended during a workout to the *tracking zone*, as shown in Figure 10.11 (right). The FE then commences to broadcast real-time data, and the watch will display and record this information.

10.3.1.6 Geocache

The Geocache profile uses the term *geocaching* to describe the situation where an end-user uses a *Global Positioning System* (GPS) receiver (**1**) to locate a hidden geocache device (**2**), as illustrated in Figure 10.12. The geocache device may contain identification characteristics, a *Short Text Message* (SMS), and other leading information as to the location of another geocache device. Similarly, the geocache device owner may also provide latitude/longitude coordinates or trailhead information to a geocaching website, database, or other service.

Figure 10.12. The GPS receiver is used to locate a geocache device.

The objective for the end-user would be to ascertain this information and commence a search for the device. The geocache device persistently transmits its ID at a low data rate that, in turn, allows the device to be easily detected by a GPS receiver. Once the end-user is in proximity to the geocache device, his or her GPS unit will request an exchange of the device's data.

10.3.1.7 Heart-rate Monitor

The heart-rate monitor is worn around the chest and measures the user's heart rate in real-time. The monitor is capable of transmitting data to a display device such as a watch, smart or cellular phone, a PC, and so on. What's more, other types of monitors may include finger sensors, earlobe sensors, or hand contact sensors mounted on fitness equipment – all these variants will have the ability to transmit ANT+ heart-rate data to a display device.

Figure 10.13. The LEV is a convenient and economic transportation vehicle ideally suited for just one person.

10.3.1.8 Light Electric Vehicle

A *Light Electric Vehicle* (LEV), such as a one-person car (see Figure 10.13 (**1**)) or an electric bicycle (**2**), typically provides convenient transportation for one person, perhaps with cargo, at a speed and cost that are sensible. The ANT+ LEV profile offers the end-user the ability to capture information such as speed, distance traveled, remaining battery life and range, and current state information (lights on/off, gear state, and travel mode). The captured information can be transmitted to a display unit, such as a bike mounted device, watch, smart or cellular phone, or a tablet device (**3**).

Some LEVs have pre-programmed *travel modes* that enable the LEV to be operated at an optimal and economical level. The operation of such modes can be controlled from the LEV itself or through a touchscreen-enabled device such as a tablet (**3**). The ANT+ LEV profile offers developers a means to provide interoperable communication between the LEV and a display unit. As such, the LEV profile uses an associated icon (as shown) to demonstrate compatibility and interoperability.

10.3.1.9 Multi-sport, Speed, and Distance

A *Multi-sport Speed and Distance Monitor* (MSM) measures the distance traveled (often with a time stamp) and calculates the speed. These types of monitors may utilize satellite positioning, radar, or other technologies to measure the distance traveled. MSMs can be used in a variety of sporting contexts to include running, cycling, skiing, skating, and so on. An MSM will also have the ability to transmit such information to a display device (a watch, smart or cellular phone, or a tablet device) and/or store data. The MSM profile uses two icons to convey to the end-user support and capabilities. The speed "SPD" icon, as shown, informs the end-user that the device will be capable of transmitting or receiving speed and distance information, whilst the "GPS" icon, as shown, will offer the end-user the ability to transmit or receive, and track the end-user's location during an activity.

10.3.1.10 Stride-based Speed and Distance

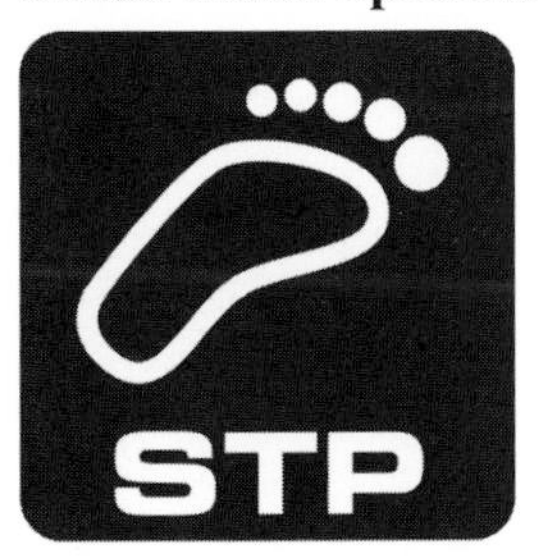

A *Stride-based Speed and Distance Monitor* (SDM) measures the number of strides taken, the speed at which the user is traveling, and/or the distance covered based on stride measurements and associated calculations. An SDM is body-worn or is close to the body, and may include foot-worn pods (**1**) (see Figure 10.14) that sit on or in the shoe. Other SDMs can be pedometers that are worn around the waist or elsewhere around the body. The information gathered by an SDM can be transmitted to a watch, smart or cellular phone, or any other ANT-capable display device.

10.3.1.11 Weight Scale

The ANT+ *Weight Scale* (WS) profile offers the end-user the ability to measure his or her weight – the end-user typically stands on the scale (see Figure 10.15 (**1**)) for a relatively short interval, normally between 10 and 30 seconds per measurement. Some weight scale devices may utilize an integrated display (**2**), where the end-user can optionally enter

Figure 10.14. An SDM is body-worn, and may include foot pods, which sit either on or in the shoe.

profile information. Likewise, a weight scale device may retrieve profile information from a watch (**3**) or other capable device (**4**) – see Figure 10.15. Profile information may include data about gender, age, and height, and this, in turn, allows the weight scale to calculate, for example, body composition as a percentage of fat/hydration or muscle/bone mass. The weight scale profile distinguishes between two types of display, namely *stationary* and *mobile* displays. A stationary display device is characterized by its static nature in that it may be integral to the scale itself or can be wall-mounted at eye level, so allowing multiple users to use the weight scale device. A mobile display device, such as a watch, smart or cellular phone, tablet, and so on, may contain additional profile information related to the end-user, and is intended to be used only by the specific end-user.

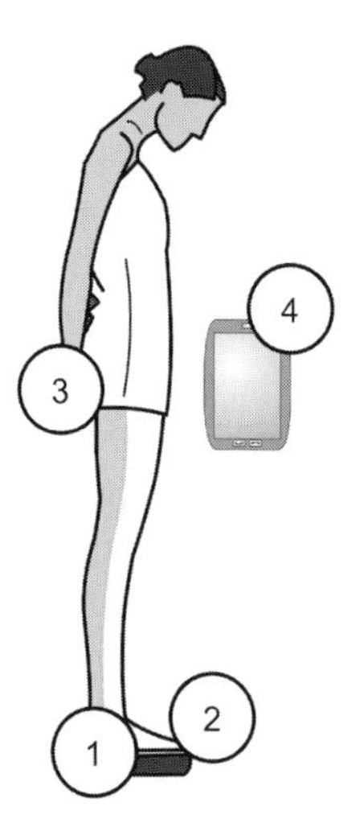

Figure 10.15. The weight scale profile enables end-users to measure weight and enter profile information to capture other data.

A weight scale device uses an associated icon (as shown) – the icon informs the end-user that it offers the ability to measure data, as well as to store and display such information.

In the sections to follow, we explore the various specifications that demonstrate ANT's acumen and foresight within this market space. The ANT+ Alliance has developed a protocol, along with numerous supportive device profiles, as already discussed, which collectively address a requirement to satisfy the hunger of consumers' needs and expectations to monitor and improve their fitness and overall performance. However, in the meantime, we'll take a look at ANT's competition within this market sector.

10.4 ANT Wireless and its Competitors

ANT has pretty much cornered the market in the sports and fitness sectors, and now only sees potential competition from the new and revised Bluetooth v4.0 specification, and possibly other proprietary technologies. As we have already touched upon elsewhere in this chapter, the Bluetooth *Special Interest Group* (SIG) has essentially reinvented Bluetooth wireless technology with its introduction of the new Specification of the Bluetooth System: Core, v4.0. The new specification has been coined *3-in-1*, a specification which includes *Bluetooth low energy* (BLE), along with the existing *classic* and *high speed* variants. You may recall that we have already discussed Bluetooth low energy in Chapter 7, "Bluetooth low energy: The *Smart* Choice." We discuss the classic and high speed variants in Chapter 14, "Bluetooth Classic and High speed: More Than Cable Replacement." Bluetooth low energy is relatively new, but it quickly found its feet within its chosen market. Since its inception, ANT has enjoyed some remarkable successes with its existing application-base, and the ANT+ Alliance continues to reinforce the technology as the de facto standard for the sports, fitness, and consumer health industries. The ANT+ Alliance hasn't unnecessarily rested on its laurels either, as it continues to evolve the technology for what may become an aggressive battle for dominance. In fact, Nordic Semiconductor has developed[2] a new converged chipset, namely the nRF51 Series, a multi-protocol "combo" solution in which ANT+ and BLE are integrated onto a small *System-on-Chip* (SoC), affording ANT+ technology an opportunity to dominate in what will become an aggressive and lucrative market sector.

10.5 Networking Topology

ANT Wireless technology supports *point-to-point*, *star*, *tree*, and *fixed mesh* networking topologies, some examples of which are illustrated in Figure 10.16 and Figure 10.17. Moreover, ANT has been optimized for point-to-point, star, and tree networks, meaning that it can offer a capacity of up to 65,536 *slave nodes*, each communicating with

[2] ANT Wireless, "ANT+ and Bluetooth low energy Concurrent Combo Chip Solution," 2012.

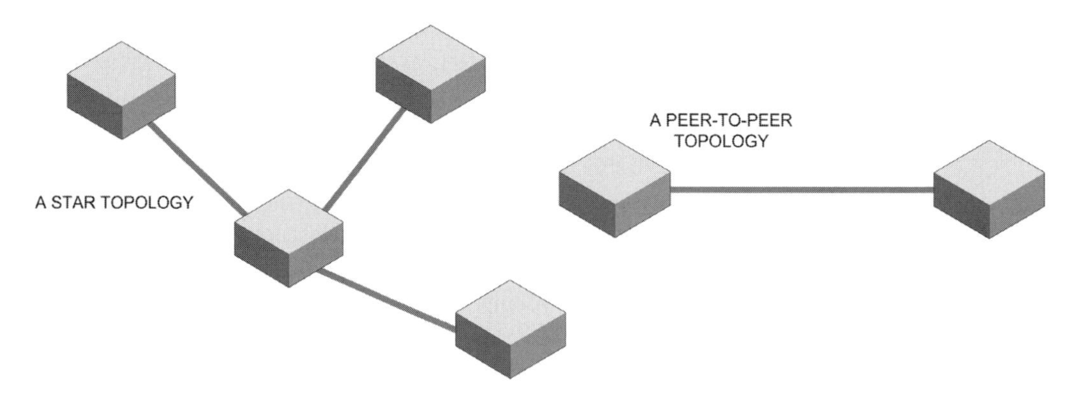

Figure 10.16. ANT Wireless supports star and peer-to-peer networking topologies.

one *master* device over a timeslot shared single channel. Most applications utilizing ANT technology can be resolved using these optimized topologies. Nevertheless, a mesh topology can also be supported, although this should be in instances where this can be justified, as a mesh network tends to increase overall complexity, increased system resources, and power consumption. The ANT managed network is composed of a series of devices that utilize the ANT radio protocol, along with ANT+ device profiles that, in turn, provide a standardization of interoperable devices that permit multiple manufacturers to coexist and interoperate successfully. Within the ANT topology, an ANT powered node can undertake either a slave or master role interchangeably. As such, nodes can behave as transmitters, receivers, or transceivers to ensure that traffic is routed

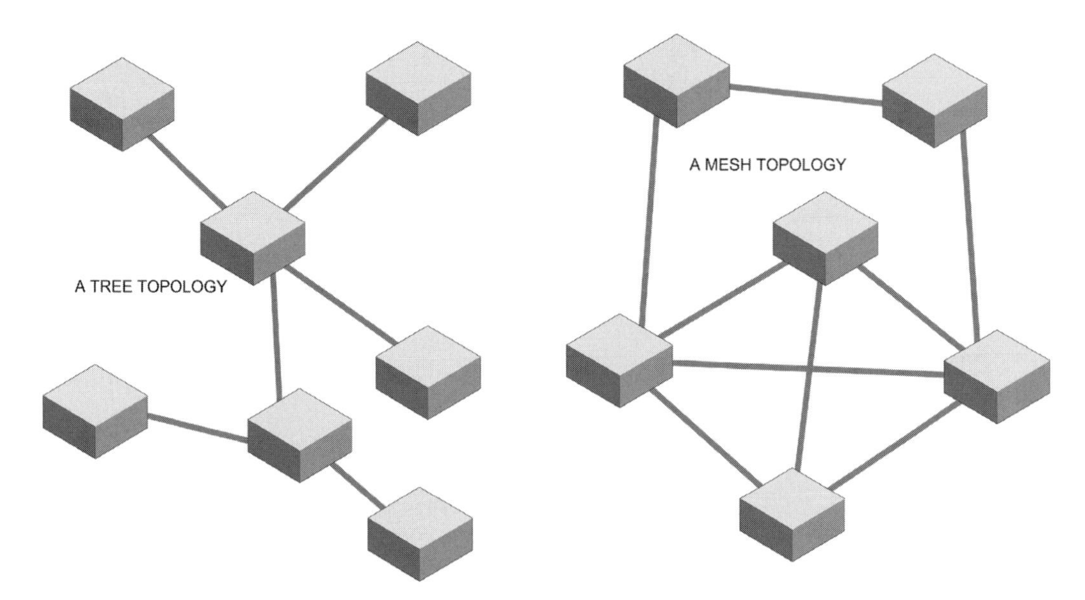

Figure 10.17. ANT Wireless also supports tree and mesh networking topologies.

efficiently. What's more, a node has the unique ability to determine the best time at which to transmit traffic based on neighboring activity, in turn eliminating the need for a *coordinator* or *supervisory* node. With ANT's ability to support an *ad hoc* structure, tens or hundreds of nodes can easily be interconnected dynamically to provide concurrent discovery and tracking, whilst possessing the ability to communicate simultaneously. Nodes also have the ability to join or leave the network, thereby reducing overhead on system resources.

10.5.1 Channel-based Communication

In this section, we'll touch upon and describe ANT's communication and its way of working within the ANT network. Successful communication is achieved through a channel-based connection, and we'll pick this up again in more detail in Section 10.6.2, "ANT Channels." For now, as it's relevant under "networking topologies," we'd just like to mention the numerous scenarios that ANT networking provides in terms of single or multiple channel connections. As we have already mentioned, an ANT network always comprises a slave and master relationship, but these roles are interchangeable. The master is the primary transmitter (TX) and, naturally, the slave behaves as the primary receiver (RX).

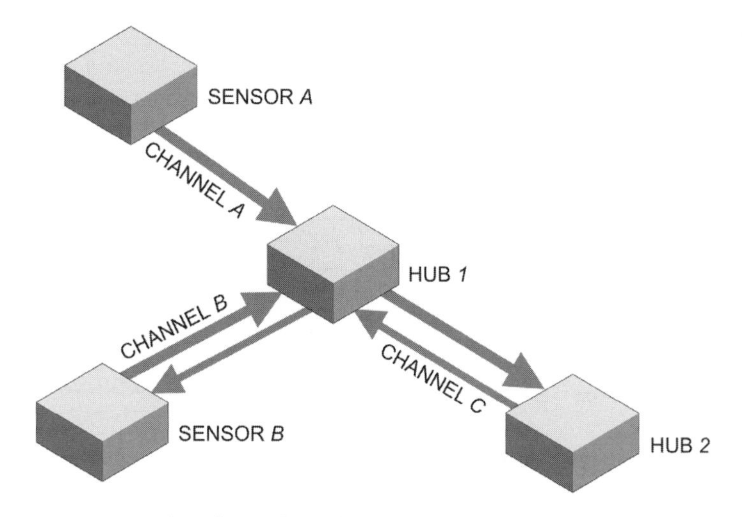

Figure 10.18. The primary flow of data from a master to a slave (thick arrows); a thin arrow represents a reverse message flow.

In Figure 10.18, we illustrate the primary flow of data from a master to a slave with "thick" arrows (*forward* flow), where the thinner arrows represent a *reverse* message flow (channels **B** and **C**, as shown). Elsewhere in the illustration, we also depict one-way communication (channel **A**), which typically utilizes low power, transmit-only devices – in Table 10.2, we summarize the master/slave relationship for each of the channels shown in Figure 10.18.

Table 10.2. A summary of master/slave status

Channel	Master	Slave
A	sensor *1* (TX-only)	hub *1* (RX)
B	sensor 2 (TX)	hub *1* (RX)
C	hub *1* (TX)	hub *2* (RX)

10.6 The ANT Architecture

In the subsections that follow, we take a closer look at the ANT stack architecture – essentially we take a top-down approach when dismantling the software stack and explore in greater detail the technology's architecture and its software and application building blocks. In Figure 10.19, we provide the key building blocks of the ANT architecture – you will note that these building blocks have been labeled numerically to ease identification and classify responsibility. Like all the protocol stacks used in wireless technology, the ANT stack models itself on the *Open Systems Interconnection* (OSI) layers, which are also shown in Figure 10.19 (left). Blocks labeled (**1**) and (**2**) are ANT responsible, while (**3**) represents a user-specific application, such as that defined by the ANT+ device profiles; see Section 10.3.1, "ANT+ Device Profiles."

You will notice from Figure 10.20 that, in a similar arrangement to that for Bluetooth wireless technology, an ANT device comprises a *host* and a *host controller* combination. As we have already illustrated in Figure 10.19 and now in Figure 10.20, blocks (**1**) and (**2**) form the host controller unit comprising the *ANT engine* – this is typically

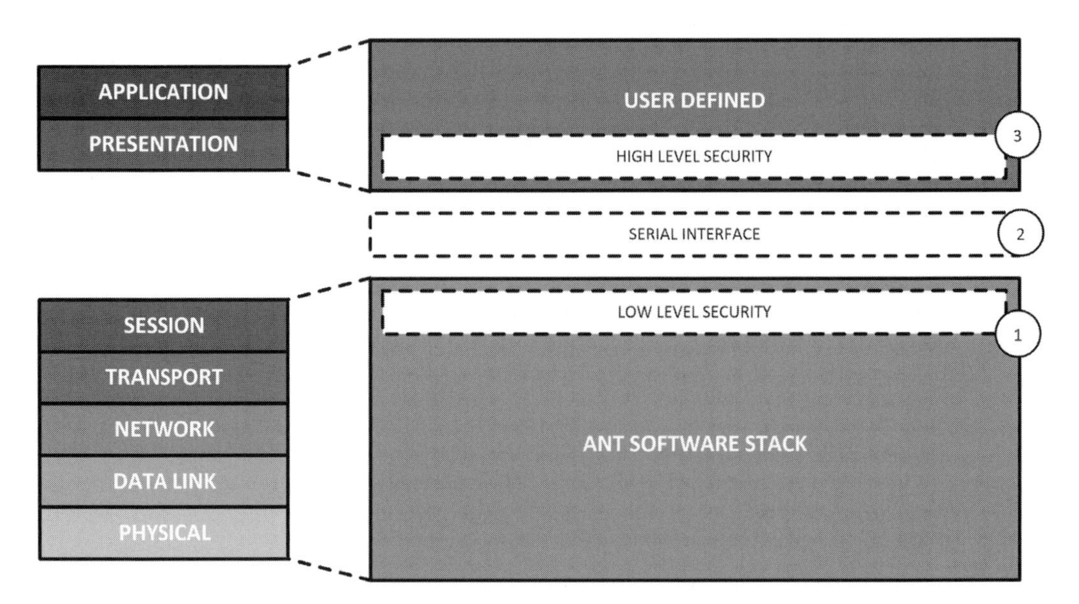

Figure 10.19. The ANT software stack architecture mapped against the OSI model.

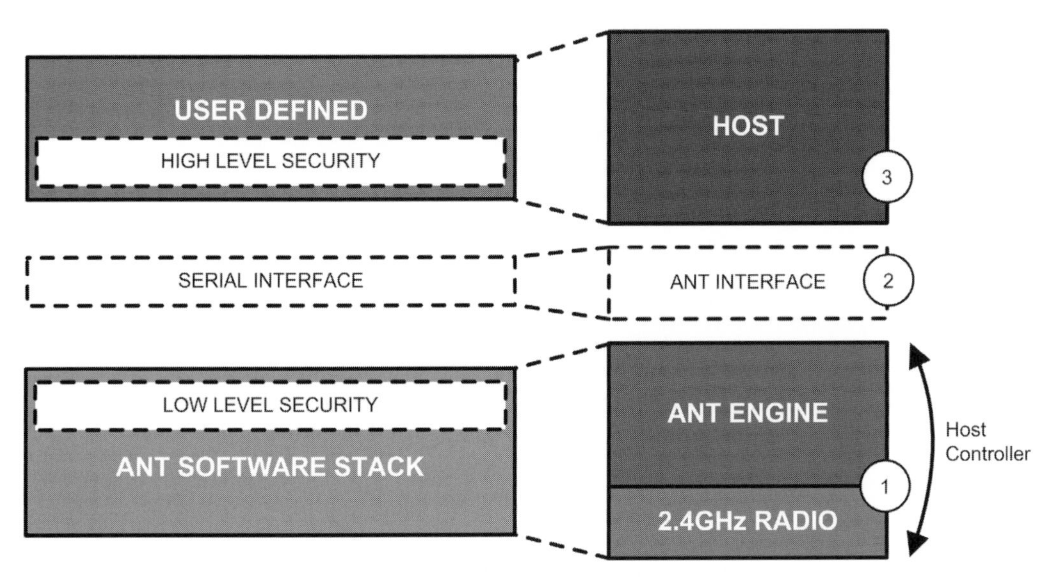

Figure 10.20. A further magnification of the ANT stack architecture reveals a host and host controller combination.

housed in a Nordic Semiconductor chip like that previously illustrated in Figure 10.4. Manufacturers can select either a first or second generation Nordic chipset or one of the Texas Instruments chipset CC257x family as their host controller. The host controller shifts the complexity away from the developer, so that development is primarily focused on the end-application. Communication between the host and host controller is managed through a serial interface, named the *ANT Interface*, which offers a simple, bi-directional, serial message-based protocol, whereby the host *Microcontroller Unit* (MCU) creates and maintains a communication pathway or *channel* (more about this later) to other ANT-enabled devices using this interface. We'll discuss the ANT interface in more detail later in Section 10.7, "The ANT Interface."

10.6.1 The ANT Node

In Figure 10.21, we take a close-up view of an ANT node (right) to form a better understanding of its composition. The node illustrated here represents a complete ANT system or device, where the end-application is present using a user-specific MCU ("host MCU," as shown). However, an ANT certified *module* is ready to be mounted onto new or existing *Printed Circuit Boards* (PCBs), which facilitate rapid integration of the ANT engine. We illustrate in Figure 10.22 an ANT module's makeup, which depicts the ANT interface, ANT engine, and the 2.4 GHz radio – essentially, the module forms the host controller components.

What's more, an ANT module can also be offered as a USB-stick, which provides access for a development PC to acquire ANT networking capabilities. The ANT

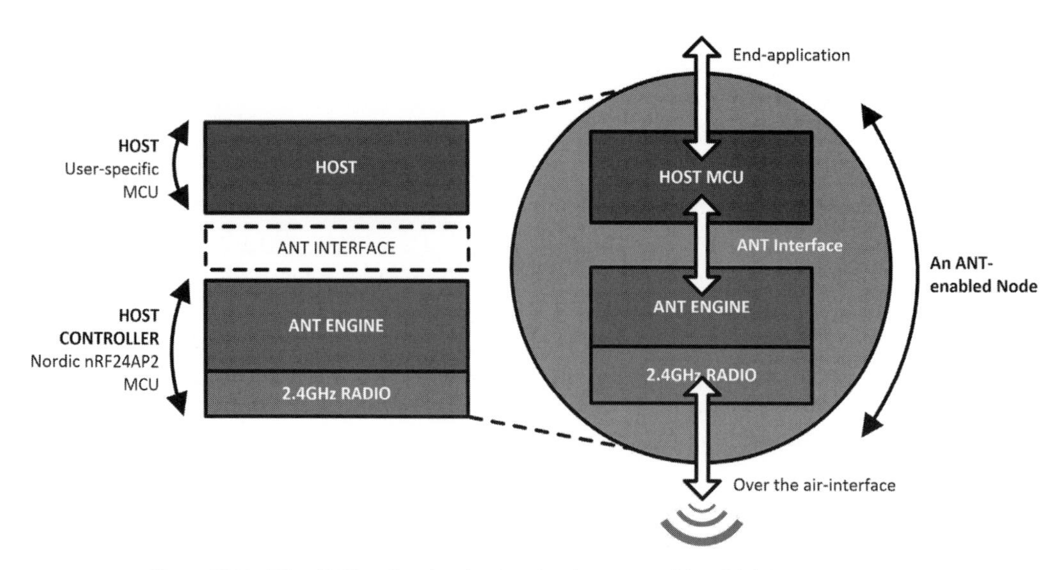

Figure 10.21. The ANT node, clearly showing its composition (right).

development kit enables third-party developers to integrate quickly ANT networking features for both an embedded and PC environment. ANT offers, for both types of environments, sample source code and device drivers to aid development, whilst providing a comprehensive PC software library.

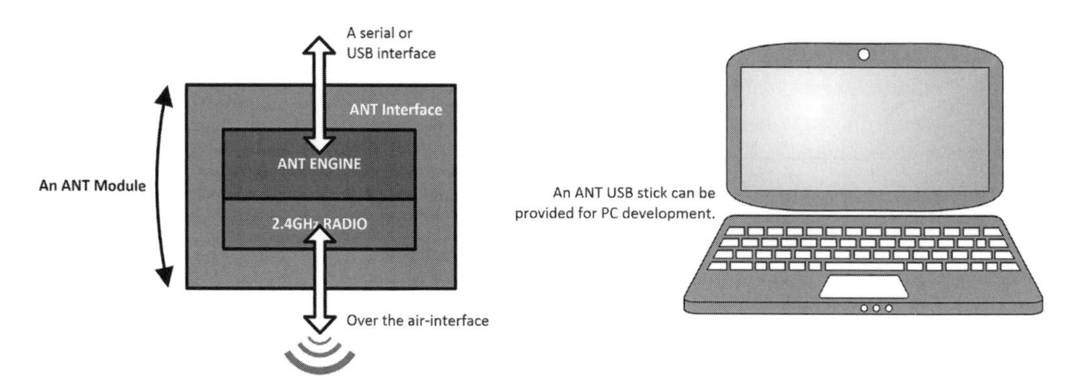

Figure 10.22. ANT PC development kit environment enables third-party developers to develop products to market quickly.

10.6.2 ANT Channels

We've already touched upon the way of working within an ANT network. We revealed earlier that to achieve successful communication a channel-based connection has to be established between ANT nodes, as we discussed in Section 10.5.1, "Channel-based Communication." So, a channel between two nodes must always be established such that, as a minimum, the network topology will consist of a master node and a slave node.

In this section, we'll explore in further detail ANT channel use and types, configuration, and device pairing. We've already hinted at communication direction types, insofar as master to slave represents a forward direction and slave to master represents a reverse direction.

Furthermore, there are three ANT data types, namely *broadcast*, *acknowledged*, and *burst*, which govern the type of communication between two nodes on its channel. We explain these data types further in Table 10.3. This ensures that when a host (the end-application) sends a message to the host controller (the ANT engine) for transmission over the air-interface, the data type and the message payload are specified.

Once the payload arrives at the ANT engine and the channel has been established and opened, the message is transmitted on each channel at the designated channel period, T_{ch}, as shown in Figure 10.23 – this behavior is shown as mandatory for the master node, whereas it is optional for the slave device to respond.

Table 10.3. The fields that make up the channel configuration

Type	Direction	Description
Broadcast	Forward.	The default data type, in which a broadcast message is sent during every timeslot that will be retransmitted if no new data have been received from the master host MCU.
	Reverse.	Only sent if specifically requested by the slave host MCU.
Acknowledged	Forward.	Sent on the next channel timeslot, if requested.
	Reverse.	Same as forward.
Burst	Forward.	A burst transfer commences at the start of the next timeslot, if requested.
	Reverse.	Same as forward.

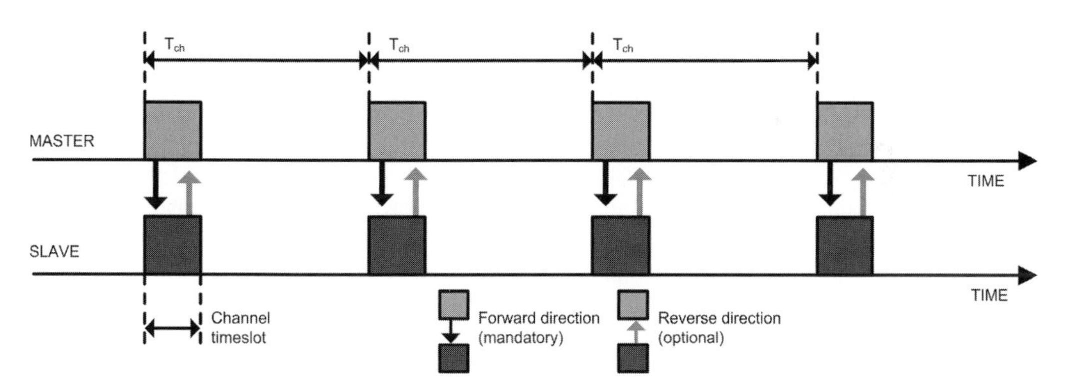

Figure 10.23. A master device will transmit its payload on each channel at the designated channel period.

10.6.2.1 Channel Configuration

The ANT system utilizes two types of channels, namely *independent* and *shared*. The independent channel type is the most commonly used between any two nodes. Independent channels provide full configurability for master or slave, network, frequency, and

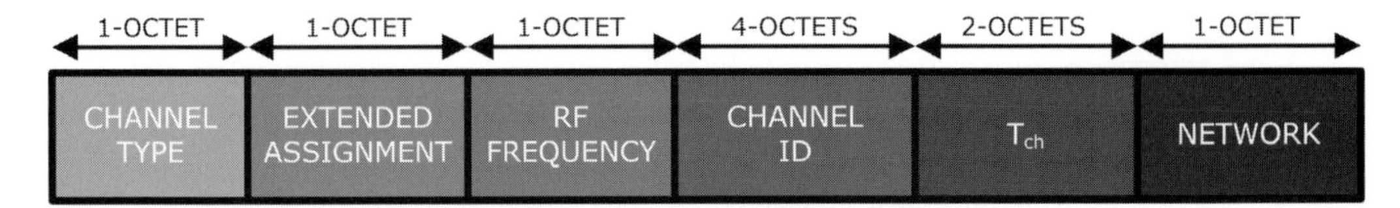

Figure 10.24. The structure of the channel configuration payload.

Table 10.4. The fields that make up the channel configuration

Name	Description	Size
Channel type	This 8-bit field specifies the type of communication on the channel. We list the common types available in Table 10.5.	8
Extended assignment	This optional 8-bit field enables various ANT features, including frequency agility and background scanning channel, to be enabled.	8
RF channel frequency	The 8-bit RF frequency field is set from `0x00` to `0x7C`, where the value represents the offset in 1 MHz increments at 2400 MHz (maximum 2524 MHz).	8
Channel ID	The most basic descriptor of an ANT channel is this 32-bit field. The channel ID comprises three further fields and is explained in more detail in the subsection entitled "Channel ID."	32
Channel period (T_{ch})	This 16-bit field represents the basic message rate of data packets transmitted by the master node.	16
Network	The 8-bit network field defines the network number. We discuss the network number and its associated network key in more detail in the subsection entitled "Network."	8

period, fully independently of all other channels assigned on the same node. However, shared channels have the advantage of being able to support more connections than the maximum number of connections possible using just independent channels. Most ANT nodes support up to eight independent ANT channels, but can support up to 65,536 slave connections derived from a single shared master channel. Shared channels require further address assignment of all participating shared slave devices, and all shared channel slave devices must exist on the same network, frequency, and channel period as the shared master channel.

As we explained earlier, we understand that a channel must be established between two nodes, that is, a master and slave device, to enable successful communication; as such, a common channel configuration must be agreed upon. More specifically, a master device will define operational parameters for a channel, and, likewise, a master node may manage and maintain multiple channels, each with its own specific parameter configuration. Moreover, once a channel has been established with its initial set of parameters, the configuration is malleable whilst the channel is open. In Figure 10.24, we illustrate the format of the channel configuration payload, and in Table 10.4 we describe the fields that make up its structure.

Table 10.5. The common channel type values

Value	Description
`0x00`	Bi-directional slave channel.
`0x10`	Bi-directional master channel.
`0x20`	Shared bi-directional slave channel.
`0x40`	Slave receive-only channel.

Channel ID

The channel ID field is regarded as the most significant parameter when used during device pairing – more about pairing later in Section 10.6.3, "Device Pairing." What's more, only devices with matching channel IDs can communicate with each other. Similarly, a slave device uses the channel ID parameter to identify which master device to communicate with. The slave device may set any of the transmission device types, and/or set the device number to `0x00`, which represents a wildcard value that allows the slave device to discover different master devices. For example, by setting the device type to heart-rate monitor and the device number to `0x00`, a slave device can discover any heart-rate monitor in proximity.

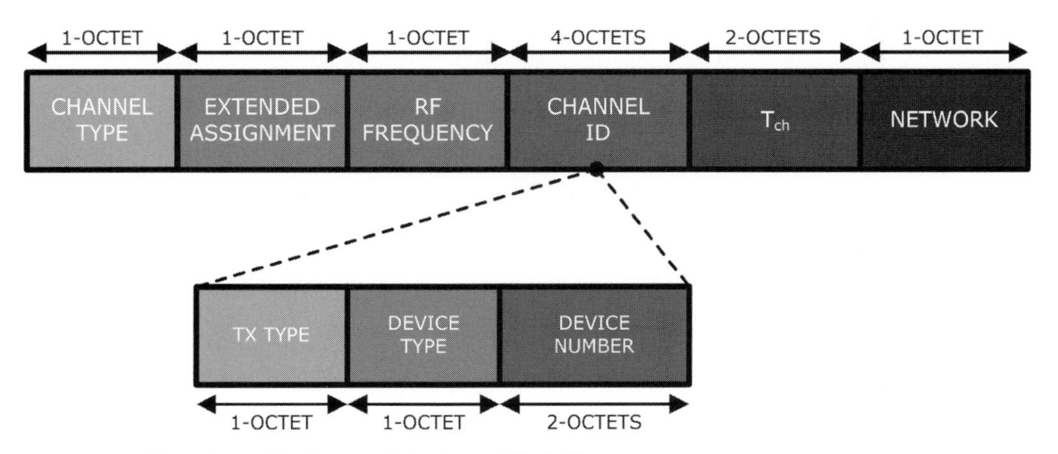

Figure 10.25. The format of the channel ID field.

As we mentioned previously in Table 10.4, the channel ID is made up of three further fields, namely *transmission* (TX) *type*, *device type*, and *device number*. We illustrate in Figure 10.25 the format of the channel ID, and explain these fields further in Table 10.6.

Network

The ANT topology supports various unique public, private, and managed networks, where each network type may impose certain characteristics for participating nodes. As we have already mentioned, only nodes participating in the same network can successfully communicate with each other. As such, this creates a network that becomes public

Table 10.6. The fields that make up the channel ID

Name	Description	Size
Transmission (TX) type	This 8-bit field specifies certain transmission characteristics of a device. The upper 4-bits of this field can optionally be used to extend the range of the device number to 20-bits.	8
Device type	This 8-bit field is used to distinguish between device types, which will inform the device as to how to interpret its payload.	8
Device number	A 16-bit value used to identify a device uniquely within an ANT network.	16

or shared between multiple vendors, thereby achieving an *open* system of interoperability. A private network, by its very nature, implies network privacy, such that restricted access is intended by certain devices.

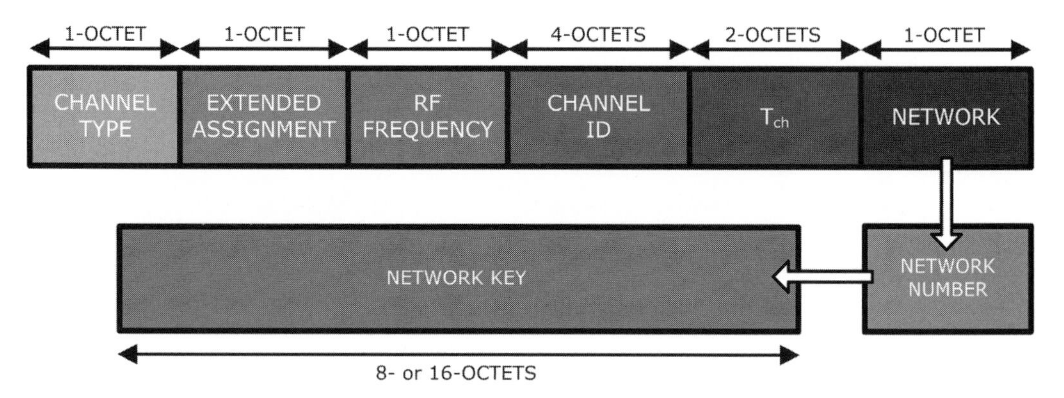

Figure 10.26. The format of the network field.

A managed network specifies behavioral characteristics that dictate its operation. This type of network affords manufacturers interoperability with numerous ANT+ products. The network number identifies the network the channel will use, and each network number is configured with a separate 8- or 16-octet network key. We illustrate the format of this field in Figure 10.26 and offer a description in Table 10.7.

10.6.3 Device Pairing

The process of pairing a slave node with a master node requires a relationship to be formed whereby communication can be established – the relationship may be *permanent*, *semi-permanent*, or *transitory*. The pairing process is initiated when a slave node acquires the unique channel ID of the master device. If the pairing relationship is to become permanent, then the slave device will save the master's ID in a permanent or non-volatile memory, allowing the slave device to open a channel based on the ID for all future

Table 10.7. The fields that make up the network

Name	Description	Size
Network number	This 8-bit field identifies which network the channel will use.	8
Network key	A 64- or 128-bit key used to enable and assign a particular network to a specific network number. The network key is configured independently from the channel configuration and is used for security and access control purposes.	64/128

communication. In a semi-permanent process, the relationship is maintained for as long as necessary – once the communication has timed out, the pairing relationship ceases. Lastly, in a transitory relationship, the connection is established for as long as necessary to receive data.

Table 10.8. The fields that make up the ANT interface

Name	Description	Size
Sync	This 8-bit field has a fixed value of `0xA4` or `0xA5` and is used to synchronize the message over the serial interface.	8
Message length	The message length is an 8-bit field containing the number of bytes contained in the data payload.	8
Message ID	This 8-bit field contains a data type identifier (`0x00` is invalid).	8
Data	The message payload.	
Checksum	XOR of all bytes, to include the sync field.	8

10.6.3.1 Other Pairing Features

ANT technology provides two additional pairing methods, which include *inclusion/exclusion lists* and *proximity search*. Only some devices support inclusion/exclusion lists, but, nevertheless, this feature enables the storage of up to four channel IDs in a slave node's "list." As such, a slave will only pair and connect with a given master ID contained in the slave's predefined list. Similarly, if an ID has been listed in an exclusion list, the slave device will not pair or connect with that master ID. Using the proximity search pairing method, channel IDs are sourced according to the distance between two devices.

10.7 The ANT Interface

We have already discussed in earlier sections that a host and host controller use a serial protocol interface to communicate with each other, and in this section we describe the

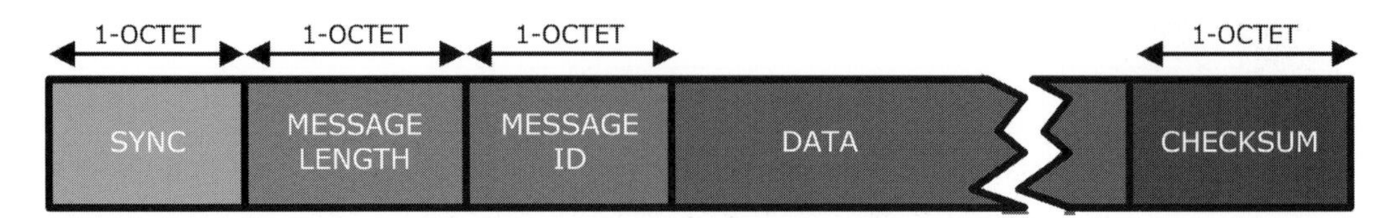

Figure 10.27. The message format used at the ANT interface.

ANT interface. You may recall from Section 10.6.1, "The ANT Node," that a host may be an embedded MCU or a PC (via a USB-stick). In Figure 10.27, we illustrate the message format used between the host and controller, and we describe the associated fields in Table 10.8.

Part III

The Classic Personal Area Network

11 Introducing the Classic Personal Area Networking Technologies

Mainstream wireless personal area networking capable consumer electronic products (excluding *Infra-red* (IR)) began to appear in the late 1990s and during the early 2000s the market witnessed the emergence of the first Wi-Fi and Bluetooth products. The ability to use products wirelessly was a completely new experience, as IR products required devices to be in line-of-sight. The TV remote control, of course, is the classic example of an IR device still commonly used today, although ZigBee, Bluetooth wireless technology, and a few others are looking to evolve this classic 40+ year old product. Nevertheless, the Motorola Xoom 2 (motorola.com) included IR capability to offer a *universal remote*, which can control multiple IR capable devices through a dedicated *App*, such as Dijit (dijit.com), a universal remote control application for smartphones and tablets. Another similarly classic wireless technology is the key-fob for vehicular access – it remains popular and doesn't require a line-of-sight operation. Some early wireless products that utilized Wi-Fi and Bluetooth were perhaps a cumbersome introduction to the wireless era, as many consumers grappled with early versions of these wireless-enabled products. The growth and development of these wireless devices as they have evolved has been exponential. The last decade or so has afforded consumers the ability to acclimatize to and increasingly become familiar with a range of wireless-enabled products and their associated idiosyncrasies, ensuring their future success. To demonstrate consumers' over-familiarity with wireless technology, Wi-Fi has become synonymous with the ability to connect to the Internet; in fact, some consumers incorrectly regard Wi-Fi *as* the Internet.

11.1 It's Never as Simple as Just Cutting the Cable!

The technologies that feature in Part III, "The Classic Personal Area Network," have been coined *classic* – primarily due to their endurance over the last decade or so and their familiarity to a generation of consumers who continue to embrace the technology. Undoubtedly, all these technologies can be recognized in the majority of consumer electronic products today. In Part III, we discuss some relative newcomers which are set to diversify connectivity scenarios for another generation of products. What's more, the majority of consumers nowadays readily identify technologies such as Bluetooth, Wi-Fi, and *Near Field Communications* (NFC), whereas upcoming technologies, such as the *Wireless Home Digital Interface* (WHDI), which is also featured in part of the book, have yet to reach mainstream popularity. Moreover, in Chapter 16, "Future and Emerging Technologies," we provide a further glimpse of other technologies that are expected to emerge, impact, and simplify connectivity scenarios. As we have already mentioned, it has taken time to enable consumers and manufacturers to overcome the awkwardness and difficulties imposed by early wireless products, and arguably this is still an ongoing process. Remember, it's never as simple as just cutting the cable!

In Chapter 1, "It's a Small Wireless World," we discussed the need to ensure that our products remain simple and transparent to all consumers by introducing an *Interoperability, Coexistence, and Experience* (ICE) model when embarking upon new wireless product development.

Figure 11.1. The Jabra HALO2 Bluetooth stereo headset – one of the many iconic Bluetooth stereo headset products that have emerged over the last decade, becoming the mainstream technology diet for so many consumers. (Courtesy of Jabra.)

In Figure 11.1, Figure 11.2, and Figure 11.3, we provide examples of three typical wireless products available today. In Figure 11.1, we show the bestselling Jabra HALO2 Bluetooth stereo headset, which affords its consumers simplicity and ease of use when listening to music, and receiving and making calls through a cell device. Figure 11.2 shows the D-Link DIR-645, Whole Home Router 1000, which provides Wi-Fi connectivity using the 802.11n standard, with *Wi-Fi Protected Setup* (WPS) and both *Wi-Fi Protected Access* (WPA) and *Wi-Fi Protected Access II* (WPA2). Finally, Figure 11.3 shows the Nokia Lumia 900 smartphone (with Microsoft's Windows Mango mobile operating system), which sports a host of personal area networking features, including Wi-Fi and Bluetooth wireless technology, which are both integrated.

These are just a few of our classic personal area networking products that have established themselves over the last decade and have become an intrinsic part of everyday use. In this chapter, we provide a short introduction to the personal area space along with a summary of what we can expect to discover about these technologies in the following chapters. You may recall that, in Chapter 2, "What is a Personal Area Network?," we discussed the origin of the local area network, which seemed to spawn a plethora of area network derivatives alluding to range, distance, and capabilities. Needless to say, a

Figure 11.2. The stunning D-Link DIR-645, Whole Home Router 1000, offering Wi-Fi connectivity (802.11n and 802.11g) with WPS and WPA2. (Courtesy of D-Link.)

Figure 11.3. The Nokia Lumia 900 (© Copyright Nokia, 2012) is representative of a product embodying classic wireless personal area networking technologies, such as Bluetooth and Wi-Fi. (Source: nokia.com/press.)

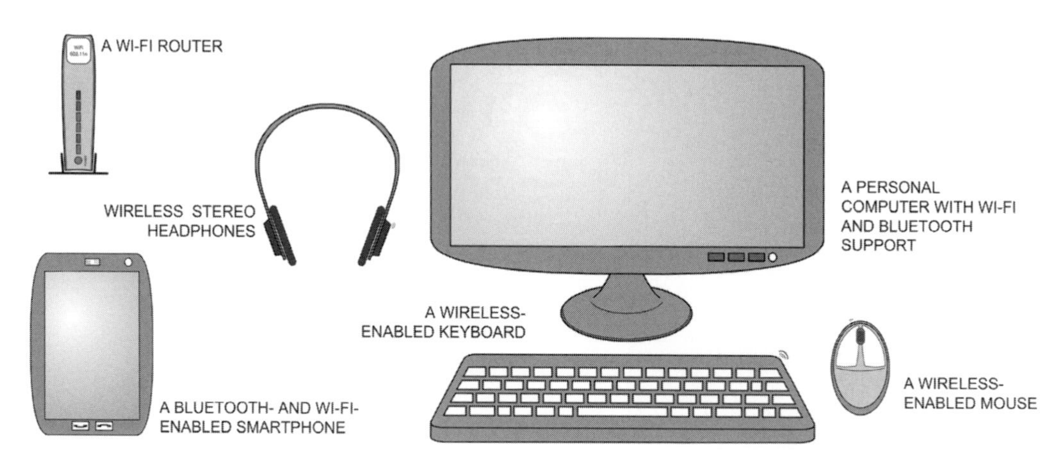

Figure 11.4. A wireless personal area network may include a PC, keyboard, and mouse, along with a smartphone and Wi-Fi router – all occupying your personal space.

personal area network is used to characterize a collection of fixed or wireless devices, such as that shown in Figure 11.4, which are personal to *you*. In fact, the *Personal Area Network* (PAN) was first coined as far back as 1996 by researcher Thomas Zimmerman. However, the definition didn't enter mainstream computing until circa 2000 with the introduction of Bluetooth wireless technology.

In Chapter 2, we also distinguished between PAN and the *Wireless Personal Area Network* (WPAN). You may recall that the devices that form your personal area network may be wireless-enabled. So, a *wireless* personal area network is a PAN in which devices permit connection through a wireless means. In other words, fixed PAN devices, such as a keyboard or mouse, may be connected to your personal computer via a cable. The definitions PAN and WPAN are occasionally used synonymously, but a WPAN, as a noun, is used to characterize the number of short-range radio technologies that may indeed enable your PAN to interconnect. For example, Bluetooth wireless technology and ZigBee are all WPAN technologies that may wirelessly enable your PAN. Therefore, a PAN represents a topology, whereas WPAN refers to the ability of that topology to interconnect wirelessly.

11.2 What Do Classic PAN Technologies Provide?

The *classic* personal area networking technologies sector is incredibly popular today, with manufacturers, innovators, and consumers alike embracing the range of consumer electronic products that embody a diverse range of wireless technologies. Most devices are not limited to one wireless technology – products nowadays integrate multiple classic technologies, all providing a range of unique and different applications. The increasing ease of use and simplicity afforded by some of these products allow consumers to continue to embrace, and likewise enjoy, the freedom that wireless technology affords. The availability of the new generation of products today is a stark reminder of how far

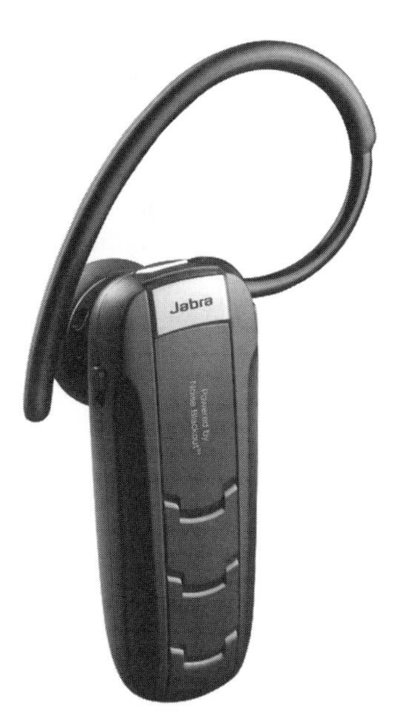

Figure 11.5. The Bluetooth wireless mono headset was one of the first classic wireless products to emerge onto the market, although the Jabra EXTREME2 pictured here is an ultra-modern style with enhanced capability when compared to early generations. (Courtesy of Jabra.)

each wireless technology within this sector has evolved. For example, in Figure 11.5, we illustrate the Jabra EXTREME2 Bluetooth mono headset, which typifies the very early generation (as far back as 2000) of Bluetooth wireless products. Indeed, one of the first Bluetooth products to emerge into mainstream consumerism was the Bluetooth mono headset. However, Jabra's EXTREME2 offers a new style and enhanced capabilities and features, which provide us with a hitherto unavailable comparison. Likewise, the D-Link DIR-645, Whole Home Router 1000 (Figure 11.2) represents a huge leap forward in Wi-Fi access point and router products that appeared over a similar timeframe to that of Bluetooth. Let's not forget that, over a decade or so ago, both Wi-Fi and Bluetooth were very much seen as competing technologies.

With Chapter 5, "Introducing Low Power and Wireless Sensor Technology," in mind, the manufacturers of wireless technologies that form Part III, "The Classical Personal Area Network," the classic section of the book, also need to be aware of the energy impact and the footprint they leave on the planet. You may recall that we touched upon the escalating cost of energy over the last decade or so – for many, the associated expense for homes and industry has become immeasurably high. So, innovators within the classic wireless technology sector must continually seek energy efficient techniques so as to optimize energy consumption within their consumer electronic products. Battery longevity, duration of use, and frequency of re-charging a product have become the overwhelming sales and marketing factors for most, if not all, consumer electronic devices. In short, allowing consumers to utilize their products for longer without the need to reach for the electricity socket to re-charge has become a unique selling point.

The chapters in Part II, "The Wireless Sensor Network," present a collection of technologies that offer varying techniques purporting low energy consumption and cost. They possess their own individual strengths and weaknesses that may suit many applications and products. Each chapter provides a structured approach to explaining each technology, whilst providing insights into specific alliances and membership structures, in turn determining the benefits of each alliance. The chapters further offer a comprehensive review of the market sector that a given technology targets, and endeavor to explain the overlap and redundancy between competing technologies.

11.3 What Should We Expect from Part III?

In Part III, "The Classic Personal Area Network," we provide a variety of classic personal area networking technologies, each purporting low energy characteristics, personable behavior, and cost effectiveness. Naturally, these technologies possess their own individual strengths and weaknesses which may lend themselves well to an abundance of applications and products. Each chapter provides a structured approach in explaining the technology, as well as insights into specific technology alliances and membership structures. As such, the benefits of each alliance may be determined. Furthermore, each chapter offers a comprehensive review of the market sector that a given technology targets, and endeavors to explain the overlap and redundancy between competing technologies. Likewise, the technology makeup is exhaustively covered, ensuring that aspects of the physical medium and software building blocks are presented, in addition to explaining the attributes of the networking topologies you can expect.

12 Just Touch with NFC

Near Field Communications (NFC) is an intuitive technology that provides an instinctive experience when it interacts with other products that are NFC-enabled. If you like, it's the magic wand of wireless technologies: a simple swish of your NFC-enabled device can gain you access to buildings, cinemas, and subways, and another swish or two can allow you to make a payment and even to simplify connectivity with other wireless technologies. We'll be discussing these very real user scenarios and much more later, in Section 12.2, "NFC's Market," and Section 12.3, "NFC's Application Portfolio."

NFC is an open standards-based, short-range technology that has its origins firmly based in *Radio Frequency Identification* (RFID) technology; in this chapter, we discuss NFC's inception and evolution. You may recall from Chapter 1, "It's a Small Wireless World," that we discussed the nineteenth-century wireless secret and revealed how RFID technology was used during World War II[1] as a means of identifying aircraft; we illustrate in Figure 12.1 how pilots would roll their planes to enable the ground forces to identify them as friend or foe. In this chapter, we'll also explore the NFC Forum, its membership benefits, and its structure. The chapter discusses the market scope, whereby initial NFC-enabled products have started to emerge in greater numbers due to the uptake of the technology in smart or cellular phones. We will, of course, look at how NFC fares against other technologies within the same market sector and how the technology has been misunderstood as a competitor to more mainstream wireless technologies such as Bluetooth. Finally, the chapter delves inside NFC and takes a first-hand look at its software architecture and protocol.

12.1 Overview

As we have already mentioned, NFC (nfc-forum.org) is an open standards-based, short-range wireless technology, supported and developed by a wealth of eclectic companies from various industries. NFC is an intuitive technology, providing a two-way interaction, passive, contactless experience through proximity, and uses a technique called *inductive-coupling*. This technique comprises "loosely coupled inductive circuits"[2] and, as such, it is able to share both data and power over a distance of a few centimeters (no more than

[1] Roberti, M., "The History of RFID Technology," 2005.
[2] NFC Forum, "How Does NFC Technology Work?," 2012.

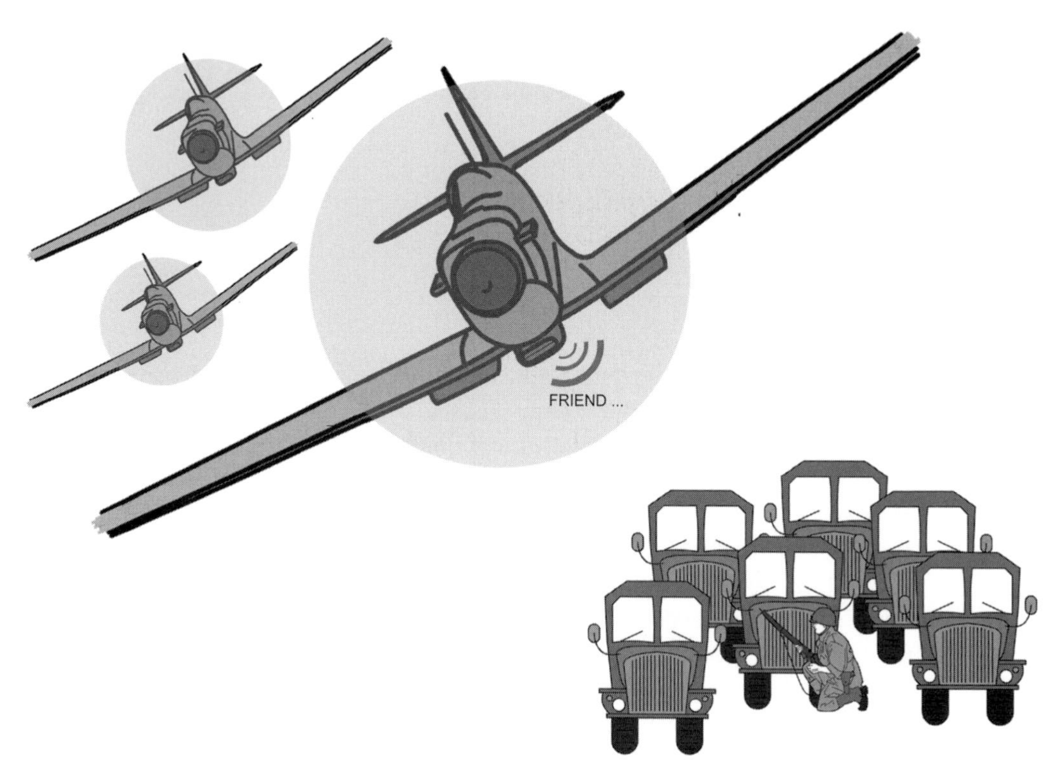

Figure 12.1. In World War II, pilots would roll their planes so that ground forces could identify them as friend or foe using RFID technology.

10 cm). The simplicity afforded by NFC is supported through proximity. What distinguishes NFC from any other technology featured in this book is the ability to exchange data or initiate a connection with a gesture of proximity; that is, it instills the capacity within proximal devices of its intention to connect. It is this intention that makes an important contribution to the innate security featured within NFC. The inherent security aspect to which we refer is the fact that communication only occurs when two NFC-enabled devices are within radio range, within a distance of no more than 10 cm for example. What's more, NFC uses the *Industrial, Scientific, and Medical* (ISM) 13.56 MHz radio frequency, and offers various data rates, from 106 kbit/s, 212 kbit/s, through to 424 kbit/s, for both *active* and *passive* communication. We will say more about NFC's communication modes later, in Section 12.6, "The NFC Architecture." Using the 13.56 MHz frequency band currently used by RFID tags and contactless smartcards, NFC is able to support a legacy base of products currently in circulation.

You may already be aware of Sony's *FeliCa*,[3] a contactless RFID smartcard technology currently used in Hong Kong, more commonly known as the *Octopus card*. It is also used across Japan, China, India, and the United States, and in Figure 12.2 we depict the product currently in use today. Furthermore, Sony's FeliCa technology was

[3] Source: sony.net/Products/felica/about.

Figure 12.2. Sony's FeliCa contactless IC card RC-S888 product. (Courtesy of Sony.)

incorporated into a number of mobile phones circa 2004. The *Osaifu-Keital*,[4] which, literally translated, means *wallet-phone* or more accurately wallet-mobile phone, was launched by NTT DoCoMo in Japan. The *Mobile Suica* offers its users comparable features to the *Suica*[5] pre-paid money card, whereby users can make purchases simply by swiping the card or device across a reader. Likewise, Philips Semiconductors (now NXP Semiconductors) also developed their own smartcard technology utilizing RFID technology in the mid 1990s, called *MIFARE* (mifare.net). Incidentally, revisions (early 2000 and beyond) of both FeliCa and MIFARE are compatible with NFC, although we won't touch upon these differentiators here; instead, we focus primarily on NFC technology. NFC uses RFID (HF) as a foundation technology and embodies a number of other key features that not only distinguish it from RFID, but also completely set it apart, as we'll discover in Section 12.1.2, "Comparing NFC and RFID." Furthermore, Sony and Philips Semiconductors (NXP) originally specified the NFC technology in 2002, when the two companies bolstered momentum within the industry to foster the new technology. NFC operates in several modes, which are based on the *International Electrotechnical Commission* (IEC)/*International Organization for Standardization* (ISO) 18092,[6] NFC *Interface and Protocol* (NFC-IP), and IEC/ISO 14443 contactless smartcard standards.

A small number of short-range technologies have wrestled with some issues surrounding initial connection, configuration, and set-up – for example, Wi-Fi, which we discuss in Chapter 13, "The 802.11 Generation and Wi-Fi," and Bluetooth wireless technology, which we discuss in Chapter 14, "Bluetooth Classic and High speed: More Than Cable Replacement," have both been criticized for what has been described by some consumers and industry pundits as employing an over complex or cumbersome connectivity method.

Later in this chapter, we will lift the lid on NFC and explore in more detail the various operational modes supported by the technology. Likewise, we'll explore the technology's features and software protocol stack, but, in the meantime, let's review the NFC Forum membership structure.

[4] Osaifu-Keital, translated as "wallet-mobile phone," incorporates Sony's FeliCa technology. (Source: http://www.nttdocomo.com/glossary/o/Osaifu-Keitai.html.)

[5] The Suica is a pre-paid e-money card, which can be used for ticket and vending machine purchases. (Source: http://www.jreast.co.jp/e/pass/suica.html.)

[6] IEC/ISO 18092.2004 Information technology "Telecommunications and Information Exchange Between Systems; Near Field Communication, Interface and Protocol (NFCIP-1)."

Figure 12.3. The NFC Forum N-Mark is the universal symbol and touchpoint for NFC. (Courtesy of the NFC Forum.)

12.1.1 The NFC Forum

We provide the NFC Forum N-Mark symbol and NFC Forum logo, respectively, in Figure 12.3 and Figure 12.4. The N-Mark is the universal symbol and touchpoint for NFC. It informs consumers that NFC functionality is available, and it is used on tags, media, devices, and software. The NFC Forum logo represents the group of companies that holistically drives the future evolution and development of the technology, as well as educates the overall consumer market. The Forum, which was formed in 2004 by Sony (sony.com), Philips (philips.com), and Nokia (nokia.com), has, to date, over 160 members, forming an eclectic group of industry leaders from financial services, manufacturers, and developers alike.

Figure 12.4. The NFC Forum logo. (Courtesy of the NFC Forum.)

12.1.1.1 Membership and the NFC Forum

The NFC Forum offers potential participants five levels of membership, as shown in Table 12.1. These five levels of membership enable companies to decide on how they wish to participate within the Forum. Like most standards-based organizations, membership is structured, so that a participant or company can decide on how much involvement they wish to have – from *Non-profit Member*, an entry level into the Forum where membership permits similar benefits as an *Associate* membership, through to *Sponsor Member*. Overall, all members, irrespective of access level, have access to a set of common benefits and privileges, which include full access to the NFC Forum community

Table 12.1. The NFC Forum offers potential participants five levels of membership, including Sponsor, Principal, Associate, Implementer, and Non-profit

Membership level	Scope
Sponsor	Sponsor level membership offers participants broad influence, with regard to the Forum itself and the overall future of the technology, in addition to a seat on the board of directors. Sponsor level participation is limited to ensure a moderate balance of industries.
Principal	At Principal level, participants can enjoy a strategic level of participation and influence; appoint voting representatives for the Technical, Ecosystem and Compliance Committees and to each working group; propose initiatives; and make contributions to the future development of NFC technology.
Associate	Associate membership allows participants to monitor NFC Forum activities, and to participate at non-voting representation within working groups. Associate members may not influence the Forum's direction, but can receive various deliverables, notifications, and comments on new working groups.
Implementer	Generally, Implementer members may monitor NFC Forum activity, and participate in and attend Ecosystem Committee working group face-to-face meetings, calls, and online working sessions.
Non-profit	A Non-profit participant receives the same benefits as Associate. However, to join as a Non-profit member, evidence must be provided to assure the NFC Forum that the organization is indeed a not-for-profit legal entity.

website; receiving all copies of published specifications; an ability to attend Plugfests; and participation in product interoperability testing and certification.

Figure 12.5. The NFC Forum Certification mark can be used on product packaging, user manuals, and other related material. (Courtesy of the NFC Forum.)

12.1.1.2 NFC Forum Certification

The NFC Forum offers manufacturers the opportunity to establish that their devices conform to the NFC Forum's official specifications, where certified devices will become registered and listed on the Forum's website. In Figure 12.5, we show NFC Forum's certification mark, which is provided to manufacturers once they have demonstrated and passed certification – a manufacturer may choose to use the mark on their product packaging, sales material, user manuals, and so on. Naturally, the certification mark is not used on the product itself; rather, the N-Mark symbol should be used, as this will convey to the consumer that the product is NFC-enabled. *Authorized Third-party Test Labs* are used to undertake certification testing of NFC-enabled products. What's more, the *NFC Forum Certification Issue Resolution Panel* (NCIRP) offers an arbitration process if an NFC Forum member disagrees with a report or test, or seeks further clarification relating to any NFC-specific specification.

12.1.2 Comparing NFC and RFID

In this section, we'll briefly touch upon some of the differentiators that set NFC and RFID apart. RFID uses several frequencies,[7] namely 120–150 kHz or *Low Frequency* (LF), 13.56 MHz or *High Frequency* (HF), 433 MHz or *Ultra-high Frequency* (UHF), and *microwave* to exchange its data for a variety of applications. Typically, LF RFID tags are used within livestock and domestic pets, where small chips can be embedded in the animal. HF RFID tags are commonly used to track miscellaneous products, such as books in a library or bookstore, for example, as well as access control for identification badges and so on. UHF RFID tags are predominately used in commercial environments, such as pallet and container tracking within trucks or an outside storage area. Finally, microwave RFID tags are used for long-range access control for a number of commercial vehicles.

As we have already mentioned, NFC borrows RFID's HF band, which is primarily used for tracking relatively small products and access control. Moreover, NFC and RFID operate identically as a tag for read/write operations at 13.56 MHz, but card emulation and peer-to-peer modes are not part of RFID – these features uniquely set NFC apart when compared with RFID. We discuss NFC's operational modes later, in Section 12.6, "The NFC Architecture." So, with the exception of HF tag read/write operations, RFID technology is largely different to NFC, as RFID provides a broader range of tracking mechanisms, which are utilized over various frequencies and distances.

12.1.3 NFC's Timeline

The infographic shown in Figure 12.6 provides a snapshot of NFC's technology timeline, from its inception to where the technology is today. NFC has been around for a decade or so, evolving from RFID, such as smartcard technology. Alas, NFC has seemingly endured a turbulent journey on its way to entering the mass market. The technology has only recently become part of mainstream consumers' diets with its integration into smart or cellular devices, despite enjoying moderate success elsewhere in the consumer industry. Sony and Philips created their respective smartcard technology, but it wasn't until 2002 that the companies combined their efforts to augment sufficient momentum to harness involvement from other companies, creating what we now understand today as near field communications.

[7] RFIDJournal.com, "What is the Difference between Low-, High- and Ultra-high Frequencies?"

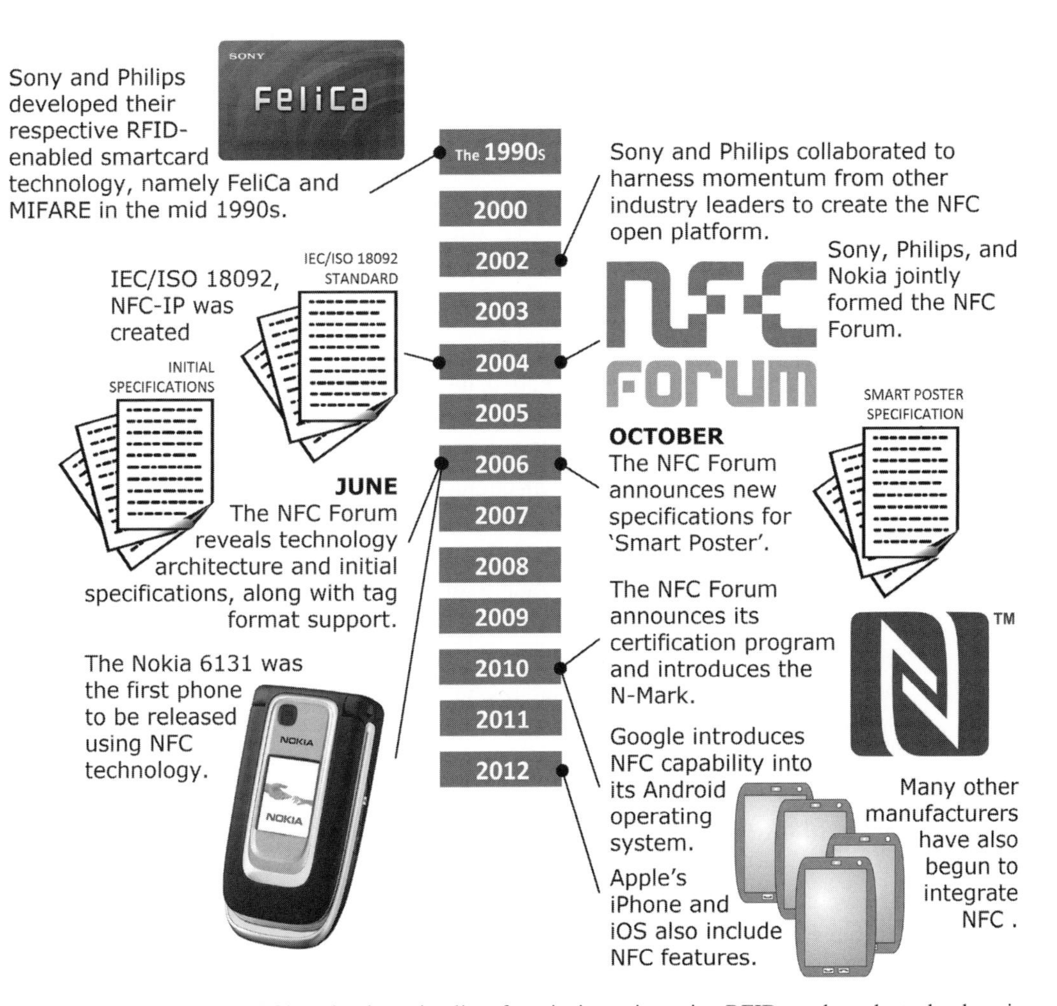

Figure 12.6. NFC's technology timeline, from its inception using RFID to where the technology is today with numerous manufacturers now integrating NFC as standard. (Nokia 6131, © Copyright Nokia; source: nokia.com/press.)

As we have already mentioned, during the 1990s Sony and Philips created their respective smartcard technologies, which ultimately led to some initial IEC/ISO standards being developed. These standards are still very much in use today in NFC technology. Of course, companies developing NFC products need to support the standards to receive product compliance certification. Following Sony and Philips' collaboration and subsequent partnership with Nokia, the NFC Forum was established in 2004. In 2006, initial specifications began to emerge – the technology architecture and the *NFC Data Exchange Format* (NDEF) and several NFC *Record Type Definitions* (RTDs) were released, as summarized in Table 12.2. We discuss both NDEF and RTD later in Section 12.9, "NFC Data Exchange Format," and in Section 12.10, "Record Type Definition."

Also in 2006, Nokia released the very first NFC-enabled phone (as pictured in the infographic, Figure 12.6), and later the NFC Forum announced a new specification for

Table 12.2. The NFC Forum provided three initial RTD specifications in which support is mandatory for compliant devices

RTD	Description
Smart poster	For posters that contain embedded tags, may contain data, or which can be read by an NFC-enabled device.
Text	Records that merely contain plain text can be read by an NFC-enabled device.
URI	The *Universal Resource Identifier* (or URI) aids an NFC-enabled reader to refer to an Internet resource.

Smart Poster. A few years later, in 2010, the NFC Forum announced their certification program and showcased their technology's N-Mark symbol (as shown in Figure 12.3), which is targeted toward consumers so that they can readily identify NFC-enabled capability within other consumer electronic products. As we have already intimated, NFC struggled with mass market uptake, but in 2010 Google (google.com) announced NFC integration within their mobile Android operating system, shortly followed by their *Google Wallet* application. Apple (apple.com) launched their new iPhone5 product in September 2012, which lacked NFC wallet and wireless charging features and capabilities, although Apple seems to suggest that one day an NFC-enabled phone could replace your wallet, but only in the medium to long term[8] (no more than 12 months or so from the time of writing).

12.2 NFC's Market

The wireless technologies that form Part III, "The Classic Personal Area Network," are typically high-end, high data rate technologies, best suited for audio-, voice-, and data-heavy-centric applications. However, NFC differs somewhat insofar as the typical data rates of the technology supported are 106 kbit/s, 212 kbit/s, and 424 kbit/s, which lends itself aptly to relatively small transactional data exchanges. NFC technology has been placed into the classic section as it is not targeted at the wireless sensor reworking domain.

NFC technology has its roots firmly embedded within RFID and, as such, NFC technology has had an extraordinary evolution spanning almost two decades. Yet, despite its longevity, it has struggled to reach the mainstream mass market. Arguably, the technology has already been used extensively across Asia under the guise of RFID (HF band), but has not reached a level of widespread recognition by everyday consumers. It has been only a recent triumph for the technology to be integrated into smart or cellular devices, and consequently we have seen some overwhelming momentum in the technology's uptake. In Figure 12.7, we illustrate NXP's PN65 NFC chipset, which has been integrated into several manufacturers' smartphones. NFC will inevitably grow in popularity and recognition as consumers come to understand its capabilities and features.

[8] Richmond, S., "iPhone 5: Price, 4G and Everything Else You Need to Know," 2012.

Figure 12.7. The NXP PN65 NFC chipset is currently used in several smartphones. (Courtesy of NXP Semiconductors.)

Moreover, the technology has been pushed and touted as an enabler for contactless purchase transactions. It's a strong first introduction for the consumer to the technology's potential, but consumers may grapple with a psychological[9] shift in disposing of their wallets and using their mobile phone as a mechanism to make a purchase – typically, NFC transactions are limited in value to, say, no more than $25. The concept isn't new, as we have demonstrated earlier with Sony's FeliCa and Philips' MIFARE smartcard technology. What's more, one of the first British banks to embrace contactless payments was Barclays (barclays.co.uk) in 2007, although Barclays' contactless product relied on RFID technology, as we illustrate in Figure 12.8. However, in 2011, *Barclaycard*[10] partnered with Orange UK[11] to offer British consumers the opportunity of making

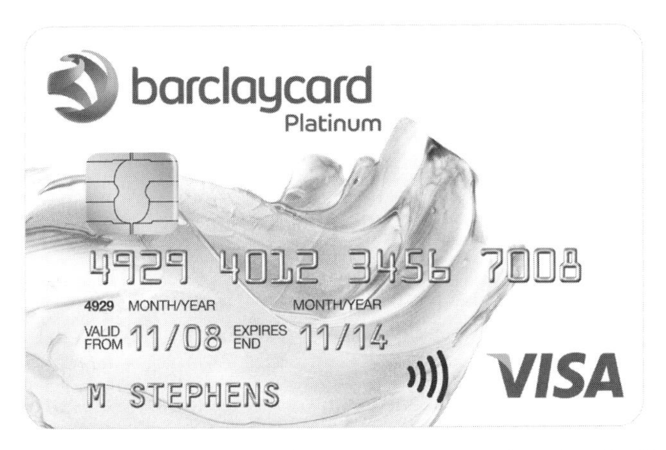

Figure 12.8. Barclays' contactless payment card was introduced in 2007 and used RFID technology. (Courtesy of Barclays.)

[9] Gratton, D. A., "Forecasting a Wireless-enabled 2012," 2012.

[10] Barclaycard is part of Barclays' retail and business banking and was one of the first credit cards to be introduced to consumers in 1996 for the United Kingdom (source: secure.barclaycard.co.uk).

[11] Orange, originally British owned, is a telecommunications corporation, and is the leading brand name of the France Télécom group providing mobile and fixed Internet and telephony services.

Figure 12.9. Barclays partnered with Orange UK to offer their consumers the ability to purchase products/goods using their mobile phone. (Courtesy of Red Consultancy.)

electronic purchases using their mobile phones, as shown in Figure 12.9. Nonetheless, NFC has a much wider scope, in terms of application potential, as we can witness from emerging specifications purported by the NFC Forum – this is something we pick up on in the section to follow.

12.3 NFC's Application Portfolio

As we've already mentioned, NFC has had a lifespan of almost two decades, and with the formation of the NFC Forum, along with the associated motivation driven by the various industry pundits that form the group, NFC has begun to populate the mobile phone market sector aggressively. In its gradual evolutionary step toward widespread popularity, the technology and its numerous applications will inevitably become instrumental in founding a proven consumer-base. All members of the NFC Forum collectively drive and share development, application, and marketing experience to develop the future of the technology. In Table 12.3, we list the broad and varied types of companies that holistically enable the NFC ecosystem.

In this section, we explore further how the technology is being used to form a better understanding of the number of applications and products supported by an NFC ecosystem. With NFC technology having been dropped into public recognition, especially

Table 12.3. The NFC Forum comprises a broad and diverse group of companies who together drive and advance the NFC ecosystem

	Market sector
1	Merchants and consumers.
2	Banking, financial sectors.
3	Telecom operators.
4	Reader manufacturers.
5	Smartcard manufacturers.
6	Semiconductor manufacturers.
7	Consumer electronics manufacturers.
8	Mobile device OEMs.
9	Education, research, and governmental sectors.
10	Testing and certification companies.
11	System integrators.

with applications such as Google Wallet (an Android app), the technology has sufficiently harvested adequate momentum for everyday use – something that undoubtedly will become abundantly innate. Consumers will increasingly become confident with using contactless payment schemes, perhaps experiencing a somewhat analogous psychological shift that occurred from using cash to using a debit or credit card. But NFC technology doesn't stop there in terms of its application portfolio. Naturally, NFC will continue to support original and existing user scenarios, such as access to subways, buildings, and so on. The industry is eager to utilize the smart or cellular phone as the active tool in diversifying NFC's application potential – it is a natural choice to place the mobile phone at the center of any use case, as the majority of consumers use their mobile phones on a daily basis. As such, the mobile phone has become the pivotal focus for all of NFC's transactional exchanges and other miscellaneous activities. In other words, the phone will be the primary device used to conduct purchases, to permit access to subways and cinemas, to allow entry into offices and buildings, and also to enable other interactions, such as those with smart posters, freeway/motorway toll payments and access, social media, and so on.

We illustrate in Figure 12.10 an example NFC ecosystem; whilst the devices that feature in the illustration are not interconnected as such, the image does portray existing applications, along with NFC's potential. NFC simply forms a *point-to-point* topology – more about this later in Section 12.5, "Networking Topology." As we can see, access to subways, trains, buses, and other forms of public transport are supported (**1**) as well as the smart poster (**2**) application that shares further information about a product. In March 2012, Orange UK partnered with EAT (eat.co.uk), a British food service chain, to launch a rewards-based or loyalty service to enable consumers with NFC-enabled handsets to receive an *EAT Treat*.[12] Future loyalty and/or social networking applications would not only allow the consumer to purchase the product directly, but also to share

[12] NFC World, "Orange UK Announces Quick Tap Treats for 200 000 NFC customers," 2012.

Figure 12.10. A typical NFC ecosystem utilizing a point-to-point topology structure.

a *tweet* through Twitter about the product, a *like* via Facebook, if the product had a Facebook or fan page, or indeed to *pin* a product image with an accompanying link via Pinterest. After all, a consumer undoubtedly has that all-important pervasive *Wide Area Network* (WAN) connection supported by our supposition of the *Lawnmower Man Effect* (LME), as discussed in Chapter 4, "Introducing the Lawnmower Man Effect."

The smart poster ethic has even been extended to graves or headstones within the memorial and funeral industries. Initially introduced as *Quick Response* (QR) codes or barcodes, this inevitably raised some concerns within Jewish, Muslim, Catholic, and

Figure 12.11. Objecs' D106, RosettaStone, microchip, which can be attached to a vertical headstone. (Courtesy of Objecs.)

other religious faiths. However, with NFC/RFID technology, there is no intrusive indicator. As such, *Objecs®* (personalrosettastone.com), based in Arizona, Phoenix, owns the RosettaStone Microchip brand for headstone technology, which aims to enhance the memorial experience (see Figure 12.11). It seems the agnostic community has embraced stone tablets or obelisks with NFC/RFID technology, whereby visitors can learn more about the deceased by using their handset to retrieve geo-tagging, photographs, and textual information. What's more, the RosettaStone NFC technology has also been accepted by one of the world's largest monasteries, *Jasna Góra*, in Poland, as well as by other smaller churches.

In Figure 12.10, the use case for a freeway/motorway toll access is supported, either through an in-built system within the vehicle (**3**), or via a mobile phone that can be swiped at the toll booth. As we move around the NFC ecosystem, we depict some of the more classic use cases portrayed by NFC. Consumers can now use their smart or cellular phones to make small purchases (**4**), and can optionally extend their feedback to their preferred social network. We have already intimated that NFC can be used as an enabler of other wireless technologies; as such, in (**5**) we see NFC being used to simplify a connection between a Bluetooth-enabled headset and mobile phone; similarly, NFC can also be used to simplify a Wi-Fi (**6**) connection set-up for a computer (**7**) and *Set-top-box* (STB) (**8**). And lastly in our NFC ecosystem it will become commonplace to own a "charger mat" (**9**) – like a mouse pad or similar surface product – which will enable some electronic products, such as the smartphone or mouse, to charge wirelessly, eliminating the need for that power cable!

NFC also allows the exchange of business or contact information or the offer of security credentials simply by a "touch" with an NFC-enabled card or handset. What's more, NFC has the potential to resurrect the notion *brand loyalty*, or indeed garner *brand awareness*.[13,14] Brand loyalty has faded somewhat due to a converged and increasingly competitive market, and consumers no longer buy in to a single brand, but rather scout for products with the best features and competitive prices. However, NFC can potentially

[13] Gratton, D. A., "Why NFC Should Keep Social Media in its Wallet!," 2011.

[14] Gratton, D. A., "Increasing Brand Awareness with NFC Technology and Social Media," 2011.

revive consumers' loyalty in brands – after all, consumers have been and remain responsible for creating and indemnifying the brands that are most recognizable today. So, the purchase transaction can be extended to the social network, where consumers may leave feedback about their purchases and experiences. Likewise, a restaurant or other food chains can harness the same ability to capture their customers' attention – providing the customer with an opportunity to leave feedback relating to the food and the quality of the experience. This is all incredibly valuable market research for any company and, moreover, it's far less intrusive than being stopped and approached by a person wielding a clipboard on the street.

The NFC topology would form part of your *Body Area Network* (BAN) since you carry your phone about your person. Likewise, if you use a smartcard to make a purchase or access the subway, for example, then your card will be inherent to your BAN. You may recall from Chapter 2, "What is a Personal Area Network?," that we discussed the divergence of the personal area networking space, which has become somewhat diluted with numerous topologies that uniquely characterize scope and function. You may also recall that we discussed and characterized the diverse number of area networking topologies that have become popular in mainstream computing. We also argued that it's still very much a PC-centric perspective, and that, despite manufacturers' attempts to steer away from PC-dependent technologies and applications, the PC still remains integral to most topologies. Nevertheless, the products portrayed in Figure 12.10 may have the ability to connect directly to the *Wide Area Network* (WAN) or Internet, in turn potentially eliminating the need to rely on a PC to source content.

12.4 NFC and its Competitors

NFC (and RFID) obviously compete with magnetic card technology – that black strip you see on the back of some debit, credit, or store cards. NFC has a very unique market sector, perhaps all to itself, since it doesn't really have any direct competition. Naturally, some other low power wireless technologies could potentially compete if the ability to offer transactional exchanges was part of that technology's portfolio. For example, Bluetooth low energy has been touted as a viable technology candidate to offer mobile payments and other similar use cases, as the technology (Bluetooth v4.0) has already been integrated into a number of consumer electronic products, such as Apple's iPhone 4S and iPad HD.[15] Nevertheless, the industry seems very much geared toward NFC as the technology to deliver the all-important user scenarios and infrastructure, as already demonstrated by Barclaycard and other industry leaders such as Google.

12.4.1 Complementary rather than Competitive

Nonetheless, many industry analysts *originally* regarded NFC as a competitor to Bluetooth wireless "classic" technology, and vice versa. Moreover, with the new Bluetooth

[15] Gabriel, C., "Apple iWallet May Favor Bluetooth Over NFC," 2012.

low energy variant introduced into the new Specification of the Bluetooth System: Core, v4.0, many pundits are wondering whether Bluetooth will venture into NFC's domain. However, whilst the feature-sets are more or less comparable, it isn't in Bluetooth's remit to offer such use cases that directly compete with NFC; and, despite rumors of Bluetooth low energy encroaching on NFC's application portfolio, it still isn't included within Bluetooth's profile to undertake such applications. We discussed Bluetooth low energy earlier, in Chapter 7, "Bluetooth low energy: The *Smart* Choice." What's more, NFC is primarily a complementary technology and remains compatible with existing contactless card infrastructures, something which we have already discussed. Likewise, in Section 12.3, "NFC's Application Portfolio," we demonstrated that NFC could ease pairing and simplify connection set-up by offering and securely sharing security credentials for both Bluetooth and Wi-Fi.

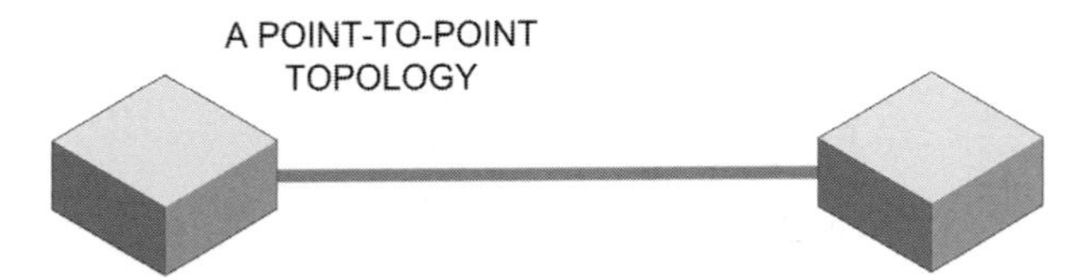

Figure 12.12. The NFC topology supports a point-to-point topology, in turn enabling bi-directional communication.

12.5 Networking Topology

NFC is a bi-directional technology supported by a point-to-point topology, as we demonstrate in Figure 12.12. NFC devices are adaptive, insofar as they have the ability to change their *mode* of operation, that is their *reader/writer*, *peer-to-peer*, or *card emulation* modes, where all modes are governed by the ISO/IEC standards, which we discussed earlier. We clarify in Table 12.4 the different operating modes available to an NFC device. An NFC *tag*, on the other hand, is a *passive* device, such as that used within a smart poster. In this instance, the tag merely contains data that can be read by an NFC-enabled device, although some tags are able to store additional information, such as secure data.

Table 12.4. NFC device modes of operation

Mode	Description
Peer-to-peer	In this mode, two NFC devices can exchange data such as a business card, photographs, and so on, as well as the link set-up parameters for a Bluetooth or Wi-Fi connection, for example.
Card emulation	The device appears to a reader in much the same way as a traditional contactless smartcard that, in turn, supports the existing infrastructure.
Reader/writer	An NFC device can read tag types, such as the smart poster tag.

12.6 The NFC Architecture

In the sections that follow, we take a top-down approach to the NFC architecture. We will unravel the technology from the top to the bottom and explore in greater detail the technology architecture along with its software and application building blocks. As we described in the preceding section, an NFC device may operate interchangeably in one of three modes and, as such, the software architectures differ conceptually depending on the device's operational mode. In the sections to follow, we'll look at each mode individually to explore its architectural makeup.

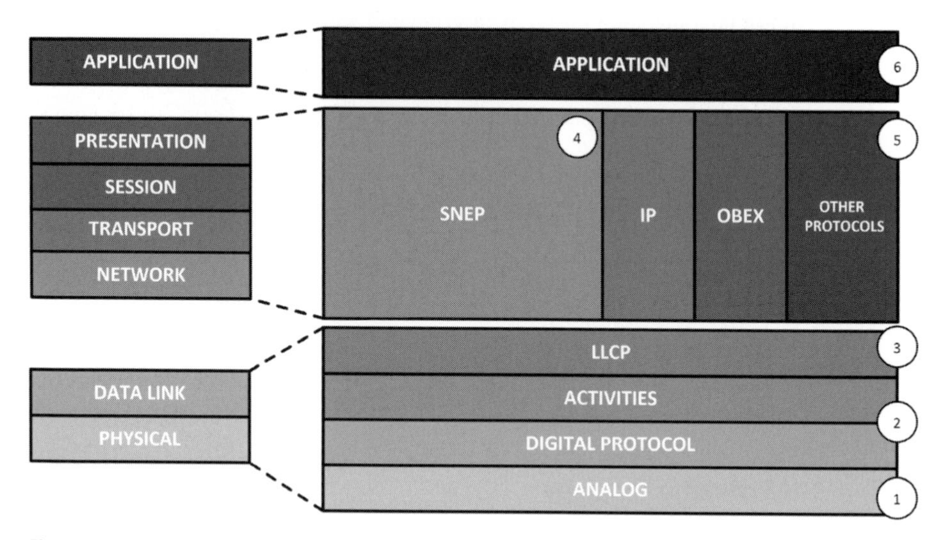

Figure 12.13. An NFC device software stack architecture mapped against the OSI model for peer-to-peer mode.

12.6.1 Peer-to-Peer Mode

The NFC stack architecture provides a number of blocks, which are layered. In Figure 12.13, we illustrate the software architectural building blocks for an NFC device operating in *peer-to-peer* mode, which is mapped against the *Open Systems Interconnection* (OSI) model (left). Furthermore, we have labeled these building blocks numerically to ease identification and classify responsibility. So, looking at our figure again, blocks labeled (**1**) and (**2**) represent the *Physical* (PHY) and *Media Access Control* (MAC) layers, which we will discuss in more detail in Section 12.11, "Activities, Digital Protocol, and Analog." In block (**3**) we see the *Logical Link Control Protocol* (LLCP), which forms the OSI data link layer. This maps and binds functionality to the MAC layer and is scoped by the NFC Forum. We discuss the LLCP later, in Section 12.8, "The Logical Link Control Protocol."

The *Simple NDEF Exchange Protocol* (SNEP) (**4**) straddles the network through to the presentation layers of the OSI model. The SNEP layer permits an application residing on an NFC-enabled device to communicate *NFC Data Exchange Format* (NDEF) messages with other devices in peer-to-peer mode. We look at SNEP and NDEF in Section 12.7, "Simple NDEF Exchange Protocol," and Section 12.9, "NFC Data Exchange Format" – both protocols are defined by the NFC Forum. In (**5**), we see a number of protocols, such as *Object Exchange* (OBEX) and *Internet Protocol* (IP), along with other protocols that bind to LLCP. And finally, the end-application (**6**) is typically defined by external specifications and does not form part of the NFC Forum's scope.

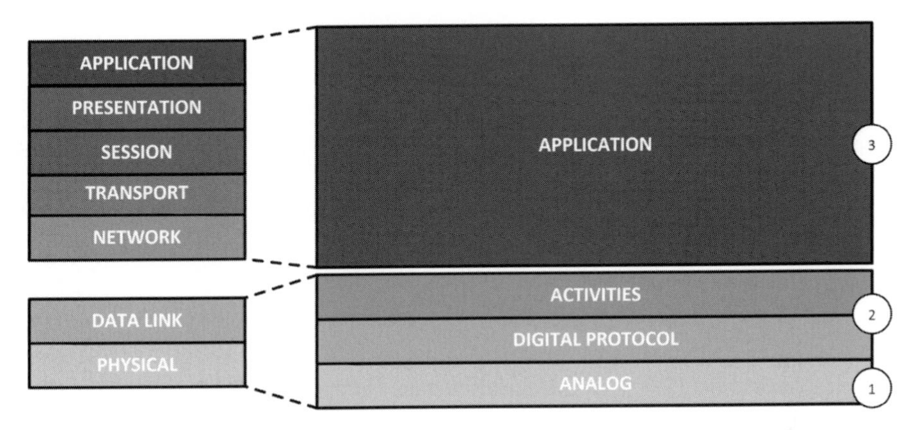

Figure 12.14. An NFC device software stack architecture mapped against the OSI model for card emulation mode.

12.6.2 Card Emulation Mode

In Figure 12.14, we illustrate the software architectural building blocks for an NFC device operating in *card emulation* mode, which is mapped against the OSI model (left). We have also labeled these building blocks numerically to ease identification and classify responsibility. Blocks labeled (**1**) and (**2**) represent the PHY and MAC layers, which we will discuss in more detail later, in Section 12.11, "Activities, Digital Protocol, and Analog." The end-application (**3**) is defined by the NFC Forum.

12.6.3 Reader/Writer Mode

The NFC stack architecture provides a number of blocks, which are layered. In Figure 12.15, we illustrate the software architectural building blocks for an NFC device operating in *reader/writer* mode, which is mapped against the OSI model (left). Furthermore, we have labeled these building blocks numerically to ease identification and classify responsibility. Blocks labeled (**1**) and (**2**) represent the PHY and MAC layers, which we will discuss in more detail later, in Section 12.11, "Activities, Digital Protocol, and Analog." In block (**3**), we see the *Tag Operations* layer, in which the NFC Forum has defined four tag types. The layer specifies the necessary commands and associated parameters needed for it to be able to read/write data to a tag, irrespective of

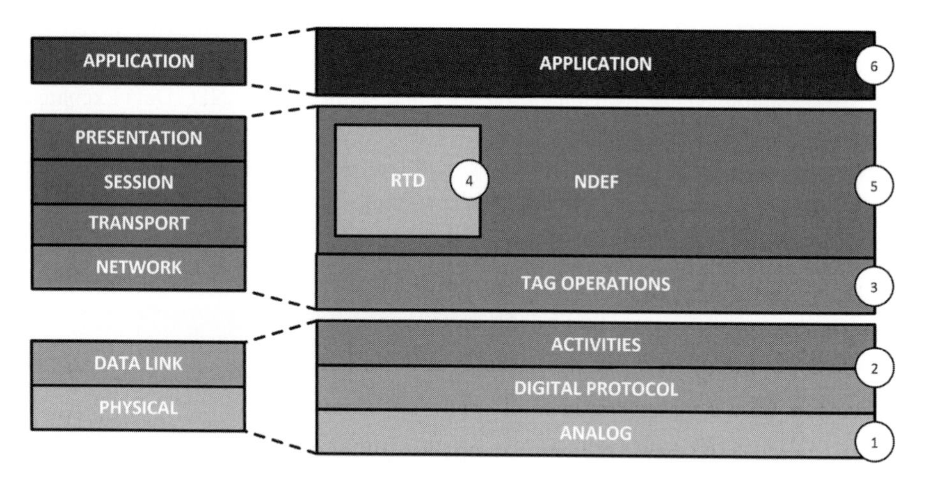

Figure 12.15. An NFC device software stack architecture mapped against the OSI model for reader/writer mode.

tag type. In (**4**) we depict the RTD, which prescribes how unique record types used by several applications are defined within an NDEF (**5**) message. We discuss NDEF later, in Section 12.9, "NFC Data Exchange Format," and RTD in Section 12.10, "Record Type Definition." The end-application (**6**) is defined by the NFC Forum.

Figure 12.16. The SNEP request and response message frame structure.

12.7 Simple NDEF Exchange Protocol

The SNEP communication protocol provides a *request/response* paradigm in which a SNEP *client* transmits a request to a SNEP *server*. Naturally, the server will execute the request and indicate a success or failure of that request. The transport protocol provides a reliable transport, which must be capable of receiving service data units of six or more octets. In Figure 12.16, we illustrate the frame format of both SNEP request and response messages and describe these fields in Table 12.5. In essence, the SNEP offers the ability to exchange an NDEF message, which is contained within the information field, as we show in Figure 12.16.

SNEP is only applicable to the peer-to-peer operation mode and sits above the LLCP layer, as we discussed earlier, in Section 12.6.1, "Peer-to-Peer Mode," and Figure 12.13. The LLCP offers a connection-oriented transport service with guaranteed delivery of service data units – typically, each data unit can accommodate no more than 128 octets.

Table 12.5. The fields that make up the SNEP request and response message frame

Field	Description	Size
Header	The SNEP header is 48-bits in length and may contain information relating to a request or a response message, which we describe further in Section 12.7.1, "The SNEP Request Message," and Section 12.7.2, "The SNEP Response Message."	48
Information	The content of the information field is subject to the value contained within the request or response field. However, the information field is not included if the length field (see Figure 12.17 and Figure 12.19) is equal to `0x00`.	

12.7.1 The SNEP Request Message

As we have already mentioned, the client sends a request to the SNEP server to invoke a particular action. We illustrate in Figure 12.17 the SNEP *request* message frame format, whilst in Table 12.6 we look at the specific fields that make up the SNEP request header, along with size of the field in bits.

Table 12.6. The fields that make up the SNEP request header

Field	Description	Size
Version	The version field is 8-bits in length and provides information relating to the format of the protocol message and how the originator understands SNEP communication. The 8-bit field is split into two further fields, as pictured in Figure 12.18and described in Table 12.7. We discuss the version field in more detail in Section 12.7.3, "Versioning."	8
Request	The request field is 8-bits in length and denotes the action to be executed by the SNEP server; we list the possible request in Table 12.8.	8
Length	The length field is 32-bits in length and specifies the total length (in octets) of the information field.	32

Table 12.7. The two fields that make up the version field

Field	Description	Size
Major	The major field is 4-bits in length and represents the *major* protocol version of the SNEP specification in use.	4
Minor	The minor field is 4-bits in length and represents the *minor* protocol version of the SNEP specification in use.	4

12.7.2 The SNEP Response Message

The SNEP response is transmitted by a SNEP server to a client – the message contains the result of the action executed by the server device. We illustrate in Figure 12.19 the

Table 12.8. Request field values

Value	Description
`0x00`	Continue (transmit outstanding fragments).
`0x01`	Get (return an NDEF message).
`0x02`	Put (accept an NDEF message).
`0x03–0x7E`	Reserved.
`0x7F`	Reject.
`0x80–0xFF`	Reserved (response field values).

SNEP *response* message frame format, whilst in Table 12.6 we look at the specific fields that make up the SNEP request header, along with size of the field in bits.

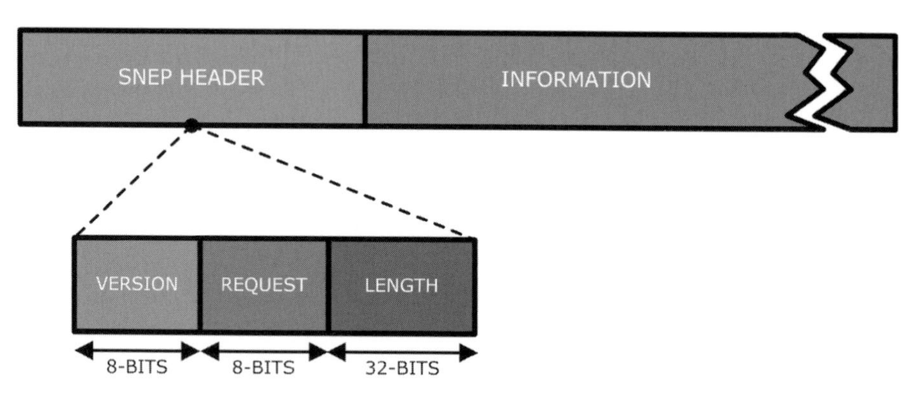

Figure 12.17. The SNEP request message.

12.7.3 Versioning

As we touched upon in Table 12.6 and Table 12.9, the request and response message header contains an 8-bit version field, which is equally split into a major (4-bit) and a minor (4-bit) field, as shown in Figure 12.18. It provides information relating to the

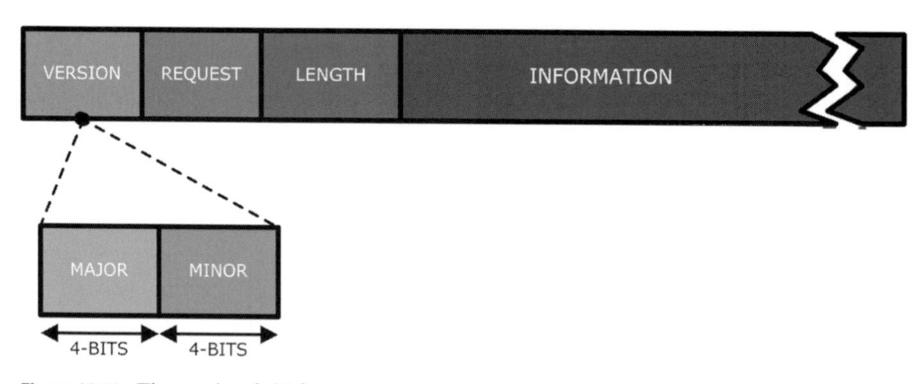

Figure 12.18. The version field format.

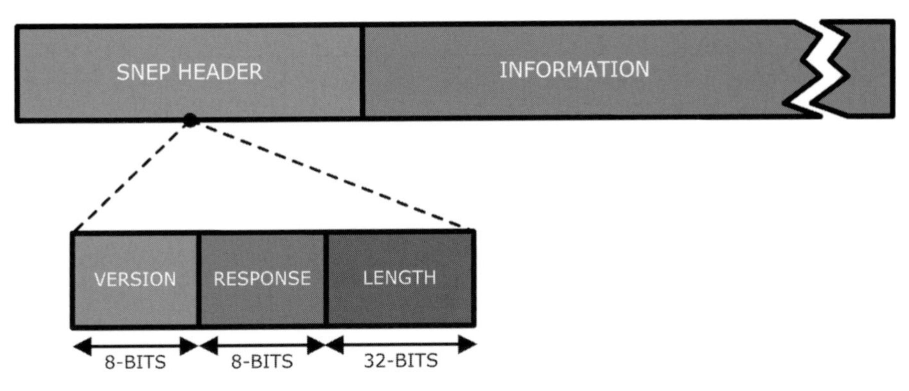

Figure 12.19. The SNEP response message.

Table 12.9. The fields that make up the SNEP response header

Field	Description	Size
Version	See Table 12.6 and Section 12.7.3, "Versioning."	8
Response	The response field is 8-bits in length and provides the results of the action that was executed by the SNEP server; we list the possible responses in Table 12.10.	8
Length	See Table 12.6.	32

format of the protocol message and how the originator understands SNEP communication. The minor field is incremented to reflect changes made to the protocol that do not have an effect on the overall message parsing algorithm, but may also indicate additional capabilities of the sender device. However, the major field is incremented to reflect a format change within the message. Furthermore, if a SNEP server receives a request message with a major version number that differs to its own, it may return an "unsupported protocol version" response, as shown in Table 12.10. In the instance where the minor field differs, both client and server agree to return an appropriate response.

12.7.4 SNEP Fragmentation

A SNEP request or response message is fragmented if the overall length exceeds the capability of the device. So, a SNEP message will be broken into fragments, as depicted in Figure 12.20, and then transmitted. For a receiver to ascertain the number of fragments to accept, the SNEP header is *always* transmitted, that is, the length field will indicate the overall size of the payload. What's more, the receiver device will inform the sender as to whether or not it can continue to receive any outstanding fragments using the continue response (see Table 12.10); however, no acknowledgement is offered.

Table 12.10. Response field values

Value	Description
`0x00–0x7F`	Reserved (request field values).
`0x80`	Continue (transmit remaining fragments).
`0x81`	Success.
`0x82–0xBF`	Reserved.
`0xC0`	Resource not found.
`0xC1`	Resource exceeds data limit size.
`0xC2`	Bad request.
`0xC3–0xDF`	Reserved.
`0xE0`	Unsupported functionality requested.
`0xE1`	Unsupported protocol version.
`0xE2–0xFE`	Reserved.
`0xFF`	Reject (stop transmitting remaining fragments).

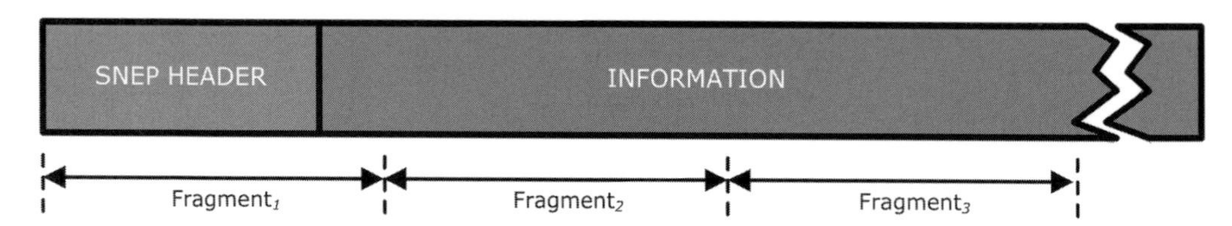

Figure 12.20. SNEP request and response message fragmentation.

12.8 The Logical Link Control Protocol

The LLCP layer occupies the OSI's data link layer, as we discussed in Section 12.6.1, "Peer-to-Peer Mode." We also discussed its use within the peer-to-peer mode operation, and in this section we explore some of the features offered by the protocol and review its architectural components. In Table 12.11, we summarize the headline features that are supported by the LLC protocol.

Table 12.11. The headline features of the LLC protocol

Feature	Description
Link activation, supervision, and deactivation	The LLCP offers features as to how two devices should behave when in proximity. Likewise, it also determines how a link is established, monitored, and deactivated.
Asynchronous balanced communication	In a client/server paradigm, the LLCP uses *Asynchronous Balanced Mode* (ABM) between two devices to provide initialization, supervision, and recovery from spontaneous errors.
Protocol multiplexing	The LLCP is capable of utilizing multiple instances of high-level protocols simultaneously.
Connectionless transport	With minimum protocol effort, this feature provides LLCP with an unacknowledged data transport, which can be used ad hoc to transfer data.
Connection-oriented transport	With an established connection, this feature affords LLCP a sequential and guaranteed delivery mechanism for data units.

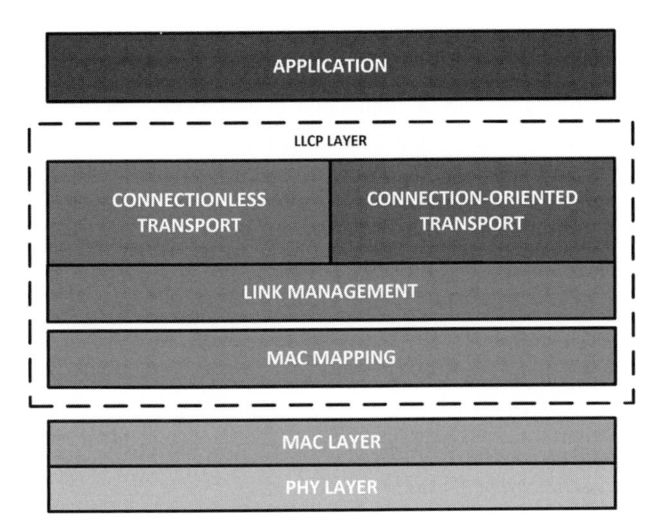

Figure 12.21. The internal building blocks of the LLCP architecture.

12.8.1 The LLCP Architecture

The LLCP has been organized logically, as shown in Figure 12.21. From the bottom up, the *MAC mapping* building block permits the existing MAC/PHY layers to be integrated fully into the overall LLCP layer. The *link management* building block provides overall responsibility for serializing connectionless and connection-oriented LLC *Protocol Data Units* (PDUs) to be exchanged and to manage smaller PDUs. We discuss the format of the LLC PDU later, in Section 12.8.2, "The LLC PDU Format." The *connectionless transport* building block is responsible for the management of *unacknowledged* data with reduced complexity, and is used when guaranteed delivery is not critical. The connectionless transport mechanism doesn't utilize a procedure for connection establishment or termination and, as such, ad hoc communication may occur without the overhead of preamble.

On the other hand, the *connection-oriented* transport is responsible for managing all connection-oriented data, along with connection set-up and termination, and offers a sequential and guaranteed delivery of data units. However, connection establishment is required between a local and a remote device, to ensure peer-to-peer communication such that the connection remains open until either device terminates the connection or there is disruption with the open link; multiple connections can be supported by LLCP.

Table 12.12. LLCP SAP values

SAP	Type	Description
1	1	SDP
2	1	IP
3	2	OBEX
4	2	SNEP

We discussed earlier, in Section 12.7, "Simple NDEF Exchange Protocol," the SNEP protocol, which relies on LLCP to transport SNEP messages over LLCP using the connection-oriented transport. So, a connection is established between a local and a remote LLC *Service Access Point* (SAP), and SNEP messages may be transmitted over the LLC connection specifically utilizing the connection-oriented transport service. A connection SNEP request offers the active SNEP server an address for its SAP – we list some examples in Table 12.12. As such, the SNEP client and SNEP server will mutually exchange request/responses, respectively, on the open connection, and the SNEP client will also be responsible for closing the connection once it has completed its communication.

Table 12.13. The link service classes

Class	Description
1	An NFC device that only supports a connectionless transport is regarded as a *link service Class 1*.
2	An NFC device that only supports a connection-oriented transport is regarded as a *link service Class 2*.
3	An NFC device that offers both connectionless and connection-oriented services is classified as a *link service Class 3*.

Figure 12.22. The LLCU PDU format.

What's more, an NFC device can be further classified by *link service class*, as we summarize in Table 12.13. In the sections to follow, we look at the LLC elements that comprise the PDU format and associated PDU types.

Table 12.14. The fields that make up the LLCP PDU

Field	Description
Header	The LLCP header is either 16- or 24-bits in length and contains information relating to a PDU type, which we describe later in this section.
Payload	The payload field may or may not be present and is subject to the PDU type.

12.8.2 The LLC PDU Format

We illustrate in Figure 12.22 the LLC PDU format, which describes how the SAP addresses, PDU type, and the associated payload are formatted. In Table 12.14, we look at the specific fields that make up the LLCP PDU.

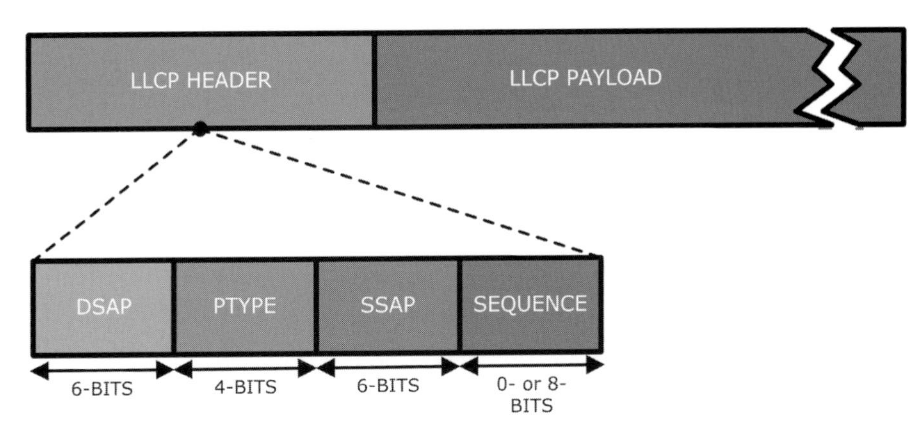

Figure 12.23. The LLCP PDU format with a view of the header makeup.

Table 12.15. The fields that make up the LLCP header

Field	Description	Size
DSAP	The DSAP is 6-bits in length and describes the destination of the intended LLCP payload. We list in Table 12.16 the possible addresses that can be used.	6
PTYPE	The PDU type field is 4-bits in length and describes the LLC payload; we discuss the PTYPE later in this section.	4
SSAP	The SSAP is 6-bits in length and describes the origin of the payload. We list in Table 12.16 the possible addresses that can be used.	6
Sequence	The sequence field is subject to the PDU type field used and is described further in Table 12.18. If the sequence field is present, then it requires 8-bits, as shown in Figure 12.23.	

Table 12.16. DSAP and SSAP address values

Value	Description
`0x00–0x0F`	Well-known SAPs, also listed in Table 12.17.
`0x10–0x1F`	A local service, which is advertised by the local SDP.
`0x20–0x3F`	A local service, which is not advertised by the local SDP.

In Figure 12.23, we illustrate the LLCP header format, whilst in Table 12.15 we look at the specific fields that make up the LLCP header, along with the size of the field in bits. As we show in Figure 12.23, the LLCP header comprises two address fields, namely the *Destination Service Access Point* (DSAP), which represents the destination of the intended LLCP payload, and the *Source Service Access Point* (SSAP), from where the LLCP payload originated.

Both the DSAP and SSAP may use any of the address values listed in Table 12.16. Incidentally, the value `0x00` always represents the LLC link management entity and

Table 12.17. Well-known LLCP SAP values

SAP	Description
1	SDP
4	SNEP

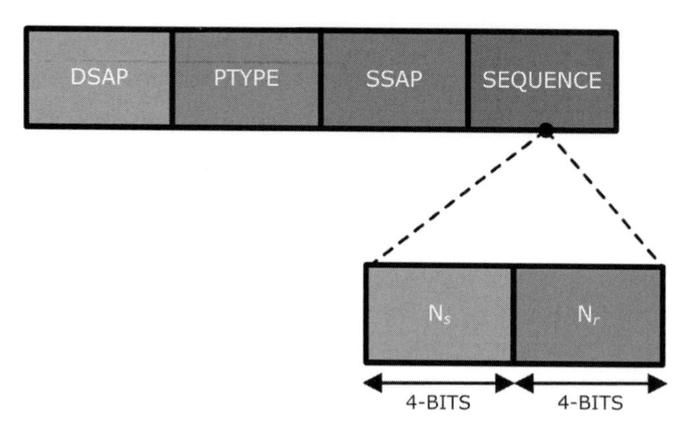

Figure 12.24. The sequence field within the LLCP header is divided into two further fields.

`0x01` will always designate the well-known SAP for *Service Discovery Protocol* (SDP), as shown in Table 12.17.

The sequence field also listed in Table 12.15 is not used if it isn't present as part of the PTYPE field, as we will show in Table 12.18. The sequence field is divided into two 4-bit fields, as shown in Figure 12.24, namely N_s and N_r, where N_s represents the *send* sequence number and N_r portrays the *receive* sequence number.

Every LLCP PDU contains a PTYPE description, and in this section we'll explore in more detail the syntax and semantics of the field. A PTYPE will use one of the values listed in Table 12.18 – we also provide a description of the valid LLC PDU. It is worth noting that a value of `0x1111` is reserved for future extension of the LLCP header.

12.9 NFC Data Exchange Format

NDEF offers a lightweight encapsulation format for messages or payloads to ease communication between NFC-enabled devices. An NDEF message may comprise one or more *record(s)*, and in Figure 12.25 we illustrate a typical NDEF message construction. Each record retains a description of its payload, which we'll discuss in the following section.

You will notice that the initial record in Figure 12.25 (left) has been labeled $\texttt{RECORD}_b$, and subsequent records are labeled $\texttt{RECORD}_2$, $\texttt{RECORD}_3$, $\texttt{RECORD}_4$, and finally $\texttt{RECORD}_e$ (right). In the record header of the first record ($\texttt{RECORD}_b$), the *Message Begin* (MB) flag is set (1) to indicate "NDEF message begins"; conversely, in the record

Table 12.18. The PTYPE value, along with its description

Value	PDU type	Description
`0x0000`	SYMM	The *Symmetry* (SYMM) PDU is used to align communication; sequence and payload fields are not present.
`0x0001`	PAX	The *Parameter Exchange* (PAX) PDU is used to exchange data concerning the link management entity; no sequence field is present, but the payload is used.
`0x0010`	AGF	The *Aggregated Frame* (AGF) is also used to exchange data concerning the link management entity; no sequence field is present, but the payload is used.
`0x0011`	UI	The *Unnumbered Information* (UI) PDU is used to transfer data units to a peer LLC, where no connection establishment is required; no sequence field is present, but the payload is used.
`0x0100`	CONNECT	The *Connect* (CONNECT) PDU is used to establish a connection between a source and destination SAP; no sequence field is present, but the payload is used.
`0x0101`	DISC	The *Disconnect* (DISC) PDU is used to terminate an established connection between a source and destination SAP; no sequence or payload fields are present.
`0x0110`	CC	The *Connection Complete* (CC) PDU is used to acknowledge successful establishment of a connection between a source and destination; no sequence field is present, but the payload is used.
`0x0111`	DM	The *Disconnected Mode* (DM) PDU is a used to indicate that the logical connection has been disconnected between source and destination, as identified by SSAP and DSAP, respectively; no sequence field is present, although the payload is used to describe the reason for disconnection.
`0x1000`	FRMR	The *Frame Reject* (FRMR) PDU is used to report an invalid PDU received over the data connection; no sequence field is present, although the payload is used to describe the reason for rejection.
`0x1001`	SNL	The *Service Name Lookup* (SNL) PDU is used to ascertain which services are available to it on a remote LLC; no sequence field is present, but the payload is used.
`0x1010`	Reserved.	
`0x1011`	Reserved.	
`0x1100`	I	The *Information* (I) PDU is used to transfer PDUs generically across an established connection. The sequence field is present: N_s will represent the sequence number corresponding to the I PDU and N_r will reflect the success of transmitting a PDU, which is akin to an acknowledgement. The size of the I PDU is determined by the *Maximum Information Unit* (MIU) size. The default value is 128 octets.
`0x1101`	RR	The *Receive Ready* (RR) PDU is used to acknowledge an I PDU and to indicate that further I PDUs can be received. The sequence field is present: N_r will reflect the success of transmitting a PDU, akin to acknowledgement. The payload field is not present.
`0x1110`	RNR	The *Receive Not Ready* (RNR) PDU is used to inform the LLC that it is temporarily unable to process subsequent I PDUs. The sequence field is present: N_r will reflect the success of transmitting a PDU, akin to acknowledgement. The payload field is not present.
`0x1111`	Reserved.	

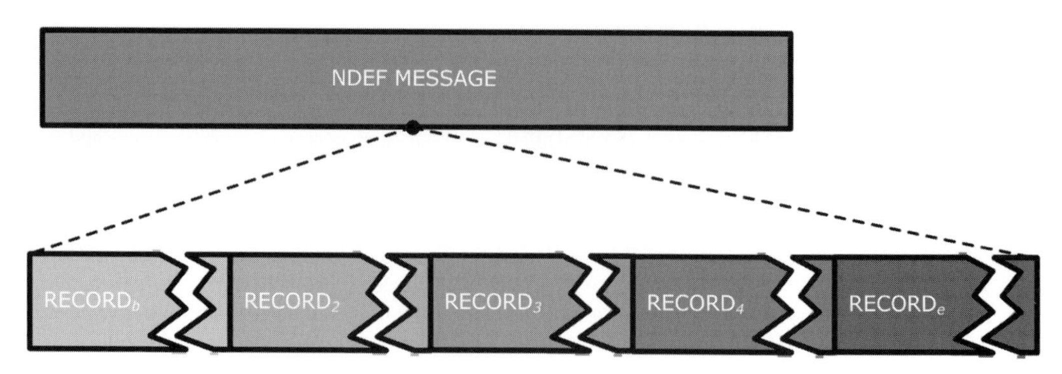

Figure 12.25. An NDEF message format.

header of the last record (RECORD$_e$) the *Message End* (ME) flag is set (1) to indicate "NDEF message ends." We describe the record header and its format in more detail in the following sections.

12.9.1 The Record Format

As implied earlier, in Figure 12.25, records are variable in length, with a specific format, as illustrated in Figure 12.26; we discuss the various fields in Table 12.19.

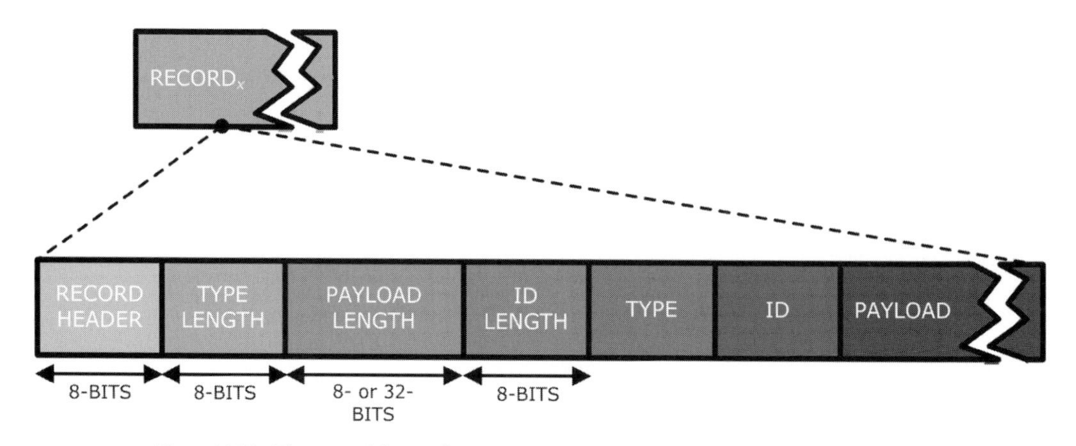

Figure 12.26. The record frame format.

12.9.2 Chunking

As we have intimated earlier, a record can be *chunked*; this, in turn, alleviates outbound buffering and can be used to portion dynamic content or to reduce large payloads into smaller records within an NDEF message. An NDEF message may comprise zero or more chunked payloads, where each payload is encoded as an *initial* chunk, which is further followed by *middle* record chunks. And finally, the sequence is ended with a *terminating* NDEF record chunk. We summarize the chunking process further in

Table 12.19. The fields that make up the record frame

Field	Description	Size
Record header	The record header field is 8-bits in length; we illustrate the record header format in Figure 12.27 and discuss its fields in Table 12.20.	8
Type length	The type length field is 8-bits in length and provides the length of the type field.	8
Payload length	The payload field is either 8- or 32-bits in length and provides the overall length of the payload. The size of the payload length field is determined by the SR flag (see Table 12.20). If the flag is cleared (0), then the payload length will occupy 32-bits, whereas when the flag is set (1), only a single octet is required to define the size of the payload.	8/32
ID length	The ID field is 8-bits in length and provides the length of the ID field; however, this field is only available if the IL (see Table 12.20) is set to 1 in the record header.	8
(Payload) type	The type field size is determined by the value contained within the type length field. It is used to characterize the type of payload, and to allow the opportunity of a user application to manage the payload accordingly. The field is made up of a string; we discuss the available types further in Section 12.10.1, "Record Types."	
(Payload) ID	The ID field is determined by the value contained within the ID length field and is optional. The identifier is a unique URI reference and provides the opportunity for a user application to identify the payload that is carried within a record.	
Payload	The intended payload for the end-application.	

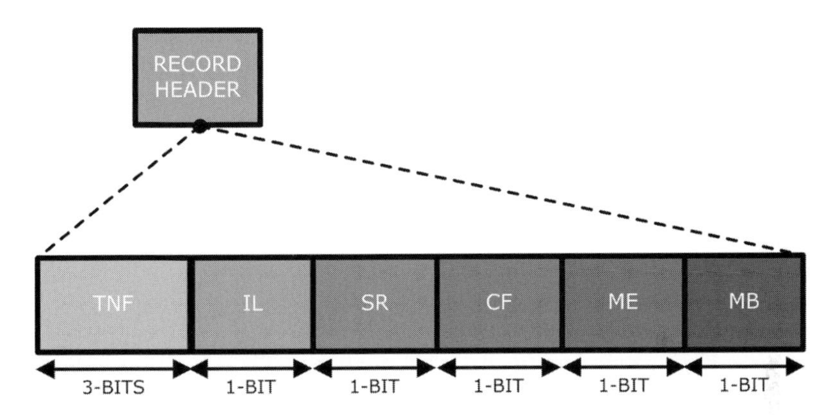

Figure 12.27. The record header format.

Table 12.22. A chunked payload may not be segmented over several NDEF messages, but rather a single NDEF message will accommodate the entire chunked payload.

12.10 Record Type Definition

As we described earlier, NDEF specifies how an NDEF message is formatted and provides the rules that govern how a record payload should be structured. It also specifies how unique record types potentially used by several applications are constructed. Furthermore, NDEF only defines the data structure used to ensure interoperability between

Table 12.20. The fields that make up the record header

Field	Description	Size
TNF	The *Type Name Format* (TNF) field is 3-bits in length and describes the structure and value of the type field – we list the possible TNF values in Table 12.21.	3
IL	The IL (`ID_LENGTH`) field is 1-bit in length; if set (1), then the ID length field (see Table 12.19) is present as a single octet.	1
SR	The *Short Record* (SR) field is 1-bit in length; if set (1), it indicates that the payload length field (see Table 12.19) is a single octet.	1
CF	The *Chunk Flag* (CF) field is 1-bit in length and indicates whether a payload is the first or continuous chunk. We discuss record chunks in Section 12.9.2, "Chunking."	1
ME	The ME field is 1-bit in length and indicates the end of an NDEF message.	1
MB	The MB field is 1-bit in length and indicates the start of an NDEF message.	1

Table 12.21. TNF field values and descriptions

Value	Name	Description
`0x00`	Empty.	In this instance, there is no associated payload with the record.
`0x01`	Well-known type.	An NFC Forum well-known type is provided; see Section 12.10, "Record Type Definition."
`0x02`	Media type.	The type field refers to a media type, as defined by the *Backus-Naur Form* (BNF); see RFC2046.
`0x03`	Absolute URI.	The type field refers to an absolute-URI BNF construct; see RFC3986.
`0x04`	External type.	The Type field refers to an NFC Forum defined external type; see Section 12.10, "Record Type Definition."
`0x05`	Unknown.	
`0x06`	Unchanged.	A value type that defines interim or terminating chunked payloads; see Section 12.9.2, "Chunking."
`0x07`	Reserved.	

NFC-enabled devices and, as such, doesn't offer how record types are defined. Therefore, the RTD is a mechanism that is able to support NFC applications and its associated framework and, in turn, offers well-known record types and third-party extension types, as we discussed earlier. In short, RTD mandates how these record types are defined in an NDEF message.

12.10.1 Record Types

In Section 12.9.1, "The Record Format," we discussed the type field (see Table 12.19) and record types; in particular, we looked at the TNF field within the record header of the record frame. (See also Table 12.23 for a list of RTD well-known types.) In this section, we discuss the actual construction of the record type as a string, referred to as the *record type name*, which is used by the NDEF application to gain a better understanding of the semantics of the record content. A record type name is postulated by the TNF within the record header (again, see Section 12.9.1, "The Record Format"), and can be defined by

Table 12.22. NDEF record chunk encoding policy

Chunk	Description
Initial	An NDEF record in which where the record header CF flag is set (1) to indicate an initial chunk. The type field will indicate the type of content within the payload.
Middle	An NDEF record in which the record header CF flag is set (1) to indicate a middle chunk. The type field will indicate the same type and identifier within the payload as the initial chunk.
Terminate	An NDEF record where the record header CF flag is cleared (0) to indicate a terminating chunk. The type field will indicate the same type and identifier within the payload as the initial chunk.

Table 12.23. RTD well-known types

Type	Description
Sp	Smart poster.
T	Text.
U	URI.
Gc	Generic control.
Hr	Handover request.
Hs	Handover select.
Hc	Handover carrier.
Sig	Signature.

either the NFC Forum or by third parties. So, the available record type names can include MIME media types, absolute URIs, NFC Forum external type names, or well-known NFC type names.

A well-known type is typically used for tags as well as for deriving other common types, but it can also be used when there is no alternative to a URI or MIME type. As we discussed in Table 12.21, a well-known type within the TNF field is set to `0x01`. There are two well-known types, as defined by the NFC Forum, namely *local* and *global* types, which are not to be modified by any third party. An external type name, similar to a well-known type, has been targeted for organizational use, where the organization can self-allocate a name space dedicated to its own specific use, and was shown earlier when an external type within the TNF field is set to `0x04`.

12.11 Activities, Digital Protocol, and Analog

In Figure 12.28, we depict the MAC and PHY layers, which we mentioned in Section 12.6.1, "Peer-to-Peer Mode," Section 12.6.2, "Card Emulation Mode," and Section 12.6.3, "Reader/Writer Mode," where we discussed the three operational modes of an NFC device. In this section, we discuss the *activities*, *digital protocol*, and *analog* building blocks, which are common to all NFC Forum devices.

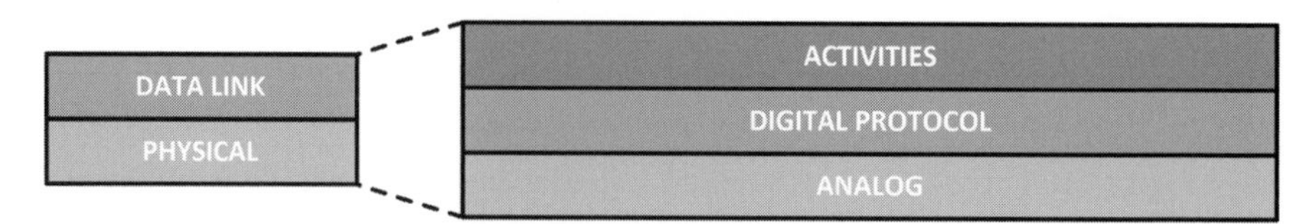

Figure 12.28. The activities, digital protocol, and analog layers comprise the data and physical layers of the OSI model.

The activity block is complementary to the digital protocol layer and prescribes how the digital protocol layer should establish its communication pathway with another device or NFC Forum tag. More specifically, the activity layer describes how to set up the communication protocol. It uses profiles, using a series of configuration parameters, to establish common behavior with NFC Forum devices. The activity block provides two specific modes, namely *listen* and *poll* mode. The digital protocol block (see Figure 12.28) resides above the analog block, which we'll touch upon in a moment. The digital protocol layer is responsible for ensuring an effective cohesion of the various ISO/IEC standards, ultimately to ensure interoperability across all NFC devices and its associated infrastructure, irrespective of manufacturer. It provides common behavioral features, which can be used by multiple manufacturers in various industry sectors. What's more, ensuring the relevant behavior of the half-duplex transmission protocol when operating in one of its modes, as well as having the ability to exchange data and bind LLCP, are the responsibilities of the digital protocol block.

Finally, the analog block provides coverage for the ISO/IEC 18092 and *European Computer Manufacturers Association*-340 (ECMA-340) NFC *Interface and Protocol*-1 (IP-1) standard, which defines[16] the communication modes for NFC using inductive coupled devices. Furthermore, the analog block also manages the modulation schemes, coding, and transfer speeds, as well as the frame format of the PHY interface.

[16] ECMA International, "Near Field Communication – Interface and Protocol (NFCIP-1)," 2008.

13 The 802.11 Generation and Wi-Fi

Let's start this chapter by distinguishing between the *Institute of Electrical and Electronic Engineers* (IEEE) 802.11 "alphabetized" standards and what we understand as Wi-Fi.[1] Consumers ordinarily shouldn't care too much about the underlying alphabet attribute that enables certain facets of Wi-Fi technology; rather, they should simply be aware that it's *Wi-Fi CERTIFIED*™ (see Figure 13.1), something we'll come back to in a moment. The inherent capability of a Wi-Fi-enabled product has been traditionally identified by a letter to represent a level of functionality, as denoted in the relevant IEEE 802.11 standard, which is the original, or legacy, specification from which all other standards have been derived. It was released as far back as 1997, and since then various enhancements to the original specification have spawned the various alphabetic variants that provide us with *Wireless Local Area Networking* (WLAN) connectivity.

Figure 13.1. Research has demonstrated that consumers identified with the Wi-Fi CERTIFIED logo rather than the a/b/g/n labeling. (Courtesy of the Wi-Fi Alliance.)

The Wi-Fi Alliance® (wi-fi.org), who are responsible for the Wi-Fi *brand*, have discovered through extensive consumer research that consumers identified with the Wi-Fi CERTIFIED logo rather than the a/b/g/n labeling. This psychological association is an accurate reflection of the Wi-Fi brand, since the majority of consumers, albeit incorrectly, associate Wi-Fi with the Internet. In fact, Wi-Fi is a WLAN technology enabling computers to interconnect wirelessly to a *Local Area Network* (LAN). We show in Figure 13.2 another branded concept from the Wi-Fi Alliance, which informs consumers that a successful connection will grant them access to the Internet via a *hotspot*; the Wi-Fi Alliance reports[2] that more than 1.4 million hotspots are available

[1] The term Wi-Fi® is a registered trademark of the Wi-Fi Alliance®.

[2] Wi-Fi Alliance, "Connect with the Wi-Fi Alliance: Membership Overview," 2012.

Figure 13.2. Hotspot logos inform consumers that they can obtain an Internet connection. (Courtesy of the Wi-Fi Alliance.)

globally and that this number is increasing! In this chapter, we discuss Wi-Fi's brand inception and evolution since making its first appearance in 1999 when the Wi-Fi Alliance was inaugurated. Furthermore, we'll explore the Wi-Fi Alliance, along with its membership benefits and structure. This chapter also discusses the market scope that Wi-Fi has already secured and explores the numerous Wi-Fi-enabled products that so many consumers have taken for granted. Naturally, we also look at how Wi-Fi holds up against other competing technologies within the same market sector. Finally, the chapter will look at how the technology underlying the 802.11 standard works by providing a detailed review of the 802.11 software architecture and protocol. In essence, the IEEE have provided a solid foundation from which to expand the classic *Internet Protocol* (IP)/*Transmission Control Protocol* (TCP)/*User Datagram Protocol* (UDP) software stack that the Internet-enabled world, for example, has become immensely dependent upon and forever connected to.

13.1 Overview

Wi-Fi technology is ubiquitous.

It is simply everywhere: in the home, the office, hotels, hospitals, shopping malls, cafés, restaurants, metros, subways, and it may even be provided in your municipality – providing connectivity for everyone, anywhere. In fact, across the globe, one in ten people[2] uses a Wi-Fi-enabled product, and over four billion products have been shipped since 2000. Essentially, Wi-Fi technology empowers everyone that has a smart device or notebook to connect wirelessly, and the Wi-Fi brand has become synonymous with an ability to connect to the Internet. No longer do we need to seek an Ethernet port to acquire our IP-fix, we can now do it wirelessly – a freedom bestowed by several innovators that simply conjectured "let's remove the cable."

13.1.1 Ethernet

Most notebook or desktop computers have an integral Ethernet port (see Figure 13.3 **(1)**). However, some compact or ultra-notebooks, such as the Apple MacBook Air, and tablet devices, for that matter, simply rely on a Wi-Fi connection to install software updates and other significant software applications (no CD-ROM required). The IEEE mandates the *Physical* (PHY) and *Media Access Control* (MAC) layers of the 802.3 standard for Ethernet, and it is still very much in use today. In essence, Ethernet (802.3) is the cabled equivalent of 802.11 – instead of a wired interface, 802.11 provides the same interface *wirelessly*. You may recall from Chapter 2, "What is a Personal Area Network?," that

Figure 13.3. An HP notebook showing a series of ports, including an Ethernet port as shown. (Courtesy of Hewlett Packard.)

we discussed how a *Personal Area Network* (PAN) represents a topology, whereas a *wireless* PAN (WPAN) refers to the ability to connect wirelessly with that topology. The definition WPAN may also refer to the number of short-range technologies that wirelessly enable your PAN to enable users to connect seamlessly, technologies such as Bluetooth, Wireless USB, and *Wireless Home Digital Interface* (WHDI). In a similar vein, WLAN is the ability to connect wirelessly to your LAN – the founding remit of the Wi-Fi Alliance.

13.1.2 The 802.11 Standards

In Table 13.1, we list the most popular set of standards that enable WLAN connectivity – these standards are maintained by the IEEE. Products using an 802.11 standard can choose to have their products certified by the Wi-Fi Alliance and can, in turn, display the Wi-Fi CERTIFIED logo. We discuss other facets of the Wi-Fi Alliance's certification program later, in Section 13.1.3.2, "Wi-Fi Certified."

13.1.3 The Wi-Fi Alliance

The Wi-Fi Alliance® is responsible for the branding and certification program for products using 802.11; we show its trademark logo in Figure 13.4. The Alliance was

Table 13.1. A set of popular standards maintained by the IEEE

Standard	Description
802.11–1999	The original, legacy specification provided by the IEEE in 1999. This now obsolete specification offered data rates of up to 2 Mbit/s.
802.11–2007	A revision containing various corrections and amendments, along with enhancements to the MAC and PHY layers of the original 802.11–1999 specification.
802.11–2012	A revision containing various corrections and amendments, along with enhancements to the MAC and PHY layers of the 802.11–2007 specification.
802.11a	Various changes and enhancements to the IEEE 802.11–1999 specification to support a higher data rate for use in the 5 GHz band.
802.11aa–2012	Enhancements to the 802.11 standard that provide robust audio and video streaming for a host of 802.11-enabled applications.
802.11b	Various changes and enhancements to the IEEE 802.11–1999 specification to support a higher data rate at the PHY layer for use in the 2.4 GHz band.
802.11e	An amendment to the 802.11 standard to improve quality of service for media-centric applications; we discuss quality issues later in Section 13.9, "Wi-Fi Multimedia."
802.11g	Various changes and enhancements to the IEEE 802.11–1999 specification to support a higher data rate at the PHY layer for use in the 2.4 GHz band.
802.11i	*Wi-Fi Protected Access*®2 (WPA2™), a subset of the 802.11i standard, was introduced in 2004; we look at WPA in Section 13.7, "Wi-Fi Protected Access."
802.11n	Various changes to both the IEEE 802.11 PHY and MAC layers to enable a higher data throughput, in turn providing up to 100 Mbit/s.
802.11ac	Further enhancements, which provide high speed WLAN connectivity (up to 1 Gbit/s) using the 5 GHz frequency band.
802.11ad	802.11ad, or WiGig, is touted to offer 7 Gbit/s using a tri-band "Wi-Fi" solution utilizing 2.4, 5 and 60 GHz spectra.
802.11ae–2012	Enhancements that offer improved network performance specifically for the prioritization of management frames across an 802.11 link.
802.11af	Wi-Fi technology used within the white space radio spectrum.
802.11ah	An early stage standard set to compete with the low energy, low power sector currently occupied by such technologies as ZigBee and Bluetooth low energy.

formed in 1999 and has 500+ members worldwide to date. The Wi-Fi Alliance is a non-profit organization, and is responsible for driving awareness and increasing the adoption of high speed WLAN alongside its primary vision to provide seamless connectivity. It also delivers exhaustive product connectivity through testing and certification.

13.1.3.1 Membership and the Wi-Fi Alliance

The Wi-Fi Alliance offers potential participants four classes of membership, as shown in Table 13.2. The four classes of membership enable companies to decide on how they wish to participate within the Wi-Fi Alliance. The organization has structured its membership from *Adopter*, an entry level where the member can access final approved versions of all documentation and specifications before they're made public, through to *Sponsor*, where a representative of a company can appoint a representative to sit on the board of directors. Essentially, membership of the Wi-Fi Alliance entitles the member to participate in the future direction of the organization and the technology. Any company

Figure 13.4. The Wi-Fi Alliance logo. (Courtesy of the Wi-Fi Alliance.)

wishing to use the Wi-Fi Alliance logo is required to join the Wi-Fi Alliance and certify their products prior to marketing or sale.

Table 13.2. The Wi-Fi Alliance offers potential participants a four-tier membership structure, to include Sponsor, Regular, Affiliate, and Adopter

Membership level	Scope
Sponsor	A Sponsor member receives all the same benefits as a Regular member, and in addition can appoint a representative to sit on the board of directors.
Affiliate	An Affiliate member has the same benefits as a Regular member, with the exception of voting rights.
Regular	A Regular member has access to final approved versions of all documentation, as well as access to working documents and deliverables for certification programs. Likewise, a Regular member has influence in group participation and is eligible for leadership roles.
Adopter	An Adopter member has access to published versions of all documentation, as well as to the certification roadmap.

13.1.3.2 Wi-Fi CERTIFIED

The majority, if not all, of the technologies featured in this book require a product to be *certified*, although some manufacturers of Wi-Fi-enabled products may choose not to certify their product. As such, manufacturers of these products are unable to use the Wi-Fi CERTIFIED logo. The Wi-Fi CERTIFIED program offered by the Wi-Fi Alliance provides a widely recognized designation of interoperability and quality, and it helps to ensure that Wi-Fi-enabled products deliver the best user experience. Products that demonstrate comprehensive testing against the IEEE standards along with successful interoperability are awarded the Wi-Fi CERTIFIED logo shown in Figure 13.1. Moreover, the Wi-Fi Alliance essentially defines *core* and *optional* programs for the certification process, and we summarize these in Table 13.3 and Table 13.4, respectively.

Table 13.3. The Wi-Fi Alliance core programs cover the following categories

Program	Description
1	Wi-Fi products based on IEEE radio standards.
2	WPA2 for wireless network security.
3	*Extensible Authentication Protocol* (EAP) used to validate the identity of network enterprise devices.
4	Wi-Fi CERTIFIED WPA2 with *protected management frames*, which provide a certain level of protection for *unicast* and *multicast* management action frames.
5	Wi-Fi CERTIFIED n, which includes *Wi-Fi Multimedia*™ (WMM®) testing; we discuss WMM later, in Section 13.9, "Wi-Fi Multimedia."

Table 13.4. The Wi-Fi Alliance optional programs cover the following categories

Program	Description
1	*Wi-Fi CERTIFIED Passpoint*™, which enables mobile devices to connect automatically and discover Wi-Fi networks in proximity.
2	Wi-Fi CERTIFIED Wi-Fi Direct™ certification, something which we discuss later, in Section 13.10, "Wi-Fi Direct."
3	*Wi-Fi Protected Setup*™ certification that implements the ability to permit easy set-up and configuration of security features, such as using a *Personal Identification Number* (PIN).
4	WMM certification, products that support the ability to prioritize traffic where content is generated by different applications using *Quality of Service* (QoS).
5	WMM-Power Save, a scheme used to place a device into sleep mode in an attempt to reduce energy consumption when using multimedia-centric applications.
6	Voice-Personal, a test for assessing the performance of voice over Wi-Fi to ensure good voice quality over the Wi-Fi connection.
7	CWG-RF, a test program for ensuring the coexistence of Wi-Fi and cellular technology.
8	Voice-Enterprise, a test for assessing a level or requirement to ensure good voice quality over a Wi-Fi connection, along with advanced WPA2 security.
9	WMM-Admission Control, an optimization program to enhance bandwidth management for voice and other multimedia traffic within a Wi-Fi network.

13.1.4 Wi-Fi's Timeline

The infographic shown in Figure 13.5 provides a snapshot of Wi-Fi's brand and associated technology timeline, from the first introduction of the 802.11 standard to where the brand and technology are today. The Wi-Fi brand has matured considerably over the last decade or so, and the Wi-Fi Alliance continues to drive its adoption of high speed WLAN connectivity ethos, as we have already discussed. The IEEE has created and released various enhancements to the original 802.11 standard over the last decade, each release evolving the technology and providing greater speed, reliability, enhancements, and modulation techniques. The Wi-Fi Alliance was formed and established in 1999 by six companies, namely 3Com, Aironet, Intersil, Lucent Technologies, Nokia, and Symbol Technologies, and the Alliance undertook the responsibility of establishing

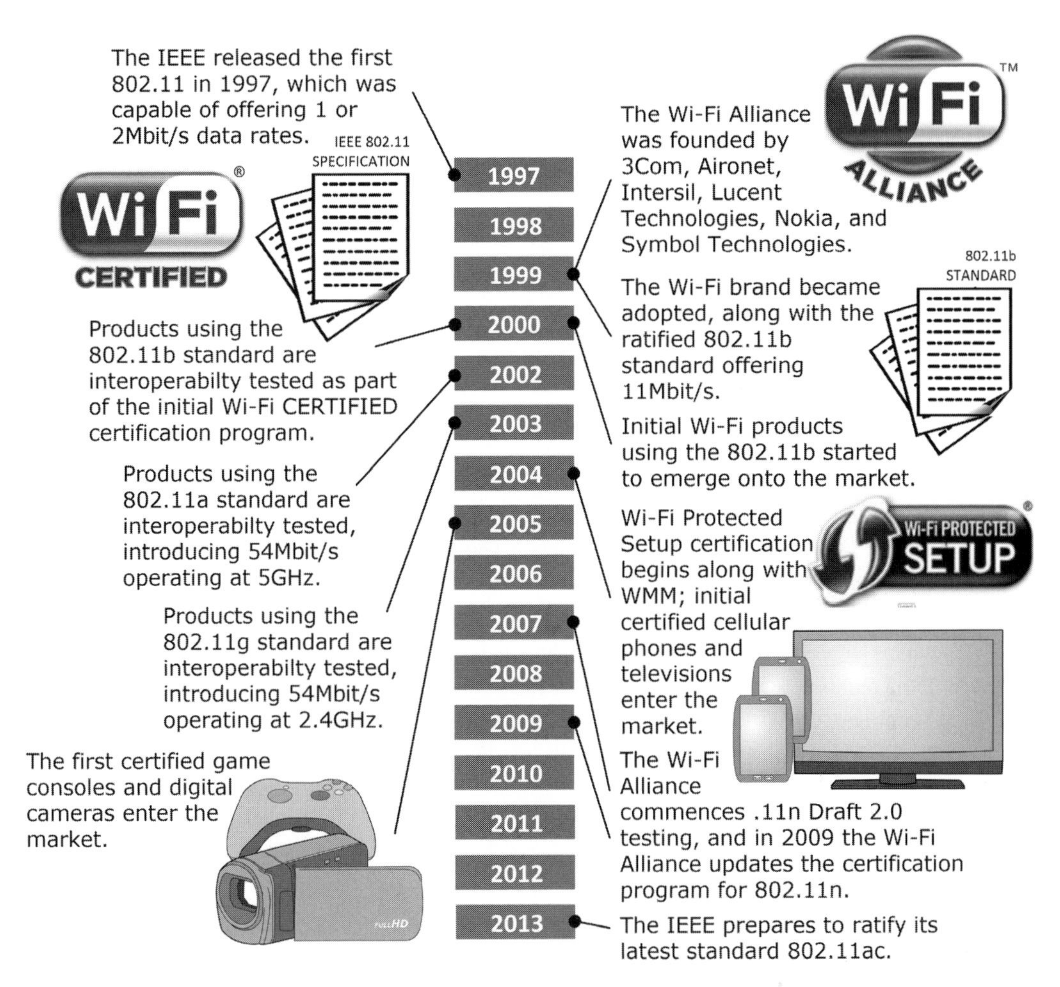

Figure 13.5. The Wi-Fi brand and technology timeline, from its inception to where the technology is today with its new 802.11ac specification.

the Wi-Fi brand. First-generation Wi-Fi-enabled products entered the marketplace circa 2000 using the ratified and popular standard, namely IEEE 802.11b, initially offering up to 11 Mbit/s using the *Industrial, Scientific, and Medical* (ISM) 2.4 GHz band. Later, in 2002, 802.11a-enabled products emerged, using the 5 GHz frequency band and boasting raw data rates of up to 54 Mbit/s. In 2003, products using the 802.11g standard were interoperability tested, as television sets and cellular phones integrating Wi-Fi technology were introduced to the market, along with *Wi-Fi Protected Setup* (WPS) and *Wi-Fi Multimedia*™ (WMM) certification. Later, in 2005, other products integrating Wi-Fi, such as game consoles and digital cameras, also entered the market. In 2007, the next generation IEEE 802.11n standard offered consumers a 300 Mbit/s raw data rate, which saw numerous products labeled as "Draft N" emerge onto the market prior to the standard becoming officially ratified in 2009. A similar route to market emerged with products being launched in mid 2012 with the latest generation IEEE 802.11ac standard.

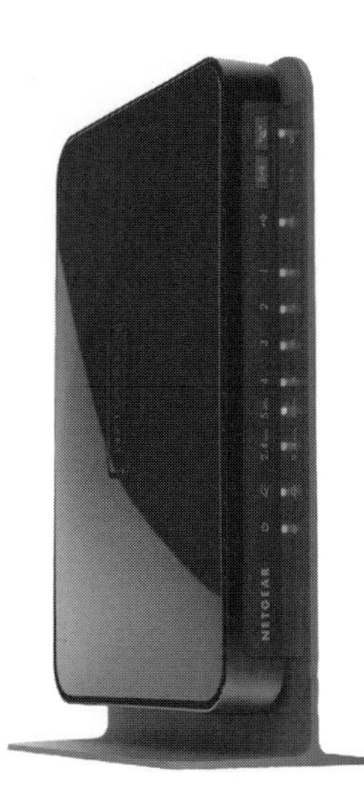

Figure 13.6. Netgear's Dual Band Wireless-N Gigabit router (WNDR3700), which provides both Ethernet and Wi-Fi connectivity within the home or office. (Courtesy of Netgear.)

This new standard is expected to be finalized in 2013, at which time products using the technology may boast raw data rates of up to 1 Gbit/s, in turn paving the way for products using the standard as a mechanism for delivering audio/video content. In the sections to follow, we review Wi-Fi's market and application portfolio.

13.2 Wi-Fi's Market

As you already know, the wireless technologies that form Part III, "The Classic Personal Area Network," are typically high-end and utilize high data rates, best suited for audio-, voice- and data-heavy-centric applications. Wi-Fi is, of course, no exception, as it provides WLAN connectivity. We may witness a new generation of 802.11-enabled products supporting a *Wireless Video Area Network* (WVAN) topology – something we touched upon earlier in Chapter 2. We illustrate in Figure 13.6 a typical WLAN product: Netgear's Dual Band Wireless-N Gigabit router (WNDR3700), which provides the consumer with the ability to connect to their broadband service and, in turn, provides both Ethernet (up to 1 Gbit/s) and Wi-Fi connectivity utilizing the 2.4 and 5 GHz frequency bands within the home or office. And, as we have already intimated, the 802.11 generation is increasingly shifting its focus to multimedia-centric content for a new generation of consumer electronic products, although no such formal announcement has been made. However, the recent introduction of the new IEEE 802.11ac standard alludes to this shift in connectivity scenarios, further compounded by two very new enhancements (something we highlighted earlier in Table 13.1), namely 802.11aa–2012 and 802.11ae–2012. The former standard specifically addresses enhancements to the 802.11 specification that provide robust audio and video streaming for a number of 802.11-enabled applications; the latter standard offers enhancements that improve network performance specifically for the prioritization of management frames across an 802.11 link. Naturally, the IEEE strategy is a clear indication of intention, as it musters up a confident standard that can undoubtedly wrestle with the likes of WHDI, WiGig, and WirelessHD. However, recently the Wi-Fi and WiGig Alliances have announced their joint effort to promote 60 GHz wireless technology.[3] You may recall from Chapter 4, "Introducing the Lawnmower Man Effect," that we discussed *media convergence* and how the delivery of content has diversified over the last few years. No longer do we have to rely on traditional forms of delivery content, as the Internet has expanded our opportunity to source content from almost anywhere. Moreover, with consumer electronic products packed with a wealth of connectivity technologies, we have no excuse but to remain connected to the wider area network. Wi-Fi is a technology that ultimately ensures we have content delivered when and wherever we want it – further solidifying our conceptual *Lawnmower Man Effect* (LME), which we discussed in Chapter 4.

Wi-Fi has a lifespan covering almost two decades and has witnessed an explosion in popularity. IEEE 802.11ac "draft" products have already begun to emerge onto the market, further driving the need for speed and connectivity robustness and reliability. The early adoption of 802.11 has proved to be incredibly successful, as initial 802.11b products have encouraged consumers to ditch the cable in favor of the seamless simplicity afforded by Wi-Fi. So much so that even the word "Wi-Fi" was introduced as part of the English language by Merriam-Webster back in 2005.[4]

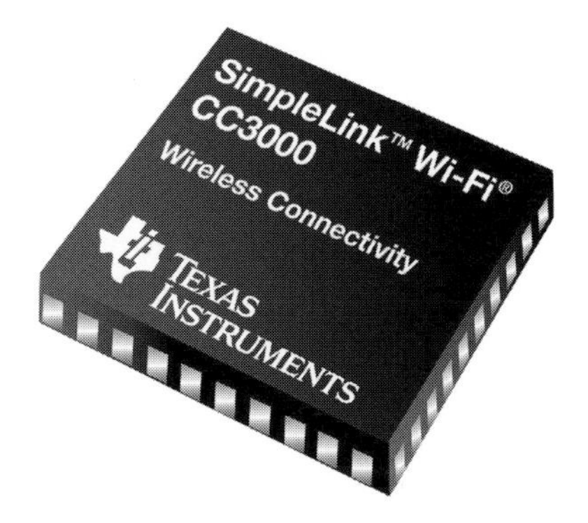

Figure 13.7. Texas Instruments' SimpleLink™ Wi-Fi CC3000, which empowers a wealth of consumer electronics products. (Courtesy of Texas Instruments.)

In Figure 13.7, we illustrate Texas Instruments' SimpleLink™ Wi-Fi CC3000 chipset, which provides 802.11b/g WLAN connectivity and is used by multiple consumer electronics manufacturers. The market potential for wireless video streaming is enormous. Products such as tablets and smartphones represent an enormous opportunity to deliver content wirelessly. Later, in Chapter 15, "One Standard, All Devices: WHDI," we look at how WHDI is starting to witness a significant increase in momentum in the professional

[3] O'Brien, T., "WiFi and WiGig Alliances Become One, Work to Promote 60 GHz Wireless," 2013.
[4] The definition is given as "Certification mark, used to certify the interoperability of wireless computer networking devices." (Source: merriam-webster.com.)

market, where there is a need for a robust, high-quality wireless system. In one particular example, applications such as orthoscopic cameras, fiber optics video filming, and vision-aided devices, to include MRI and CT scanners, potentially offer the opportunity for freedom within the medical sector. Naturally, Wi-Fi has become synonymous to connecting with the Internet and unintentionally offers its users the ability to stream content either directly from the Internet or from a PC to a media center or Wi-Fi-enabled TV. In Chapter 15, we discuss how compromises in latency and quality[5] may disrupt your viewing experience, which can, in turn, naturally impose quality of service issues on to the professional video sector. Moreover, Wi-Fi was never intended to deliver wireless streaming of audio/video content, but many consumers use the wireless technology as a streaming mechanism. Like WHDI, Wi-Fi may help with delivering content to multiple screens for public or consumer infotainment and advertising within shopping malls and airport terminals. Wi-Fi is, first and foremost, a WLAN solution, but its inadvertent penetration into a completely new market may concern other technologies all vying for the same application space.

13.3 Wi-Fi Application Portfolio

As we have already discussed, Wi-Fi's remit is WLAN connectivity; in Figure 13.8, we see a Wi-Fi access point (**1**) permitting a connection to the *Wide Area Network* (WAN) or the Internet for a number of notebooks (**2**) and tablet and smartphone (**3**) devices. Wi-Fi connectivity within the home has become a commonplace way to connect to the Internet and, as such, every household member expects to experience that all-important connection. Likewise, these same users expect to receive connectivity outdoors, where often they assume that the Wi-Fi range will inevitably extend to the remotest areas of their garden, for example. In fact, with products such as Netgear's Universal Wi-Fi Range Extender (WN2000RPT), pictured in Figure 13.9, every corner of the home and garden *can* be covered. Its simple configuration and set-up procedure allow any consumer to extend an existing Wi-Fi access point typically beyond the advertised range. The Wi-Fi ecosystem allows wireless connectivity in an enterprise context in which an Internet connection may also be available. However, the technology potentially has the opportunity to encroach into a market perhaps best suited to other wireless technologies. For example, Wi-Fi Direct, something we'll come back to later in this chapter, has been perceived by many industry pundits to compete directly with Bluetooth. What's more, with the new IEEE 802.11ac standard, Wi-Fi can also place itself into an application space from where dedicated technologies, such as WHDI, WirelessHD, and WiGig, are beginning to emerge, in order to offer wireless media-centric streaming.

[5] *Latency* and *quality* refer to the over the air-interface. If a user is streaming content using the Internet or via DLNA, then the quality of the image shown on the display is limited to the inherent quality of the originating content.

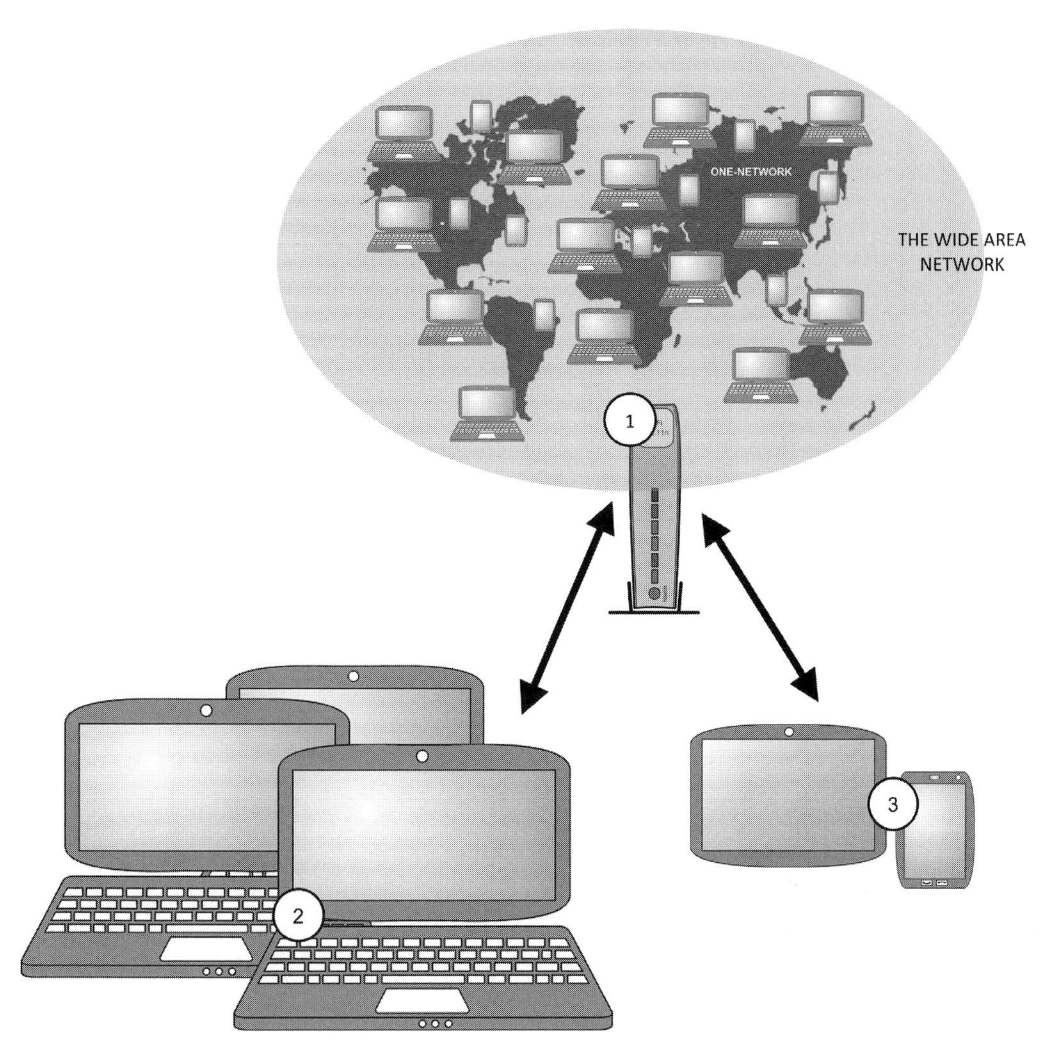

Figure 13.8. A typical WLAN ecosystem simply providing Internet connectivity for devices connected to a Wi-Fi access point.

Figure 13.9. Netgear's Universal Wi-Fi Range Extender (WN2000RPT) extends your Wi-Fi range to the whole garden and even the pool! (Courtesy of Netgear.)

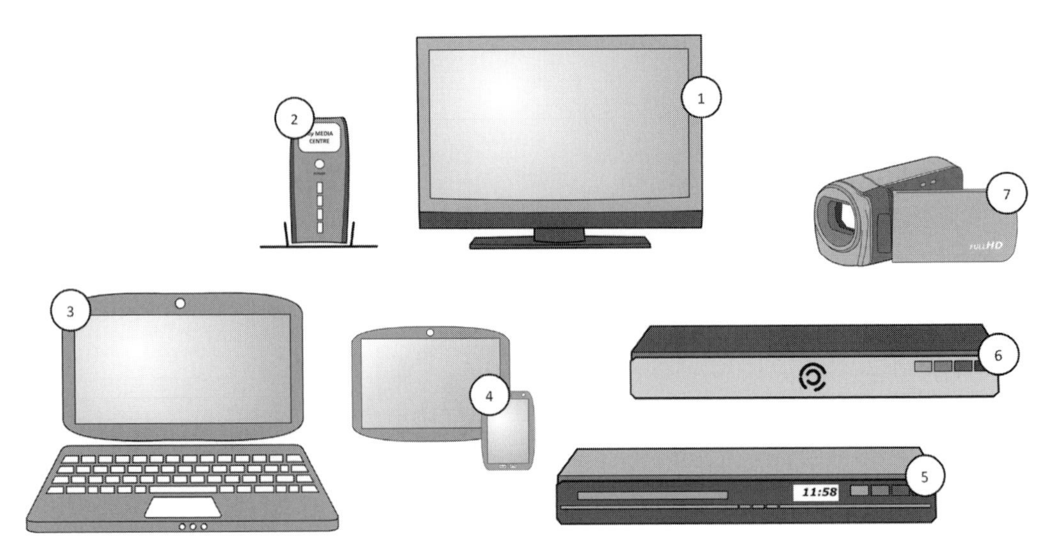

Figure 13.10. The new 802.11 generation may encroach onto the home entertainment sector, where we will begin to witness products delivering audio/video content.

The Wi-Fi Alliance has delivered a brand, as well as a certification program that ultimately empowers a new generation of devices that may revolutionize home entertainment systems by wirelessly delivering audio/video content, in turn offering almost total flexibility when interconnecting multiple audio/video sources. In Figure 13.10, we depict a series of products, such as Wi-Fi-enabled TV (**1**) or Wi-Fi-enabled media center (**2**), that may be capable of receiving audio/video content from devices such as notebooks (**3**) using *Digital Living Network Alliance* (DLNA); tablet and smart or cellular phones (**4**), Blu-ray/DVD players (**5**), *Set-top-boxes* (STBs) (**6**), or a digital camera (**7**).

13.4 Wi-Fi and its Competitors

As we have already suggested, it seems as if Wi-Fi has encroached on other wireless technologies' application space. Wi-Fi pretty much stands alone as a WLAN enabler, but it has yet to reach a market providing a low power solution, albeit that many companies tout their "specialized" take on a low power variant. Admittedly, with the Alliance's *power save* initiative, devices are encouraged to *sleep*, in turn reducing their overall power consumption. Even so, the overhead required to execute the Wi-Fi protocol stack, if you like, does require a more powerful processor and an increased memory footprint when compared with similar wireless technologies and application use cases. In Chapter 6, "Enabling the Internet of Things," we discussed the use of an IP stack on a small device to allow it to be connected to the *Internet of Things* (IoT) whilst utilizing as few resources as possible to provide an affordable and effective solution for a smart object.

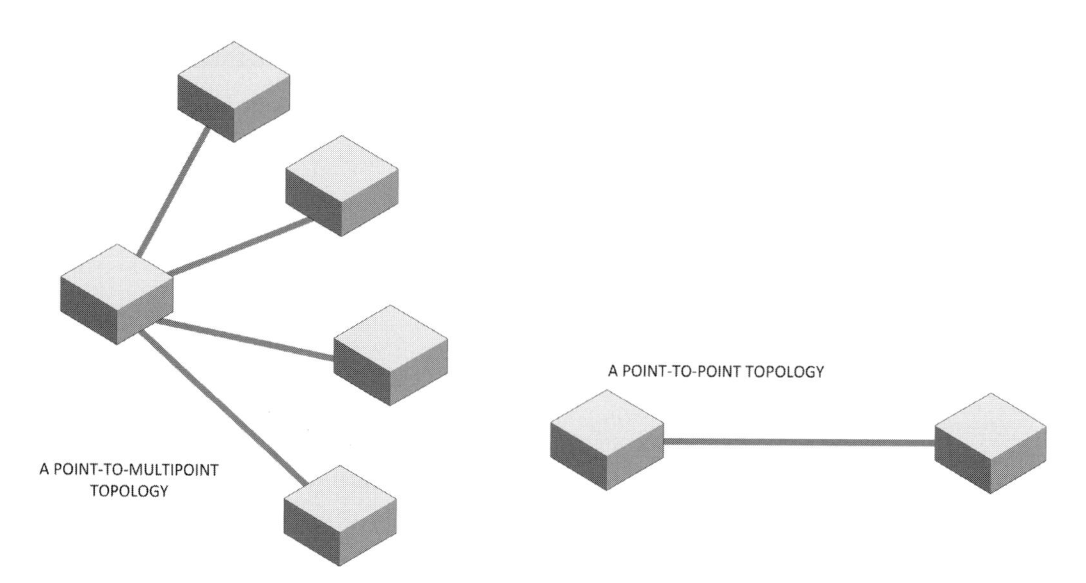

Figure 13.11. A Wi-Fi ecosystem typically supports point-to-multipoint and point-to-point topology configurations.

13.5 Networking Topologies

WLAN typically supports point-to-multipoint and point-to-point (or peer-to-peer) topologies, as illustrated in Figure 13.11; however, there are additional components within the 802.11 architecture that may specify the nature of a WLAN topology. In this section, we discuss aspects of the wireless *Station* (STA) and its associated *modifiers*, the *Basic Service Set* (BSS), the *Basic Service Area* (BSA), and the *Independent* BSS (IBSS), and we also look at the dynamics and flexibility of a *Distribution System* (DS), whilst reviewing the *Extended Service Set* (ESS).

A WLAN network, in contrast to a wired LAN, possesses varying characteristics and potential shortcomings. These factors may affect the design and overall performance of a WLAN device; for example, a WLAN device remains unprotected from other radio signals that share the same PHY medium – the same can be said of any wireless technology that features in this book. Similarly, the consistency of a WLAN medium when compared with a wired LAN may be somewhat less reliable. What's more, the dynamics of propagation and possible interference from other 802.11 networks may also affect reliability. In short, these are factors that need to be addressed when designing a WLAN device. In the sections to follow, we uncover the dynamic topologies offered by the 802.11 standard and detail the software architecture of an 802.11-enabled device.

13.5.1 Service Set Identifier

We have dedicated this section to the *Service Set Identifier* (SSID), something we have become accustomed to as Wi-Fi consumers, despite the associated complexity we face in having to name or search and recognize the default name of our access point. We

Figure 13.12. The SSID for both 2.4 and 5 GHz are broadcast, but WPA2-PSK [AES] is enabled for both networks, and the security key must be offered to gain access. (Source: Netgear Genie, WNDR4500.)

discuss this later, in Section 13.8, "Wi-Fi Protected Setup," where we see how the simplicity afforded by some Wi-Fi techniques simplifies access to a Wi-Fi network by sharing credentials such as the SSID and security passkey. Ordinarily, an SSID identifies a network or the name of a network you wish to join – you know that by joining this network successfully you will gain access to the Internet and/or LAN. In some instances, the SSID will be hidden from public broadcast as a security requirement, but in Figure 13.12 we show a configuration that has opted to advertise publically "MissGatsby" (2.4 GHz) and "MissGatsby5G" (5 GHz) for discovery. However, WPA2 (see Section 13.7, "Wi-Fi Protected Access") is in use, and any user wishing to gain access to the Wi-Fi network must offer and know the correct credentials to do so successfully. Later, in Section 13.5.3.1, "The Basic Service Set Identifier," we introduce the *Basic Service Set Identifier* (BSSID).

13.5.2 Wireless Stations

The IEEE 802.11 ecosystem identifies an STA, the addressable unit from which a message originates or where it is destined to go. Within the topology there are modifiers that are used to distinguish the type of STA; for example, *fixed* STA, *portable* STA, and *mobile* STA are used to characterize the station further. Nevertheless, a station may simultaneously exhibit multiple characteristics; for example, a station may exhibit fixed STA and *hidden* STA characteristics. Likewise, a distinction is made between a portable and a mobile station. In particular, a portable STA is a unit that is moved about a fixed

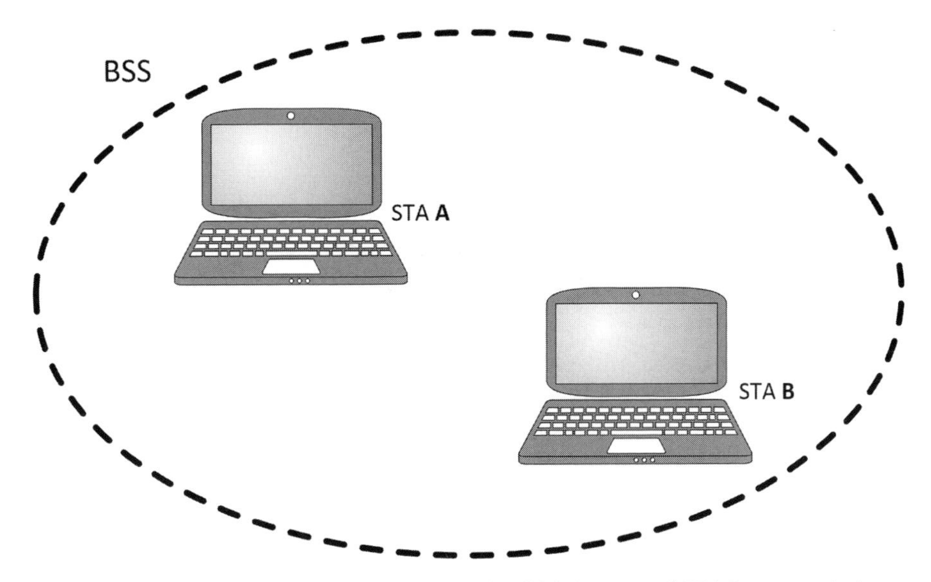

Figure 13.13. The BSS represents the radio range in which STA **A** and STA **B** can remain in point-to-point contact with one another.

location, for example a notebook computer in an office may be moved from one area to another, whereas a mobile STA is a device that accesses the LAN whilst in motion.

13.5.3 Basic Service Set

In Figure 13.13, we illustrate the BSS concept, where the elliptical region depicts the BSA or its radio range where our two STAs (**A** and **B**) are members of the BSS and are in communication with each other. If STA **A** moves out of the BSA, then it no longer has contact with STA **B**. A BSS forms the basic building block of an 802.11 LAN.

The membership of a station within a BSS can be dynamic since the STA can enter and exit radio range, and also be powered on and off. What's more, an IBSS offers a basic type of 802.11 LAN, whereas, in fact, a BSS may only comprise two STAs, such as that illustrated earlier in Figure 13.13. This type of topology, in which a station can spontaneously join or leave a network and communicate with another STA, in a point-to-point or *peer-to-peer* fashion, is referred to as an *ad hoc network*. The very nature of this "informal" connection to a network does not require the availability of an *Access Point* (AP), and uniquely shapes our IBSS.

13.5.3.1 The Basic Service Set Identifier

We discussed earlier, in Section 13.5.1, "Service Set Identifier," the characteristics of the SSID; the BSSID (or the MAC address), however, uniquely identifies a BSS, whereas the SSID may be used to morph multiple infrastructure BSSs (see Section 13.5.4, "Distribution Systems").

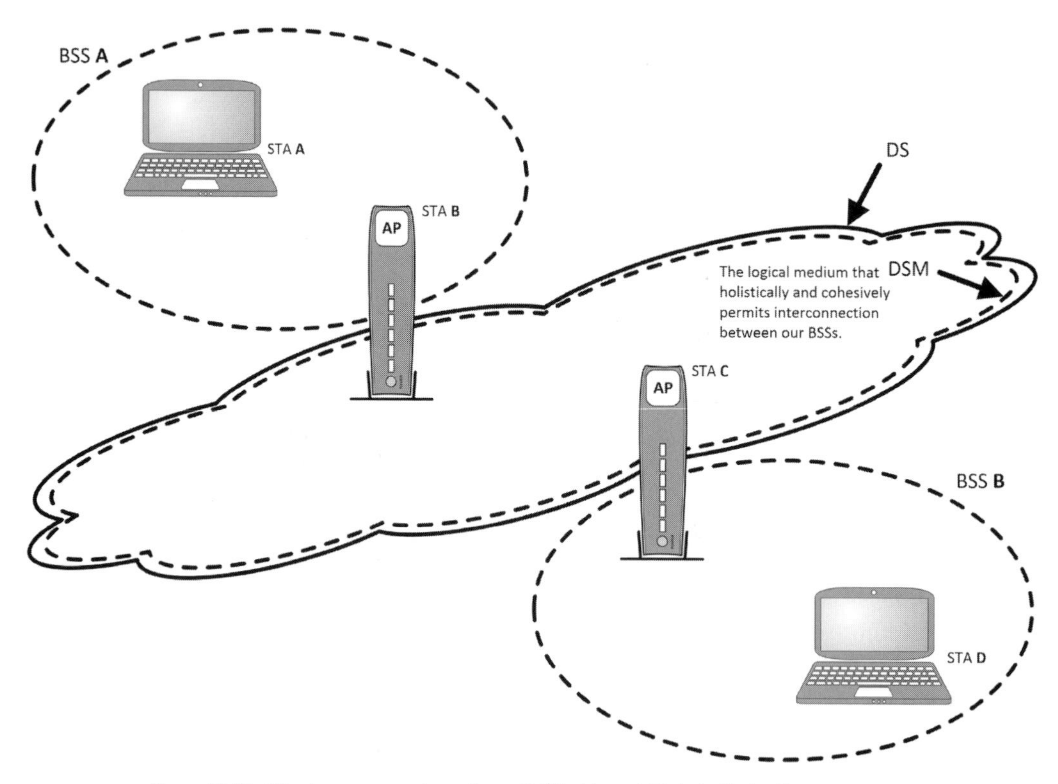

Figure 13.14. The interconnection of two BSSs (**A** and **B**) is holistically connected via two APs, along with the logical DSM.

13.5.4 Distribution Systems

The radio medium used for an 802.11 network has a fixed range, which varies from product to product. Typically, this range is adequate for most applications within the home or office; however, in other scenarios, extended coverage may be required. As such, multiple BSSs may be used to provide contiguous and seamless WLAN connectivity. A DS architectural component is used to characterize this notion, where interconnection is provided between multiple BSSs with the provision of an AP. In fact, the 802.11 standard alludes to the interconnection of large and complex wireless networks of *infrastructure* BSSs as an ESS. An ESS holistically combines infrastructure BSSs with the same SSID, all connected through the DS. A logical distinction is made between the actual wireless medium (the PHY) and the *Distribution System Medium* (DSM) – this distinction within the architecture demonstrates 802.11 LAN's flexibility as well as assures its independence from any specific implementation. In other words, the DSM behaves as a logical medium and is instrumental in interconnecting our two BSSs, as shown in Figure 13.14. In the illustration, we see an AP, which provides the same functionality as an STA and permits actual access to the wireless medium for other connected STAs. So, communication moves from a BSS to the DS through an AP, on to the interconnecting AP, and then to the destination STA.

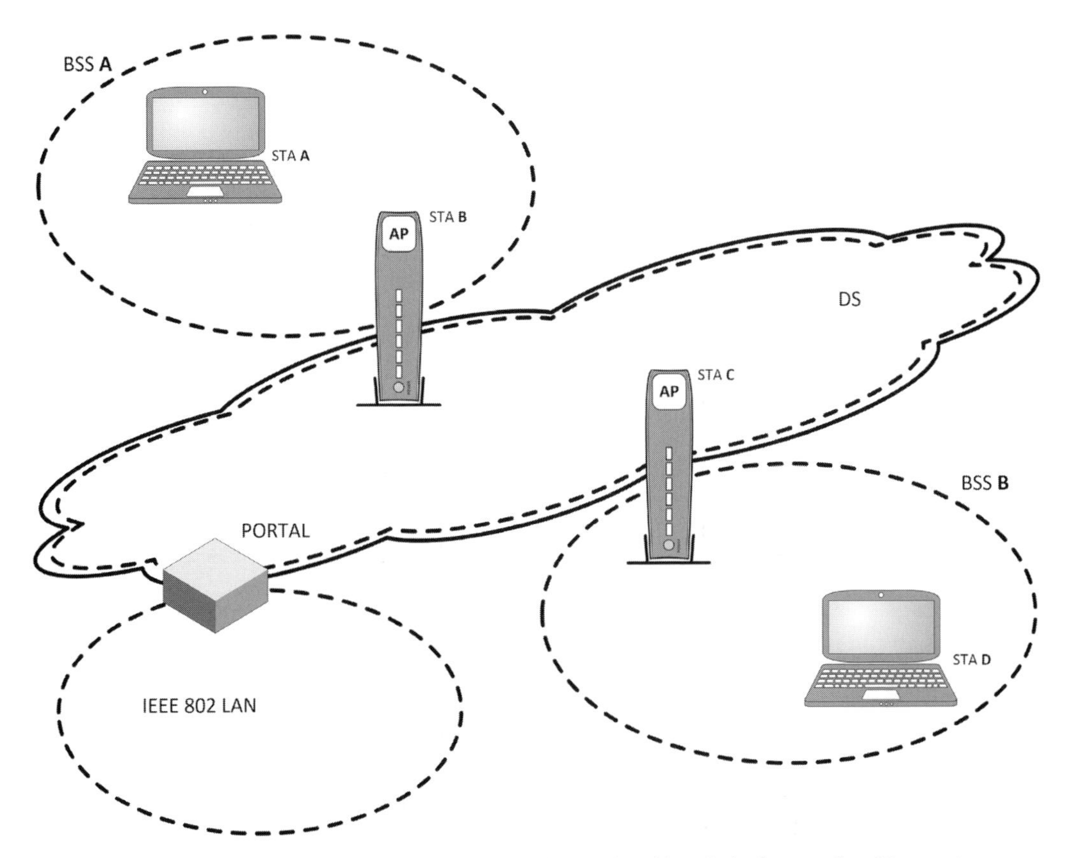

Figure 13.15. The interconnection of a fixed LAN is achieved via the portal architectural component allowing BSS **A** and BSS **B** to intercommunicate via the DS.

13.5.4.1 Integrating 802.11 with a Wired LAN

The *portal* is an architectural component used to characterize the integration of a wireless (802.11) network with a fixed LAN, as we illustrate in Figure 13.15. In essence, data from the fixed LAN enter the wireless network via this portal (and vice versa).

13.5.5 Mesh BSSs

A *Mesh BSS* (MBSS) is a wireless network that comprises a series of autonomous STAs or *mesh* stations. A station within the MBSS is capable of establishing a connection with other neighboring STAs. What's more, a *multi-hop* function is used to ensure data can be transferred between STAs that aren't necessarily in direct communication or in proximity. An STA within an MBSS may be a *source* device, a *sink*, or a *broadcaster* (perhaps to other STAs). However, a mesh STA can only cooperate with other like-enabled mesh STAs within the MBSS and will not become a member of an IBSS or an infrastructure BSS, although, as we illustrate in Figure 13.16, an MBSS can not only exist independently, but also may access the DS through yet another architectural component called the *mesh gate* (as shown). As such, data can flow between the MBSS

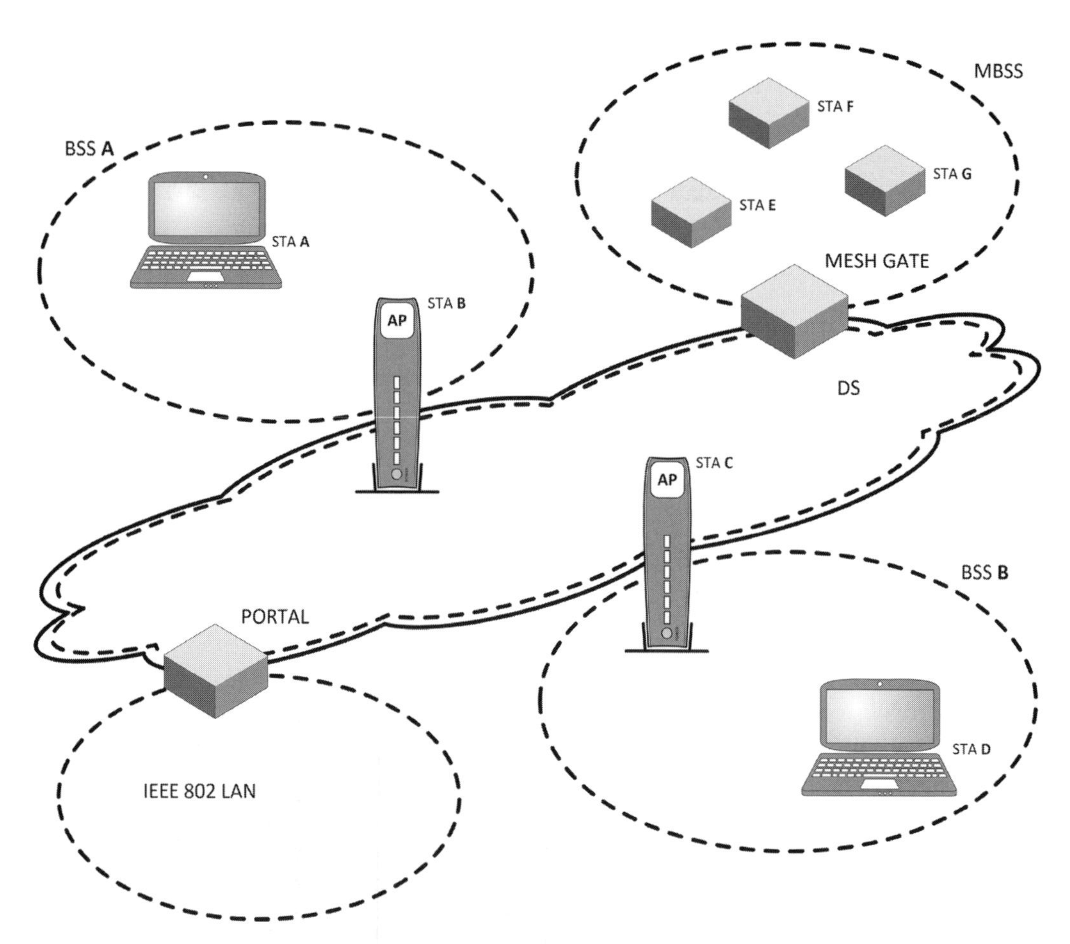

Figure 13.16. The interconnection of an MBSS is achieved via the mesh gate architectural component, allowing the fixed LAN, BSS **A**, and BSS **B** to intercommunicate via the DS.

and the DS, allowing it to communicate with fixed LANs and/or infrastructure BSSs, for example.

13.6 The 802.11 Software Architecture

The IEEE 802.11 standard provides the MAC and PHY layers, as we have already discussed, and essentially these layers provide the wireless interface equivalent for Ethernet (802.3) – see Section 13.1.1, "Ethernet." In Figure 13.17, we map the 802.11 MAC and PHY layers against the *Open Systems Interconnection* (OSI) model.

The *Internet Protocol Suite*, commonly known as the *TCP/IP*[6] *model*, is a set of protocols used for the Internet and other similar networks, such as the PAN, LAN, WAN,

[6] The *Transmission Control Protocol* (TCP)/*Internet Protocol* (IP) should not ordinarily need expanding as it is so commonplace within mainstream computing.

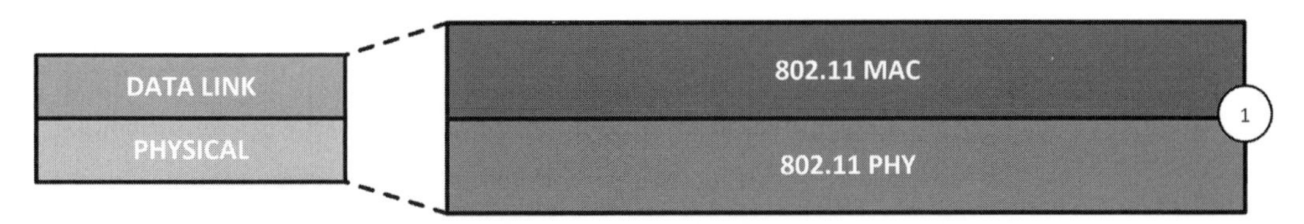

Figure 13.17. The MAC and PHY layers form part of the data link and physical layers within the OSI model.

and so on. The Internet protocol suite's "migrational" fit to the OSI model has caused some confusion within mainstream computing; nevertheless, what we have is a four-layered reference paradigm used to conceptualize and portray the layered components that occupy the Internet architecture, as shown in Figure 13.18. Naturally, the TCP/IP model is used to facilitate communication over both fixed and wireless LANs, and it remains identical to both platforms. Looking again at Figure 13.18, we can identify four layers within the model, namely *network access* (**1**), *Internet* (**2**), *transport* (**3**), and *application* (**4**). You may recall from Chapter 6, "Enabling the Internet of Things," that we discussed how the TCP/IP model was used in smaller devices to enable smart objects to remain interconnected.

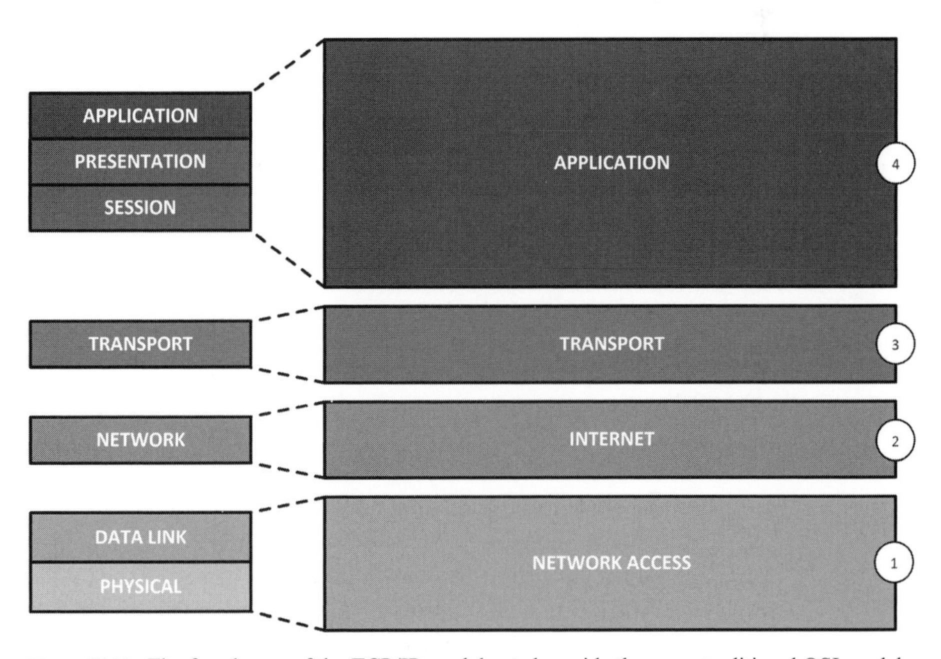

Figure 13.18. The four layers of the TCP/IP model set alongside the more traditional OSI model.

We illustrate in Figure 13.19 the complete protocol stack as used within a fixed or wireless environment, which accurately reflects the inclusion of the 802.11 MAC and PHY layers as the data link and physical layers of the OSI model. The application layer of the TCP/IP model, and the application layer shown in Figure 13.19 for that matter,

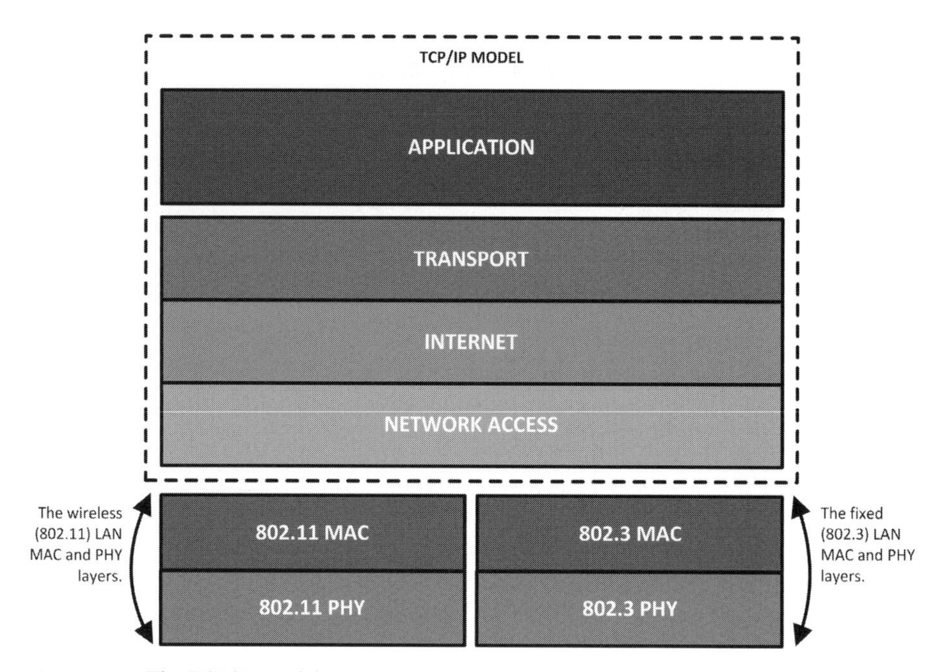

Figure 13.19. The TCP/IP model *in situ* above the 802.11 (wireless) and fixed (Ethernet) MAC and PHY layers, respectively.

may support Telenet, the *File Transfer Protocol* (FTP), *Simple Management Network Protocol* (SMNP), *Domain Name System* (DNS), and the *Hypertext Transfer Protocol* (HTTP), just some possible examples of applications supported by both the TCP and *User Datagram Protocol* (UDP).

13.6.1 MAC Data Service

The MAC data service provides an opportunity for the *Logical Link Control* (LLC) layer to exchange *MAC Data Service Units* (MSDUs) with its peer entity. In Figure 13.20, we illustrate the MAC and LLC sub-layers, along with the *MAC Layer Management Entity* (MLME), which we'll come back to in a moment. Naturally, this service is supported by the asynchronous and connectionless transport service that is provided at the PHY layer, which we discuss later in Section 13.6.2, "PHY Services." The PHY-level services transport the MSDUs to a MAC peer entity and, in turn, the peer LLC. So, all STAs support data transport on a best effort basis, but STAs that utilize *Quality of Service* (QoS) distinguish between MSDUs and look to prioritize traffic based on type of content; QoS is something we discuss later in Section 13.9.1, "Empowering Wi-Fi Networks with QoS."

As we show in Figure 13.20, the MLME provides management responsibilities, whereby layer management service functions can be invoked. Each STA has a *Station Management Entity* (SME) which executes layer-specific and generic management

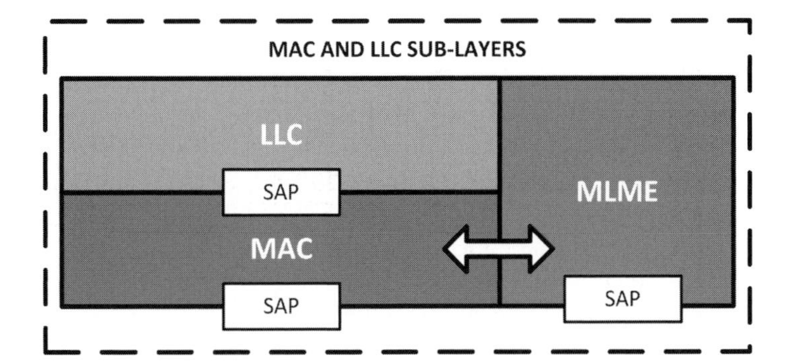

Figure 13.20. The relationship between the LLC and MAC sub-layers, also showing the interaction with the MLME.

for both the MAC and PHY layers. As such, certain primitives are defined and exchanged across the specific *Service Access Point* (SAP), such as `GET` and `SET` operations for the MLME and SME, amongst others. In Section 13.6.2, "PHY Services," we discuss the PHY sub-layers and the associated management entity, whilst in Section 13.6.6, "Layer Management," we provide the complete interworking reference model of the MAC and PHY sub-layers.

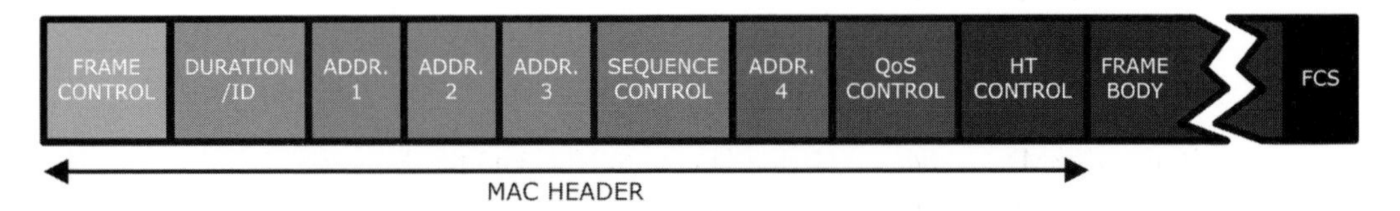

Figure 13.21. The generic MAC frame format.

13.6.1.1 The Generic MAC Frame Format

We illustrate the generic MAC frame format in Figure 13.21. The initial three fields comprise the *frame control*, *duration/ID*, and *address1*, where the last field in the structure contains the *Frame Check Sequence* (FCS) – these fields are common and present in all frame formats. However, fields *address2*, *address3*, *sequence control*, *address 4*, *QoS control*, *High-throughput* (HT) *control*, and, lastly, the *frame body* or payload are only present in certain frame types. In Table 13.5, we look at the specific fields that make up the MAC frame format, along with size of the field in bits.

13.6.2 PHY Services

As you may already know, various PHYs are defined, for example, for use with 802.11a, 802.11b, 802.11g, 802.11n, 802.11ac, and so on. Furthermore, in Section 13.6.3, "The 2.4 GHz DSSS System," Section 13.6.4, "The OFDM PHY System," and Section 13.6.5, "The HT PHY," we touch upon these various PHY systems that are used. Typically, each PHY will comprise two functions. (1) A *convergence* function supports

Table 13.5. The fields that make up the generic MAC frame

Field	Description	Size
Frame control	The frame control field is 16-bits in length and is broken down further in Figure 13.22 and Table 13.6.	16
Duration/ID	The duration/ID is also 16-bits in length and its content varies with frame type and subtype.	16
Addresses 1 to 3	Looking back again at Figure 13.21, we identified four address fields (1 to 4) in our generic MAC frame structure, and these are used to identify the BSSID, *Source Address* (SA), *Destination Address* (DA), *Transmission* [STA] *Address* (TA), and a *Receiving* [STA] *Address* (RA); as we explained earlier, some frames may or may not contain these fields.	
Sequence control	This 16-bit field consists of two further sub-fields and is discussed in Figure 13.23 and Table 13.8.	16
Address 4	See "Addresses 1 to 3" above.	
QoS control	The QoS control field is 16-bits in length and identifies the associated MAC frame, that is, *Traffic Category* (TC) or *Traffic Stream* (TS), along with other QoS information.	
HT control	The high-throughput field is persistently present in a control wrapper frame and is also present in QoS type data and management frames.	32
Frame body or payload	A variable length field, which contains information relevant to frame and sub-types.	
FCS	A 32-bit field containing a *Cyclic Redundancy Code* (CRC), which uses the MAC header and frame body to calculate its value.	32

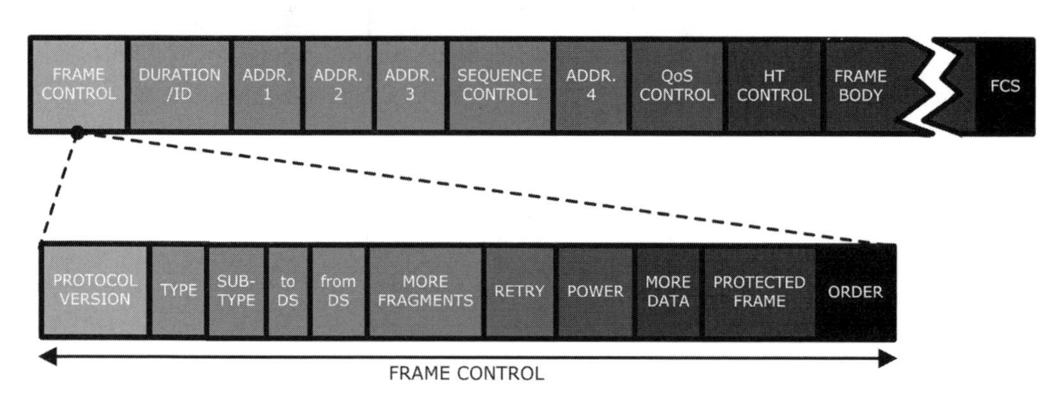

Figure 13.22. The format of the frame control field of the generic MAC frame.

the *Physical Layer Convergence Procedure* (PLCP). The PLCP provides a method for which *MAC Protocol Data Units* (MPDUs) can be mapped into frames that are suitable for transmission using the relevant *Physical Medium Dependent* (PMD). (2) A PMD system defines the features and methods for transmitting and receiving data through the wireless medium between devices. In Figure 13.24, we illustrate the PHY sub-layers, where the PLCP provides as much independence as possible from the underlying PMD and offers a simplified interface with the MAC SAP. The PLME provides management of the immediate PHY operation in cooperation with the MLME. The PMD sub-layer,

Table 13.6. The fields that make up the frame control field

Field	Description	Size
Protocol version	This 2-bit field contains the value `0x00`; other values are reserved for future use.	2
Type and sub-type fields	The type field is 2-bits in length, whereas the sub-type field is 4-bits in length. The fields jointly identify the function of the frame, where the three frame types available are *control*, *data*, and *management*.	6
To and from DS fields	The combination of these values is explained in Table 13.7.	2
More fragments	A 1-bit field that is only set to 1 if data or management type frames have been fragmented, and is set to 0 with all other frame formats.	1
Retry field	The retry field is also a 1-bit field in length and is set to 1 if a data or management frame is repeated; it is used to alleviate duplicate frames.	1
Power (management)	The power management field is 1-bit in length, and is used to indicate the power management mode of an STA.	1
More data	The more data field is 1-bit in length and is used to instruct an STA that is in PS mode that additional *Buffered Units* (BUs) are buffered at the AP.	1
Protected frame	The protected frame field is also 1-bit in length and is set to 1 if the frame body or payload contains information that has been processed by a cryptographic capsulation algorithm.	1
Order	The order field is 1-bit in length and is set to 1 for a QoS data or management frame.	

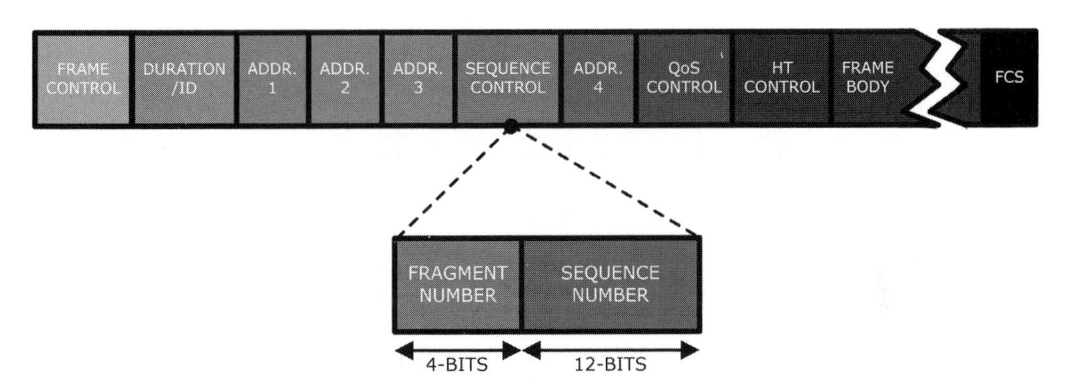

Figure 13.23. The sequence control field structure.

on the other hand, offers a transmission interface, which is ultimately used to transmit and receive data over the air-interface between devices.

As we show in Figure 13.24, the PLME provides management responsibilities, whereby layer management service functions can be invoked. Each STA has an SME, which manages layer-specific and generic management for both the MAC and PHY. As such, certain primitives are defined and exchanged across the specific SAP, such as `GET` and `SET` operations for the PLME and SME, among others. In Section 13.6.6, "Layer Management," we provide the complete interworking reference model of the MAC and PHY sub-layers. However, in the sections that immediately follow, we look at specific

Table 13.7. The "to" and "from" fields that make up the frame control field

Description	To	From
These values represent a payload from one STA to another within the same IBSS; a payload from a non-AP to another non-AP within the same BSS; or a payload that is outside a BSS, along with management and control frames.	0	0
A payload which is destined for the DS or sent by an STA associated with an AP on a port access entity for that AP.	1	0
A payload that exits a DS or is sent by the port access entity in an AP or a group, which has been addressed mesh data frame, with mesh control field present.	0	1
A payload that uses the four-address MAC header format, which is a combination used in a mesh BSS.	1	1

Table 13.8. The two fields that make up the sequence control field

Field	Description	Size
Fragment number	This 4-bit field denotes the number of each fragment of an MSDU; it is set to 0 for the first or only frame and is incremented for each successive fragment frame.	4
Sequence number	This 12-bit field indicates the sequence number of an MSDU.	12

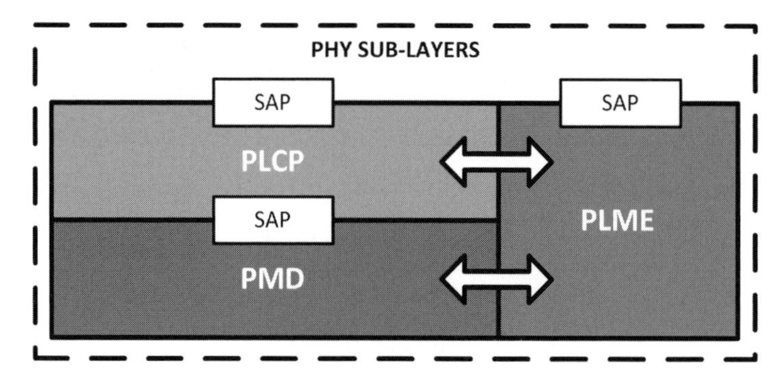

Figure 13.24. The relationship between the PLCP and PMD sub-layers; the interaction with the PLME is also shown.

PHY implementations and functions for 2.4 GHz, 5 GHz, and high-throughput PHY implementations for WLAN connectivity.

13.6.3 The 2.4 GHz DSSS System

The original and popular standard 802.11b uses the *Direct Sequence Spread Spectrum* (DSSS) modulation technique. The PHY system comprises two specific protocol functions. (1) A PHY *convergence* function supports the PLCP. The PLCP provides a method for which MPDUs can be mapped into frames that are suitable for transmission between two or more STAs using the relevant PMD system. So, the PHY will exchange a *PLCP Protocol Data Unit* (PPDU) which, in turn, contains a *PLCP Service Data Unit* (PSDU) that corresponds with an MPDU. (2) A PMD system defines the features and methods for transmitting and receiving data through the wireless medium between two or more STAs using the DSSS system. See Figure 13.24 for an illustration of the 2.4 GHz DSSS PHY architecture.

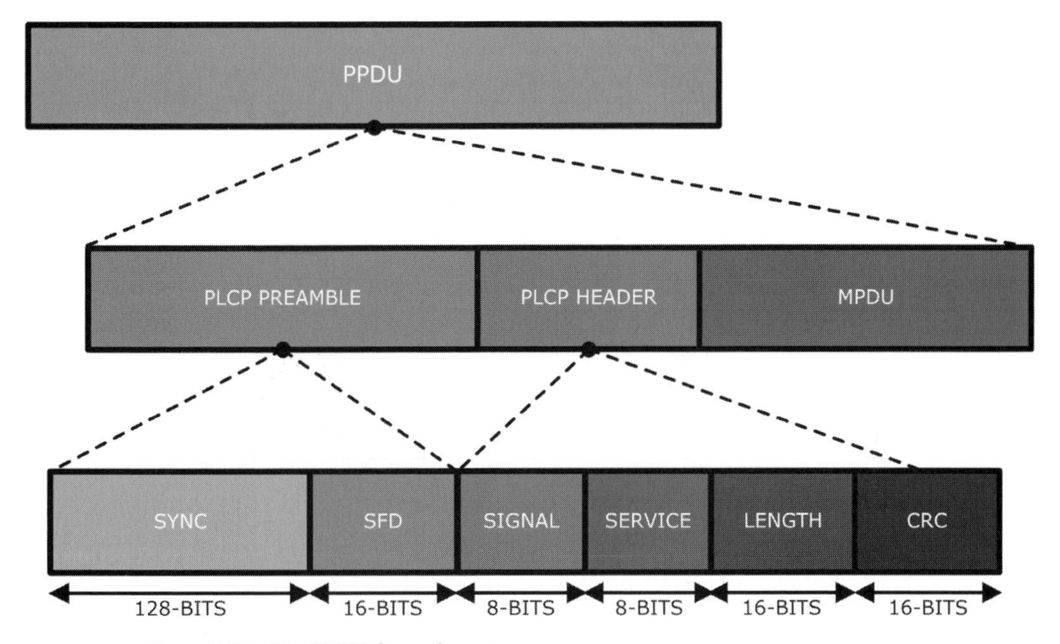

Figure 13.25. The PPDU frame format.

In Figure 13.25, we illustrate the PPDU frame format, which includes the DSSS PLCP preamble, the PLCP header, and the MPDU. Prior to transmission, the MPDU is prefixed with the PLCP preamble and header, which form the PPDU, and, of course, at the receiving entity, the PLCP preamble and header are processed accordingly to deliver the MPDU. In Table 13.9, we describe the fields that form the PPDU preamble, whilst in Table 13.10 we list the fields that form the PPDU header.

Table 13.9. The fields that make up the PPDU preamble

Field	Description	Size
SYNC	This 128-bit field comprises a series of scrambled "1s," and this is used to aid synchronization.	128
SFD	The *Start Frame Delimiter* (SFD) is 16-bits in length and is used to indicate the start of PHY-specific parameters within the preamble.	16

In Figure 13.26 we illustrate the PMD sub-layer reference model, which depicts the relationship of the entire DSSS PHY. The PMD sub-layer receives PLCP sub-layer

Table 13.10. The fields that make up the PPDU header

Field	Description	Size
Signal	The signal field is 8-bits in length and indicates what type of modulation should be used for transmission and reception of the MPDU.	8
Service	Reserved (for future use).	8
Length	An unsigned 16-bit integer alluding to the number of microseconds required to transmit the MPDU.	16
CRC	This 16-bit value protects the integrity of the signal, service, and length fields of the PPDU header using the ones' complement.	16

service primitives via its SAP (as shown), which provides the mechanism by which data shall be sent or received from the air-interface. The PLCP sub-layer supports a number of primitives that support PLCP peer-to-peer interaction and support other interactions at high-level sub-layers.

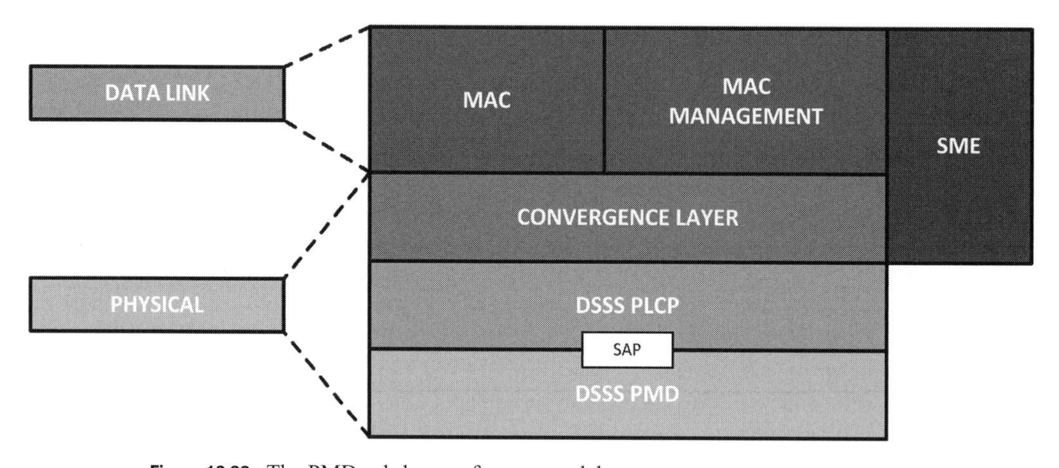

Figure 13.26. The PMD sub-layer reference model.

13.6.3.1 HR DSSS PHY

In this section, we discuss the *High-rate* (HR) extension to the PHY for the 2.4 GHz DSSS system, namely 802.11b (up to 11 Mbit/s), which builds upon the existing capabilities (see Section 13.6.3, "The 2.4 GHz DSSS System"). More specifically, an eight-chip *Complementary Code Keying* (CCK) is used as the modulation scheme, where the rate is 11 MHz. The HR/DSSS essentially utilizes the same PPDU preamble and header as discussed earlier (see Figure 13.25) and, as such, both PHYs are able to coexist within the same BSS. A shorter PPDU header may be used to increase data throughput, which is referred to as HR/DSSS/short, whereby the short preamble can coexist in certain contexts and/or perhaps on another channel.

13.6.3.2 Extended Rate PHY

Building further on the original DSSS system, the *Extended Rate PHY* (ERP) provides data rates of up to 54 Mbit/s using the 2.4 GHz band. The ERP also provides two functions, namely the convergence function and the PMD system, as previously discussed. The ERP uses the basic reference model that we illustrated in Figure 13.26.

13.6.4 The OFDM PHY System

The *Orthogonal Frequency Division Multiplexing* (OFDM) PHY system provides data rates of up to 54 Mbit/s and is used by 802.11g (2.4 GHz), 802.11a (5 GHz), 802.11n (2.4 and 5 GHz), and 802.11ac (5 GHz). The OFDM system also uses sub-carriers that are modulated with *Binary* or *Quadrature Phase Shift Keying* (BPSK or QPSK) or instead uses *16-* or *64-Quadrature Amplitude Modulation* (16-QAM or 64-QAM). The OFDM PHY system comprises two specific protocol functions. (1) A PHY *convergence* function supports the PLCP. The PLCP provides a method for which MPDUs can be mapped into frames that are suitable for transmission between two or more STAs using the relevant PMD system. (2) A PMD system defines the features and methods for transmitting and receiving data through the wireless medium between two or more STAs using the OFDM system.

13.6.5 The HT PHY

The HT OFDM PHY system is capable of sending and receiving frames for both 2.4 GHz (20 MHz channel width) and 5 GHz (20 MHz channel width) frequency bands. The HT OFDM PHY uses four spatial streams operating at 20 MHz, as well as 40 MHz, which, in turn, supports data rates of up to 600 Mbit/s, that is, 40 MHz bandwidth with four spatial streams. The sub-carriers used are modulated using BPSK, QPSK, 16-QAM, or 64-QAM. The OFDM PHY system comprises two specific protocol functions. (1) A PHY *convergence* function supports the PLCP. The PLCP provides a method for which PSDUs can be mapped into a frame format (PPDU) suitable for transmission between two or more STAs using the relevant PMD system. (2) A PMD system defines the features and methods for transmitting and receiving data through the wireless medium between two or more STAs using the OFDM system, where an STA may support a mixture of HT PHY, DSSS PHY, OFDM PHY, HR/DSSS PHY, and ERP.

13.6.6 Layer Management

In Figure 13.27, we illustrate the complete reference model of the MAC and PHY, along with their associated sub-layers and management entities. You will note that the SAPs have been labeled numerically to aid identification. The sub-layer SAPs are MAC_SAP (**1**), PHY_SAP (**2**), and PMD_SAP (**3**), whilst the management SAPs within this model are the MLME-PLME SAP (**4**), MLME-SME SAP (**5**), and PLME-SME SAP (**6**).

As we have already mentioned, each STA has an SME, which affords it layer-independency. The SME coordinates its effort with the MLME and PLME and provides various functions on behalf of the system; the relationship between these entities is depicted in Figure 13.27. The SAP permits the exchange of service primitives, which include the `GET` and `SET` operations between the management entities.

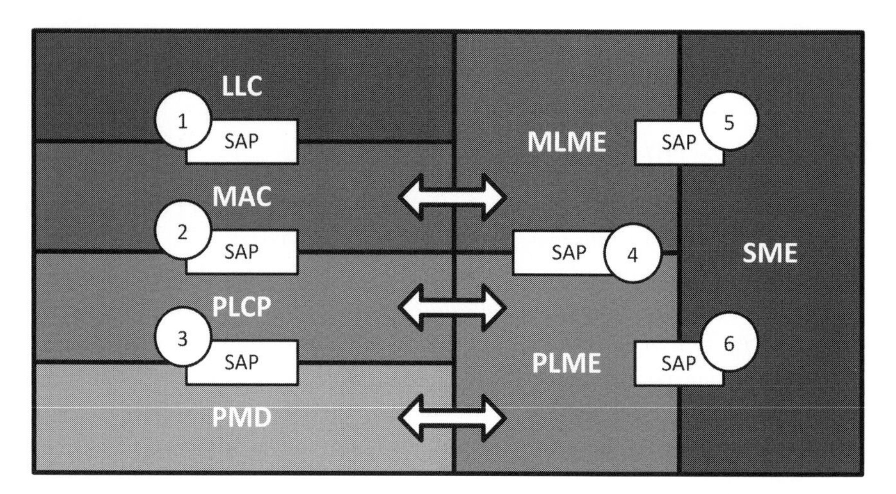

Figure 13.27. The interworking reference model for the MAC and PHY layers of the 802.11 standard.

13.7 Wi-Fi Protected Access

In a joint collaboration, the Wi-Fi Alliance worked alongside the IEEE to formulate a security specification; the result was *Wi-Fi Protected Access®* (WPA™). It provides enhanced and interoperable security for over the air-interface data protection within a secured Wi-Fi network. WPA was quickly introduced to overcome many of the shortcomings identified with the original security standard, namely *Wired Equivalent Privacy* (WEP), which faced a lot of criticism when it was introduced in early 1999. WPA2™, a certification program from the Wi-Fi Alliance, was introduced in 2004, and in 2007 WPS (see Section 13.8, "Wi-Fi Protected Setup") simplified the use of WPA2 within residential contexts. WPA2 is nowadays supported in current Wi-Fi CERTIFIED devices and relies on two-key protocols: the 128-bit *Advanced Encryption Standard* (AES) and IEEE 802.1X. AES is a trusted encryption protocol, which is currently used by the United States and other government agencies to secure WLANs and other environments. IEEE 802.1X, mostly used in enterprises, provides robust authentication and complex network access control features. Mutual authentication is achieved with a *Pre-shared Key* (PSK) in *personal* mode, whilst 802.1X/*Extensible Authentication Protocol* (EAP) is used in an *enterprise* mode.

13.7.1 Security Features with WPA2

WEP lacked effective authentication procedures, but WPA uses an enhanced encryption scheme, namely *Temporal Key Integrity Protocol* (TKIP), along with 802.1X/EAP authentications, as already mentioned. TKIP, in fact, uses a key hierarchy system, which offers better protection. Likewise, TKIP includes a *Message Integrity Check* (MIC)

Table 13.11. Comparing WEP and WPA security features

	WEP	WPA
Encryption	Ineffective and was cracked quickly by scientists and hackers.	Addresses all WEP shortcomings.
	40-bit keys.	128-bit keys.
	The same static key was used by everyone on the network.	A dynamic session key is used for each user and each session, and per packet keys.
	A manual distribution of keys, which is hand-typed into each device.	WPA uses an automatic distribution of keys.
Authentication	Ineffective since the WEP key itself was used for authentication.	WPA now uses strong authentication using 802.X/EAP.

to offer additional protection against packet forgeries. We compare WEP and WPA in Table 13.11.

13.8 Wi-Fi Protected Setup

As we discussed in Section 13.7, "Wi-Fi Protected Access," Wi-Fi security was strengthened with the introduction of WPA and later WPA2. However, entering a hexadecimal number to gain access to a Wi-Fi network can be a little cumbersome, so the Wi-Fi Alliance introduced WPS, which offers several convenient ways to access a network. WPS,[7] which was released in 2007, supports an ease-of-use philosophy that enables consumers to configure, set up, and connect with their Wi-Fi product(s) easily. In Figure 13.28, we show the WPS logo which is used on certified products, in turn ensuring interoperability and compliance with the *Wi-Fi Simple Configuration Specification*.

Figure 13.28. The Wi-Fi Protected Setup logo, which is used by certified products. (Courtesy of the Wi-Fi Alliance.)

Manufacturers can choose to use WPS, a certification program from the Wi-Fi Alliance. The program is primarily targeted for use within the *Small Office/Home Office* (SOHO) rather than an enterprise. It alleviates the user from having to delve into certain parameters which are typically required to connect with an AP, such as the SSID and (WPA2) security key – the user no longer has to sift through what can be overwhelming security choices.

[7] Gratton, D. A., "A Touch of Genius: Wi-Fi Protected Set-up," May 2008.

13.8.1 Set-up Options

Products that are WPS-enabled may provide one of two options to users to set up and configure their equipment, namely the *Personal Identification Number* (PIN) or the *Push Button Configuration* (PBC). This means that an AP must be capable of supporting both options, and that a client device must provide the PIN set-up.

13.8.1.1 PIN Method

A PIN (which may be placed or stuck onto the device) is provided for each device that wishes to join the network; a dynamic PIN can be produced if the product has a suitable display or interface. The user enters the PIN into the *registrar* (see Section 13.8.2, "The Registrar") via the AP or through a management pane, which may be presented onscreen by another device that is already connected to the network. The PIN is primarily used to ensure the intended product is to be included within the network and removes any opportunity for unauthorized access.

13.8.1.2 Push Button Configuration Method

The PBC method provides the user with the opportunity to connect their device to the network by pushing buttons on both the AP and client devices or on a device that acts as gatekeeper in a Wi-Fi group context. (We'll come back to this shortly.) A fixed time window, say two minutes, is opened from the moment the button is pressed on the AP and is closed when the button on the client device is pushed. If the user fails to push the button on the client device within this two minute window, then the opportunity to connect the device is lost and the button on the AP has to be pushed again. This provides a further security mechanism to prevent unauthorized device access.

13.8.2 The Registrar

A *discovery* procedure, which is common to all Wi-Fi devices irrespective of manufacturer, simplifies connection within a network by using a registrar to offer credentials to devices wishing to connect – APs using WPS have the capability of being a registrar, and multiple registrars are supported within a WPS-enabled network. However, a registrar may reside on the AP itself (referred to as an *internal* registrar), whereas a registrar that sits elsewhere within the Wi-Fi network is known as an *external* registrar. As we have already discussed, a user wishing to join a network uses one of the two methods, PIN or PBC, and this, in turn, launches the configuration wizard. As such, WPS commences an exchange of credentials, namely the SSID and security key, between the device that wishes to join and the registrar. The process is usually undertaken just once, as the AP and device will remember their credentials and will continue accessing the network.

13.9 Wi-Fi Multimedia

As we have discussed, Wi-Fi technology is prevalent and integrated into a range of consumer products – most, if not all, smart or cellular phones, cameras, TVs, and other entertainment systems include Wi-Fi. Wi-Fi has been widely adopted for non-PC devices, and it has seen an enormous shift into delivering multimedia content. In addition, teenagers, families who live apart, and residences and businesses embracing no/low-cost telephony have all taken advantage of *Voice over Internet Protocol* (VoIP) services such as Skype (skype.com) and Google Hangouts (google.com) to speak with loved ones and colleagues. What's more, Wi-Fi is confidently expected by the consumer to provide video and music streaming, and to support real-time audio and video content, along with other ancillary services. So, a level of *quality* has to be assured by Wi-Fi technology to deliver and accommodate the demands of these and other similar media-centric applications. In essence, Wi-Fi Multimedia™ or WMM offers Wi-Fi technology a QoS for Wi-Fi-enabled networks using the IEEE 802.11e standard.

Table 13.12. WMM allows a user to set four priority levels

AC	Priority
Voice	Highest.
Video	Second highest.
Best effort	Third highest: typically representing applications such as browsing, file-transfer, and email.
Background	Low: for applications that are not latency-sensitive.

13.9.1 Empowering Wi-Fi Networks with QoS

With the diversification and use of Wi-Fi technology, QoS is required to support media-centric applications and services. WMM provides several features to improve overall quality and subsequent user experience in which it prioritizes traffic demands from various applications over the Wi-Fi network. As such, a user can define four priority levels for certain types of traffic or *Access Category* (AC), as shown in Table 13.12. For example, voice traffic over a Wi-Fi network has the highest priority and therefore the voice call is less likely to degrade. The best effort and background priorities typically represent traffic from legacy devices or applications that do not require or support QoS.

13.9.1.1 WMM Prioritization

WMM provides an enhancement to the MAC sub-layer and is also an extension to the legacy *Carrier Sense Multiple Access* with *Collision Avoidance* (CSMA/CA) mechanism – we discussed the MAC sub-layer earlier in Section 13.6.1, "MAC Data Service." In short, all devices have the same priority based on best effort (see Table 13.12). Therefore, each device will wait a random *backoff* period prior to transmission and will also utilize the Listen Before Talk (LBT) algorithm to ensure that no other devices are transmitting. However, when there are high traffic demands, whereby a network can become overloaded, quality and performance affect all devices. Utilizing WMM traffic prioritization allows data packets to be categorized by traffic type (Table 13.12) – data packets that are not assigned a type are placed into the best effort category. In Figure 13.29, we illustrate the mechanism that is used within a WMM-enabled device where data

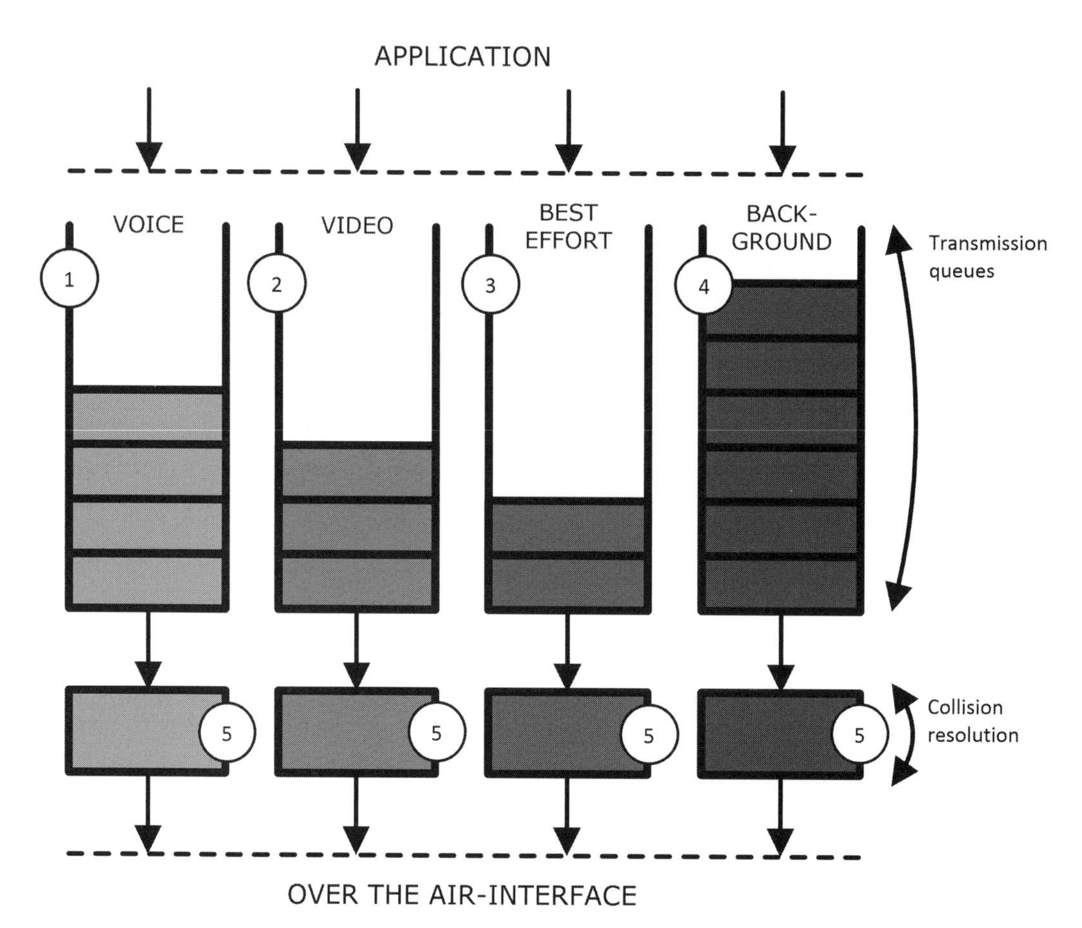

Figure 13.29. A WMM-enabled device, where data packets are placed into one of four transmission queues for voice, video, best effort, or background.

packets are placed into one of four transmission queues, that is, voice (**1**), video (**2**), best effort (**3**), or background (**4**). Furthermore, the WMM device will use a collision resolution mechanism (**5**) to overcome collisions with different queues, and will also manage collisions prior to transmission over the air-interface – ultimately these mechanisms determine which will be offered the *Opportunity to Transmit* (TXOP).

The collision resolution mechanism is probabilistic and uses two timing parameters that differ across the four types of transmission queue – these timing values are inherently smaller for high priority traffic. The first timing parameter is the minimum inter-frame space or the *Arbitrary Inter-frame Space Number* (AIFSN); the second parameter is the *Contention Window* (CW), or random backoff period, which we mentioned earlier. For each traffic type, a backoff value is calculated based on the sum of the AIFSN, along with a random value from zero to CW, which varies through time. When a collision occurs, the CW value is doubled until a maximum value is reached – once transmission occurs, the CW is reset to its initial value associated with the AC. As such, the traffic type with the smallest backoff period is offered TXOP and is permitted to transmit for a predetermined period, which is dependent on the AC and the PHY rate. This effective transmission "burst" technique offers an efficient mechanism for high data rate traffic for audio/video streaming.

13.9.2 WMM-Power Save

We discussed in Chapter 5, "Introducing Low Power and Wireless Sensor Technologies," how we are encouraged to think of the "bigger picture" and to understand the repercussions of our decisions today, which ultimately may have an impact on our future. As such, manufacturers are seeking alternative energy solutions which either extend battery life or harness ambient resources to energize new products. WMM-*Power Save* (PS) is a certification program defined by the Wi-Fi Alliance to encourage manufacturers of battery powered equipment to reduce overall power consumption, which, in turn, extends battery life. The WMM-PS program is especially suited to smart or cellular phones, tablets, mice, keyboards, sensors, and other small portable products that ultimately rely on batteries. As we have already mentioned, a device that is WMM CERTIFIED is encouraged to enter a "doze" state. In this state, a device will consume less power, improving overall performance and minimizing transmission latency.

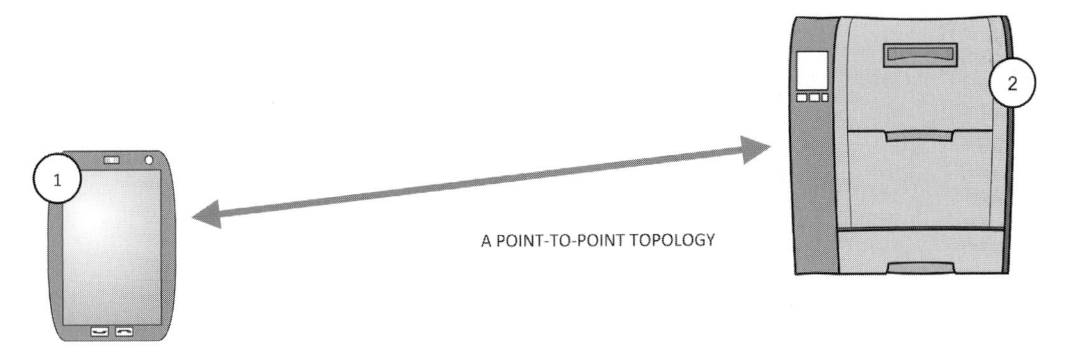

Figure 13.30. The technology empowering Wi-Fi Direct provides the opportunity to create a one-to-one connection.

13.10 Wi-Fi Direct

In Section 13.5, "Networking Topologies," we covered the essential architectural components that typically form a Wi-Fi-enabled network. Wi-Fi Direct™, on the other hand, allows Wi-Fi-enabled devices to connect directly with each other. As such, Wi-Fi Direct offers users the convenience to transfer, share, print, and synchronize content without the need to join a home, office, or hotspot network, for example. The topology supported in this context is point-to-point and point-to-multipoint, as illustrated earlier in Figure 13.11, and as we illustrate in Figure 13.30 and Figure 13.31, respectively. However, this is unlike the legacy ad hoc (or IBSS) mode; it is simply an extension to the infrastructure mode of operation where there is no requirement for a dedicated AP. The technology

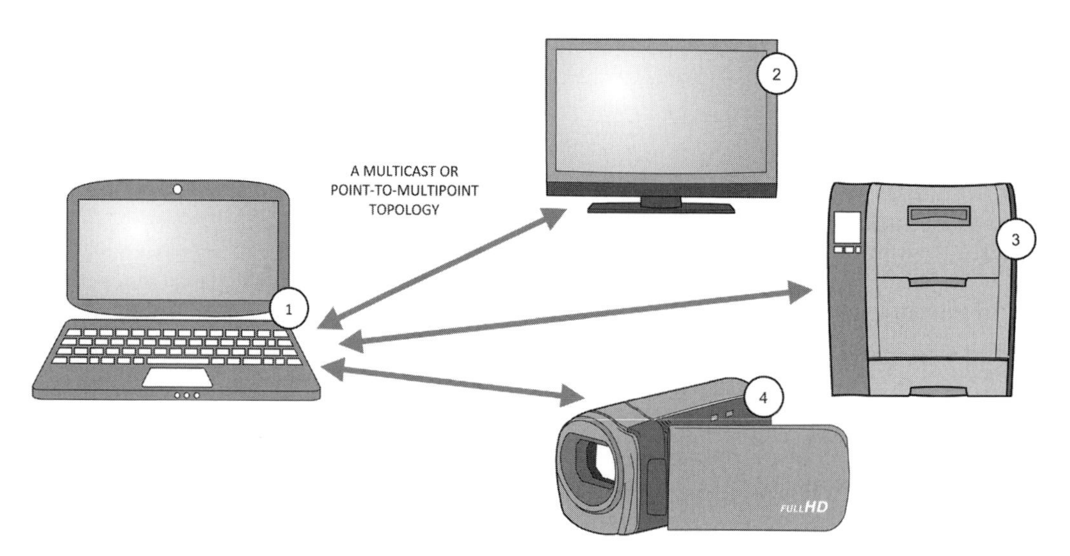

Figure 13.31. The technology empowering Wi-Fi Direct also provides the opportunity to create a one-to-many connection.

empowering Wi-Fi Direct differs in that a Wi-Fi Direct-certified device can concurrently maintain a connection to other Wi-Fi Direct devices, as well as to an infrastructure network, which is not possible with ad hoc.

You may recall from Chapter 4, "Introducing the Lawnmower Man Effect," that we discussed the dynamics of media convergence; in particular, we highlighted how "We carry our content with us and it's become pocketsize!"[8] The Wi-Fi Alliance has identified this trend with consumers and has extended Wi-Fi technology to encompass flexibility and portability of users' content across all their devices. It ultimately empowers users of the technology not only to connect with each other's devices spontaneously, but also to share pictures and videos, and even to display this content on a Wi-Fi-enabled TV, all in all offering a sense of immediacy and convenience.

13.10.1 Wi-Fi is More Than Internet Connectivity

As we mentioned previously, Wi-Fi is commonly associated with the ability to connect to the Internet, and the Wi-Fi Alliance has extended the technology's use by establishing it to perform tasks that would ordinarily be accomplished with a cable. In direct competition with Bluetooth wireless technology, Wi-Fi Direct certified devices can be used by consumers, who can embrace existing traits of Wi-Fi technology, such as throughput, range, and security. Wi-Fi Direct will notify certified devices in proximity that it's capable of establishing a connection and, as such, either users can view the available devices and request a connection, or a device may offer an invitation to connect to a Wi-Fi Direct certified device. Once two or more Wi-Fi Direct certified devices are connected, they form a Wi-Fi Direct *group*. As we have already mentioned, Wi-Fi Direct devices can

[8] Gratton, S., *Follow Me! Creating a Personal Brand with Twitter*, John Wiley & Sons, 2012.

connect either on a one-to-one basis, as shown in Figure 13.30, or as a group, as in Figure 13.31. A typical scenario for Wi-Fi Direct devices connecting in a one-to-one manner would be as in Figure 13.30, where we show a smart or cellular device (**1**) connecting to a printer (**2**). Likewise, Wi-Fi Direct devices may connect with multiple devices to form a unique group, in which one device will act as the *gatekeeper*, which becomes responsible for sending invitations to other devices in proximity. The gatekeeper device will also be responsible for determining whether a device is allowed to connect and will provide authorization. In Figure 13.31, we see that our notebook (**1**) has established a connection with a TV or monitor (**2**), a printer (**3**), and a digital camera (**4**).

13.10.2 Wi-Fi Direct in an Enterprise Context

Within an enterprise context, Wi-Fi Direct devices will still be identifiable to an infrastructure AP as "direct" capable devices. As such, a Wi-Fi Direct certified device has the capability of joining an infrastructure network as an STA (see Section 13.5.2, "Wireless Stations"). As you already know, within a Wi-Fi Direct group, a gatekeeper controls which devices may participate within the group – the gatekeeper device will appear as an AP to other legacy clients and will offer similar services typically provided by an infrastructure AP. An infrastructure AP can prevent a Wi-Fi Direct device from connecting to it, or may decide to disconnect the device completely; it may even configure Wi-Fi Direct device's parameters. A Wi-Fi Direct device must be capable of taking responsibility and negotiating which device takes on the role of gatekeeper when interacting with other Wi-Fi Direct devices. Incidentally, since Wi-Fi Direct-capable devices do not provide full functionality of an infrastructure AP, a standard AP will often be the advisable choice for satisfying the requirements within a home, hotspot, or business.

Table 13.13. The Wi-Fi Alliance peer-to-peer specification offers several underlying mechanisms

Mechanism	Description	Required
Device discovery	A mechanism to discover other Wi-Fi Direct devices within proximity and to exchange device information.	Mandatory.
Service discovery	A mechanism used to ascertain higher-layer services, which can be undertaken prior to connection establishment.	Optional.
Group formation	A mechanism used to determine who is responsible for the group.	Mandatory.
Invitation	A mechanism that allows a Wi-Fi Direct device to invite another device to participate within the group.	Optional.
Client discovery	A mechanism that allows a Wi-Fi Direct device to establish which Wi-Fi Direct devices are within an existing group.	Mandatory.
Power management		
P2P-PS and P2P-WMM-Power Save	Several adapted mechanisms that provide additional power save and WMM-PS features for Wi-Fi Direct devices.	Mandatory.
Notice of absence	A technique used to offer further power consumption by communicating a scheduled absence.	Mandatory.
Opportunistic power save	A group device may enter a doze state if other devices within the group are also dozing to help reduce power consumption.	Mandatory.

Table 13.14. The Wi-Fi Alliance peer-to-peer specification offers several optional important capabilities

Mechanism	Description
Persistent groups	A mechanism that allows a previously established group to be reinstated.
Concurrent connection	A mechanism that allows a Wi-Fi Direct device to maintain simultaneous connections to a group and/or to a WLAN.
Multiple groups	A mechanism that provides a Wi-Fi Direct device to maintain membership in multiple groups.
Cross-connection	Offers a Wi-Fi Direct device that's in charge of a group the opportunity to provide infrastructure access for other devices within the group.
Managed device	A feature that allows a Wi-Fi Direct device to undertake management direction from an AP reference coexistence, channel selection, power management, and so on.

What's more, within the enterprise context it is expected that devices, such as smart or cellular phones, tablets, and other portable equipment, will temporarily connect for a short period to undertake small tasks such as file-transfer; other scenarios may include sharing access to a Wi-Fi-enabled projector or printer.

13.10.2.1 Peer-to-Peer Mechanisms

The Wi-Fi Alliance has defined several mechanisms in its peer-to-peer specification that underlie Wi-Fi Direct's capabilities, and in Table 13.13 we summarize these mechanisms along with their responsibilities; other important features within the specification are summarized in Table 13.14.

Table 13.15. The Wi-Fi Alliance's series of certifications tests for a Wi-Fi Direct device

Test	Description
1	Baseline: a device must support, as a minimum, 802.11g and WPA2-Personal.
2	WMM and WPS: these are mandatory features of a Wi-Fi Direct device.
3	All mandatory: represents the minimum functionality that a Wi-Fi Device must support.
4	Optional features: devices that support optional features must be tested for that feature.

13.10.3 Certification

We discussed the Wi-Fi Alliance's certification program earlier in Section 13.1.3.2, "Wi-Fi." You may recall that the Wi-Fi CERTIFIED program offered by the Wi-Fi Alliance encourages compliance and interoperability of Wi-Fi products and ensures an improved user experience – only products that have been certified can use the Wi-Fi CERTIFIED logo shown in Figure 13.1. However, devices wishing to become Wi-Fi Direct certified will be able to bear the mark, "Wi-Fi CERTIFIED Wi-Fi Direct" upon completion of certification. As such, a Wi-Fi Direct device must pass a series of certifications tests, as summarized in Table 13.15.

14 Bluetooth Classic and High speed

More Than Cable Replacement

Both Bluetooth *classic* and *high speed* are included in the Specification of the Bluetooth System: Core, v4.0, along with the recent addition of Bluetooth low energy, which we discussed earlier in Chapter 7, "Bluetooth low energy: The *Smart* Choice." The inclusion of low energy capabilities enables a new generation of Bluetooth *smart* devices, which are capable of operating for months and possibly years just utilizing coin-cell batteries. Bluetooth wireless technology, like Wi-Fi, has been enabling our *Wireless Personal Area Network* (WPAN) for almost two decades; in fact, these two technologies were regarded as competitors. A modicum of time has afforded both technologies to tread their own respective applications paths, find their feet so to speak, and establish two very different user scenarios and applications.

Bluetooth was originally touted as a cable replacement technology, looking to replace RS232-like cabling and instead offer the convenience of wireless connectivity. Bluetooth still very much retains its original objective, but the technology itself has such a rich and diverse market and application-base that it has become more than just a cable replacement technology.

Figure 14.1. The Bluetooth logo. (Courtesy of the Bluetooth SIG.)

14.1 Overview

Bluetooth wireless technology now comprises two systems, namely *Basic Rate* (BR) and, of course, *Low Energy* (LE), where both systems provide capabilities that enable device discovery, connection establishment, and other mechanisms. The LE system, which we discussed earlier in Chapter 7, is derived, to some extent, from the BR system. However, the LE system has several capabilities that ultimately offer lower cost, low power, and provide reduced complexity when compared with the BR system, and it is designed for products and applications which utilize low data rates. The LE system provides data throughput of around 1 Mbit/s, but it is not designed to provide file transfer, for example; rather, it is ideally suited to applications where only small packets of data are exchanged. On the other hand, the BR system provides optional *Enhanced Data Rate* (EDR), *Alternative Media Access Control* (MAC), and *Physical* (PHY) or AMP extensions, in turn offering synchronous and asynchronous connections with data rates of up to 721 kbit/s as a minimum, whilst 2.1 Mbit/s can be achieved with EDR and 24 Mbit/s for high speed operation using 802.11 AMP. Bluetooth devices that implement both systems are, of course, capable of communicating with other like-enabled devices and have the opportunity to provide a greater range of applications. What's more, a product implementing both systems may also communicate with other devices that implement either system.

Figure 14.2. The Bluetooth Special Interest Group logo. (Courtesy of the Bluetooth SIG.)

14.1.1 The Bluetooth Special Interest Group

The Bluetooth *Special Interest Group* (SIG) is responsible for the evolution and future of the technology, as well as for the promotion of the brand and its awareness. The Bluetooth SIG also provides a product qualification program. We show the SIG's trademark logo in Figure 14.2. The Bluetooth SIG was formed in 1998, just one year prior to the Wi-Fi Alliance, and has an overwhelming 17 000+ member companies worldwide to date. The Bluetooth SIG is a private, non-profit organization, and is responsible for driving awareness and encouraging the adoption of Bluetooth wireless technology.

14.1.1.1 Membership and the Bluetooth SIG

The Bluetooth SIG offers potential participants three classes of membership, as shown in Table 14.1. The three classes of membership enable companies to decide on how they wish to participate within the SIG. The organization has structured its membership from *Adopter*, an entry level where the member can access many Bluetooth resources, to *Promoter*, where a representative of a company can appoint a member to sit on the board of directors. Essentially, membership of the Bluetooth SIG entitles the member to participate in the future direction of the technology, as well as to use the Bluetooth word mark and logo. Any company wishing to use the Bluetooth word mark and logo is required to sign up with the SIG and qualify their products prior to marketing or sale.

14.1.1.2 Bluetooth Qualification

The majority, if not all, of the manufacturers of technologies featured in this book require a product to be certified, or *qualified* in Bluetooth's case. Qualification is mandatory if a company wishes to obtain a Bluetooth intellectual property license and apply the Bluetooth trademarks to a design or product. As such, once these products are qualified,

Table 14.1. The Bluetooth SIG offers potential participants a three-tier membership structure to include Promoter, Associate, and Adopter

Membership level	Scope
Promoter	The Bluetooth SIG has a minimum number of Promoter companies, including Apple, CSR, Ericsson, Intel, Lenovo (formerly IBM), Microsoft, Motorola, Nokia, Nordic, and Toshiba. Each company has one representative sitting on the board of directors.
Associate	An Associate member has the same benefits as an Adopter, but has early access to Bluetooth specifications and the opportunity to participate with working groups and committees, whilst assisting with the future development of the technology.
Adopter	An Adopter member has access to Bluetooth resources, as well as the use of Bluetooth specifications and the license to use the Bluetooth word mark and logos.

they have to become listed on the *End Product Listing* (EPL) – a rich source from where consumers and manufacturers may locate all qualified Bluetooth-enabled products. The Bluetooth qualification process comprises four stages, as we illustrate in Figure 14.3 and summarize in Table 14.2.

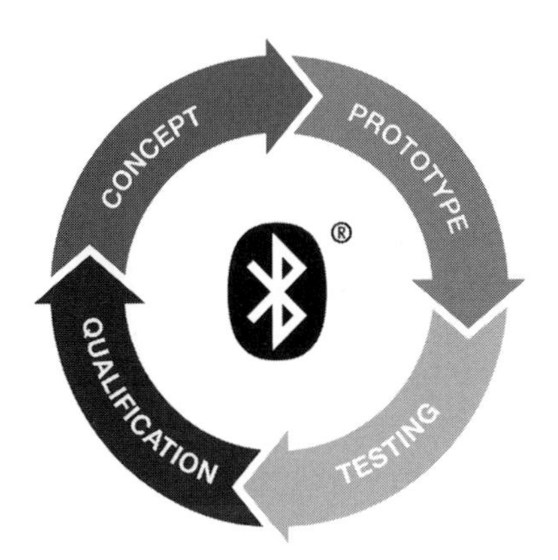

Figure 14.3. The qualification cycle. (Courtesy of the Bluetooth SIG.)

14.1.2 Bluetooth's Timeline

The infographic shown in Figure 14.4 provides a snapshot of the Bluetooth classic and high speed technology timeline, from its inception to where the technology is today. Bluetooth wireless technology was conceived as far back as 1994 by Ericsson, and the Bluetooth SIG was established soon after, in 1998, after garnering support from some prominent promoter companies. However, first-generation Bluetooth-enabled products didn't start to enter the market until 2000, as the technology was heavily polluted with industry hyperbole. With the furor surrounding its initial capabilities and features, marketers from numerous manufacturers and early advocates were guilty

Table 14.2. The Bluetooth qualification process comprises four stages

Stage	Description
Concept	The concept stage is used to initiate the qualification process whereby a member receives their *Qualified Design Identifier* (QD ID).
Prototype	The member company at this stage identifies specific core and profiles features that are utilized in their product and, in turn, generates a test plan.
Testing	The testing stage requires the member to commence testing their prototype against the test plan, which was generated at the prototype stage. The member company submits their test report to the Bluetooth SIG.
Qualification	The member company begins to detail product features and so on and to list their product on to the EPL. The *Declaration of Compliance* (DoC) and *Supplier Declaration of Conformity* (SDoC) documents are signed, which declare the product compliant with the various Bluetooth specifications.

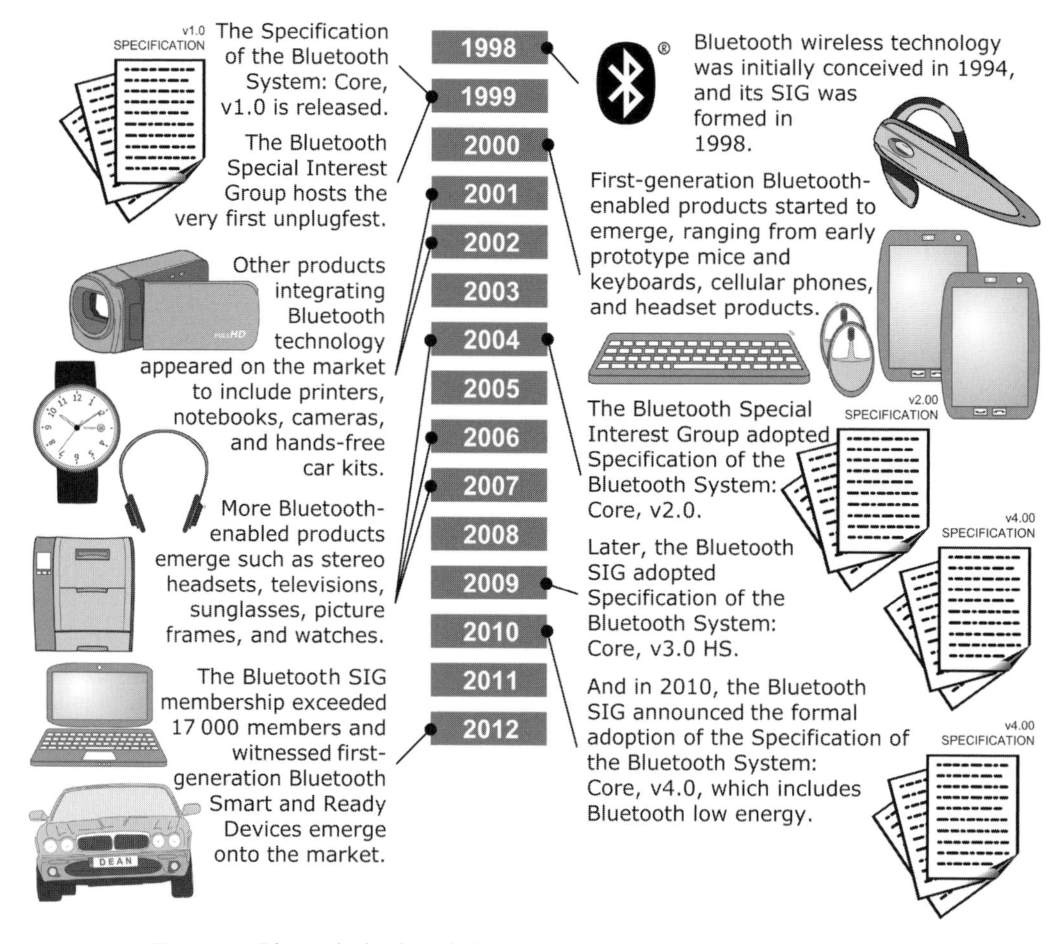

Figure 14.4. Bluetooth classic and high speed technology timeline, from its inception to where the technology is today with its Specification of the Bluetooth System: Core, v4.0. (Bluetooth logo courtesy of the Bluetooth SIG.)

of overselling a technology that needed an opportunity to walk before it could run. Nonetheless, once this initial hype dissipated, Bluetooth rapidly become widely adopted by a number of leading manufacturers, and a plethora of consumer electronic products integrating Bluetooth as standard was embraced by consumers. In 2000, the market witnessed a surge in first-generation Bluetooth products to include cellular phones and headsets, along with prototype mice, keyboards, and USB dongles. Later, in 2001 and 2002, the consumer electronics sector became populated with Bluetooth technology integrated within notebooks, printer, and camera products, as well as hands-free car kits, which included speech recognition. In 2004, the Bluetooth SIG adopted a new core specification, namely v2.0, which included new enhanced data rate capabilities offering up to 2.1 Mbit/s data throughput. With increasing popularity, Bluetooth technology was integrated into a vast range of products, such as watches, sunglasses, TVs, and even picture frames. On celebrating its tenth anniversary in 2008, the SIG welcomed its 10 000th member company, and in 2009 the SIG adopted core specification v3.0 with high speed technology. In 2010, the Bluetooth SIG took a leap forward with its technology by adopting core specification v4.0 with Bluetooth low energy capability, something which we discussed earlier in Chapter 7, "Bluetooth low energy: The *Smart* Choice."

In 2011, the Bluetooth SIG announced Bluetooth Smart and Smart Ready brand extensions, and Microsoft announced support for v4.0 in its new Windows 8 operating system. Likewise, Apple announced that its iPhone 4S would also support v4.0 and became the first Bluetooth Smart Ready phone. In 2012, the Bluetooth SIG announced that its member companies had exceeded 17 000 members, and first-generation tablets and music players were launched onto the market all branding Bluetooth Smart Ready.

14.1.3 Shaping the Personal Area Network

A *Personal Area Network* (PAN) is a term used to describe a particular networking topology and is the subject matter for this book. You may recall from Chapter 2, "What is a Personal Area Network?," that we characterized the *Wireless Personal Area Network* (WPAN) as a diverse collection of short-range radio technologies that have had an influence on the dynamics of the personal area network. The PAN is a relatively new term to the computing industry, although it was initially coined as far back as 1996 by researcher Thomas Zimmerman.[1] However, the definition arguably didn't enter mainstream computing until circa 2000 with the introduction of Bluetooth wireless technology. The devices that form your PAN may or may not be wireless-enabled. Nonetheless, a *wireless* personal area network is a PAN in which devices permit connection through a wireless technology, such as Bluetooth, *Wireless Home Digital Interface* (WHDI), and ZigBee.

14.2 Bluetooth's Market

The wireless technologies that are shared in Part III, "The Classic Personal Area Network," typically serve high-end, high data rate products best suited for audio-, voice-,

[1] Zimmerman, T., "Personal Area Networks (PAN): Near-field Intra-body Communication," 1996.

and data-heavy-centric applications. Bluetooth wireless technology accommodates and provides a somewhat diverse and eclectic application space, in turn offering the high-end, high data rate for audio, voice, and data-heavy capabilities. Both the classic and high speed Bluetooth ecosystems boast a healthy application-base, something we cover later in Section 14.3.1, "Bluetooth Profiles." Bluetooth classic and high speed have matured and captured a healthy market share for a wealth of consumer electronic products, including cellular or smart phones, PCs, stereo headsets, car kits, and so on. Bluetooth low energy, a recent introduction to the core specification v4.0, has been architected from the ground up; it has a new radio, software protocol stack, several new specific profiles, and is all designed to operate from a coin-cell battery.

14.3 Bluetooth's Application Portfolio

Bluetooth classic and high speed continue to enable consumer electronic products with the simplicity afforded by the premise of wireless technology, and, with new and improved profiles empowering the Bluetooth ecosystem, the technology simply goes from strength to strength. Bluetooth has been in use for over a decade and has matured to accommodate a diverse and competitive consumer electronics market sector; with the introduction of Bluetooth low energy, the market will inevitably witness hybrid products ensuring lower power consumption for a new generation of Bluetooth applications and services.

14.3.1 Bluetooth Profiles

> Bluetooth profiles enable end-user functionality by defining user and behavior characteristics. They have many defining objectives, one of which is to achieve a level of interoperability between manufacturers. (Dean Anthony Gratton, *Bluetooth Profiles*, 2003, pp. 567–568)

A Bluetooth application achieves interoperability through the use of *Bluetooth profiles* – which simply define how an application should manifest itself from the physical layer through to the *Logical Link Control and Adaptation Protocol* (L2CAP), along with any other associated protocols that support the profile. Likewise, a profile will prescribe specific interaction from layer to layer and vice versa, as well as peer-to-peer interaction. The Bluetooth system uses a core profile, which all Bluetooth-enabled devices must implement – this includes both Bluetooth systems, that is, BR/EDR and LE. The *Generic Access Profile* (GAP) specifically describes behaviors and procedures for device discovery, connection establishment, security, authentication, and service discovery, and this is something we discuss further in Section 14.6.8, "Generic Access Profile."

In Table 14.3, we provide a comprehensive list of profiles, and in Table 14.4 we list the associated protocols that have already been adopted. In the sections to follow, we provide an overview of some of the applications that have enabled a new generation of Bluetooth products and service/applications. The profiles listed here are specific to the BR/EDR system and are dependent on the GAP, which we discuss later. What's more, in

Table 14.3. A Bluetooth profile specifies behaviors and provides a common user experience when developing applications; we list the most common Bluetooth profiles used today

Acronym	Definition
A2DP	Advanced audio distribution profile.
AVRCP	Audio/video remote control profile.
BIP	Basic imaging profile.
BPP	Basic printing profile.
FTP	File transfer profile.
GAVDP	General audio/video distribution profile.
GOEP	Generic object exchange profile.
GNSS	Global navigation satellite system profile.
HCRP	Hardcopy cable replacement profile.
HDP	Health device profile.
HFP	Hands-free profile.
HID	Human interface device profile.
HSP	Headset profile.
MAP	Message access profile.
OPP	Object push profile.
PAN	Personal area networking profile.
PBAP	Phone book access profile.
SAP	SIM access profile.
SDAP	Service discovery application profile.
SPP	Service port profile.
SYNC	Synchronization profile.
VDP	Video distribution profile.

Table 14.4. The number of protocols that underpin profile functionality and behavior

Acronym	Definition
AVCTP	Audio/video control transport protocol.
AVDTP	Audio/video control distribution protocol.
BNEP	Bluetooth network encapsulation protocol.
OBEX	Object exchange.
MCAP	Multi-channel application protocol.
RFCOMM	RS232 serial emulation.

Table 14.3, we also list several profiles that may underpin other Bluetooth profiles that, in turn, offer ancillary capabilities for that profile, such as the *General Audio/Video Distribution Profile* (GAVDP), *Generic Object Exchange Profile* (GOEP), *Message Access Profile* (MAP), the *Object Push Profile* (OPP), *Service Discovery Application Profile* (SDAP), *Serial Port Profile* (SPP), and the *Video Distribution Profile* (VDP). Likewise, Table 14.4 lists a number of protocols that also underpin profile functionality and behavior, and we touch upon these dependencies in the following sections.

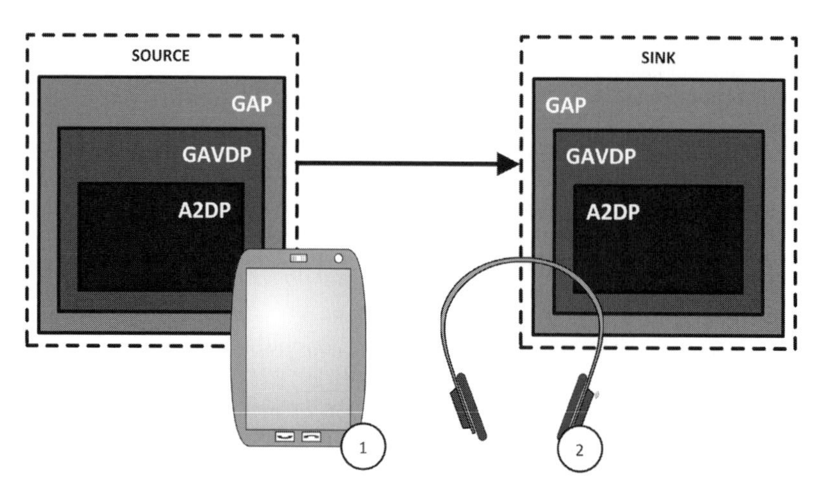

Figure 14.5. The source device (an MP3 player) streams audio content to the sink device (a stereo headset).

14.3.1.1 Advanced Audio Distribution Profile

The *Advanced Audio Distribution Profile* (A2DP) defines two roles for an audio device, namely the *source* and the *sink*. For example, such devices that might use A2DP are stereo headphones and speakers, MP3 players, and microphones. The profile permits the distribution of audio content, mono, or stereo and relies on GAP and GAVDP; A2DP supports the *Sub-band Coding* (SBC) codec and, optionally MPEG-1, 2 Audio, MPEG-2/4, *Advanced Audio Coding* (AAC), and *Adaptive Transform Acoustic Coding* (ATRAC).

In Figure 14.5, we illustrate an MP3 player (**1**) as the source device streaming audio content to the headset (**2**) – the sink device; conversely, in Figure 14.6, we see a

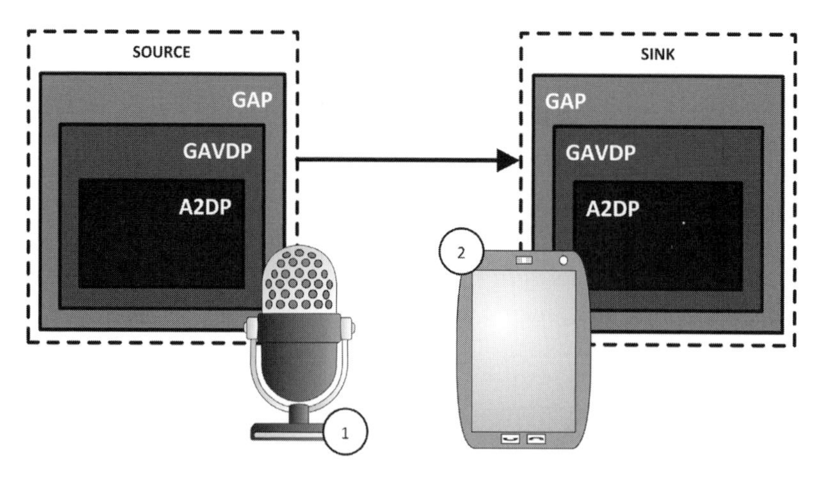

Figure 14.6. In this depiction, we see a source device (a microphone) stream audio content to the sink (a recording device).

microphone (**1**) streaming audio to the recording device (**2**) (the sink device). The illustrations also depict A2DP profile dependencies on both the GAP and GAVDP.

14.3.1.2 Audio/Video Remote Control Profile

The *Audio/Video Remote Control Profile* (AVRCP) defines two roles, a *controller* and a *target*. AVRCP is used to offer an interface to control a number of electronic devices, which may be used in combination with A2DP and the VDP. A controller device is perceived to be the "remote control," which can manipulate menu functions, play/pause/record, and adjust the volume and brightness of your television or other audio functions on your sound system, for example. A controller device may include smartphones, PCs, and headphones, whereas a target device can include an audio/video player/recorder, an amplifier, or headphones. In Figure 14.7, we illustrate a smartphone (**1**) as the controller device controlling the DVD player (**2**) – the target device. The illustration also depicts AVRCP's dependency on GAP.

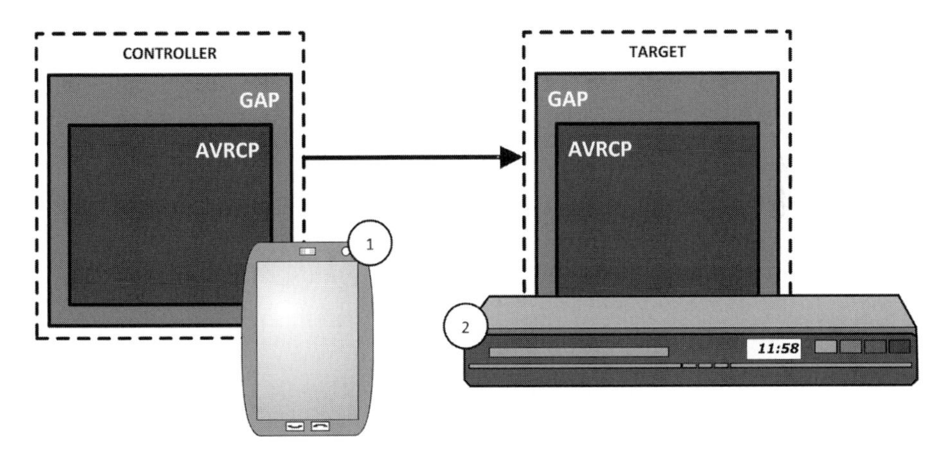

Figure 14.7. The controller device (a smartphone) controls the target device (a DVD player).

14.3.1.3 Basic Imaging Profile

The *Basic Imaging Profile* (BIP) defines two roles, an *initiator* and a *responder*. BIP allows the manipulation (such as the resizing) of images, the control of an imaging device, how such a device can print, and how images can be transferred between devices such as a storage device. BIP also provides the functions that allow devices to harmonize when transmitting, receiving, or browsing images. What's more, your smartphone may browse and retrieve images from a camera, or your PC may automatically retrieve images once your camera is in proximity, for example.

14.3.1.4 Basic Printing Profile

The *Basic Printing Profile* (BPP) defines two roles, a *sender* and a *printer*. BPP provides the ability to send any kind of printable information, such as text, an email, vCard, and images, to a printer. A PC, smartphone, or camera, for example, can connect to a printer device.

14.3.1.5 File Transfer Profile

The *File Transfer Profile* (FTP) defines two roles, namely a *client* and a *server*. FTP allows a client device to browse and retrieve content from the server device.

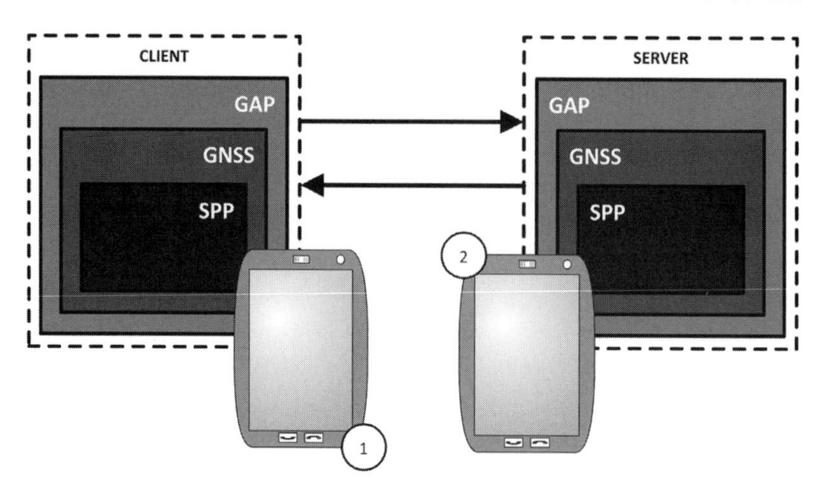

Figure 14.8. The client device (a smartphone) retrieves data from a GPS-capable device.

14.3.1.6 Global Navigation Satellite System Profile

The *Global Navigation Satellite System Profile* (GNSS) defines two roles, a *client* and a *server*. GNSS allows a PC, smartphone or other device that's capable of running navigation software to obtain positioning information from a GNSS server device; typically, these devices will not possess a *Global Positioning System* (GPS) receiver. In Figure 14.8, we illustrate a smartphone (**1**) as the client device receiving data from a GPS-capable device (**2**) – the server device. The illustration also depicts GNSS's dependency on GAP and SPP.

14.3.1.7 Headset Profile

The *Headset Profile* (HSP) defines two roles, namely an *Audio Gateway* (AG) and a *Headset* (HS). HSP allows a PC, smartphone or other device to communicate with a headset device. In Figure 14.9, we illustrate a notebook (**1**) and a smartphone (**2**) as the headset device (**3**) receives audio content from the AG. The illustration also depicts HSP's dependency on GAP and SPP.

14.3.1.8 Hands-free Profile

The *Hands-free Profile* (HFP) defines two roles, an AG and a *Hands-free Unit* (HF). Primarily, HFP allows a smartphone or other device to communicate with a hands-free unit, such as a headset or a hands-free device which has been installed in a vehicle. In the context of HFP, a headset is a hands-free unit when used within a vehicle, for example. The hands-free profile suggests that, with HFP, additional functionality is available in the HFP context but, as with HSP, the headset remains exactly the same in terms of use and operation. However, it's important to make this *conceptual* distinction, as it's

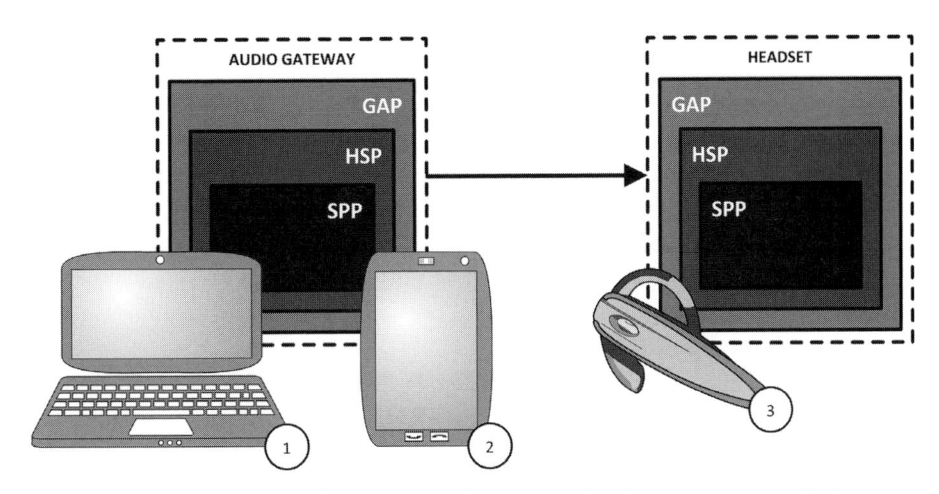

Figure 14.9. The audio gateway device (a smartphone or notebook) communicates with the headset.

often misunderstood – the headset is deemed the hands-free unit when used within a vehicle, as it does offer the driver hands-free capability. In Figure 14.10, we illustrate a headset (**2**) and a hands-free unit within the vehicle (**3**) that behaves like a headset device that, in turn, receives audio content from the AG. The illustration also depicts HFP's dependency on GAP and SPP.

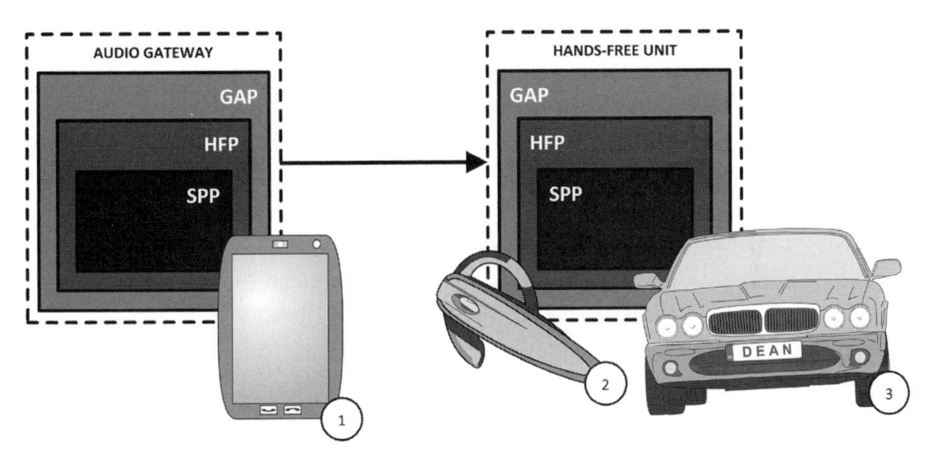

Figure 14.10. The audio gateway device (a smartphone) communicates with the headset or an inbuilt hands-free unit within the car.

14.3.1.9 Human Interface Device Profile

The *Human Interface Device Profile* (HID) defines two roles, a *Human Interface Device* (HID) and a *host*. HID allows a PC, tablet, smartphone, or other device to communicate with a mouse, keyboard, gaming device, and so on.

14.3.1.10 Personal Area Networking Profile

The *Personal Area Networking* (PAN) *profile* defines two roles, namely a *Network Access Point* (NAP) and *NAP service*, and a *Group ad hoc Network* (GN) and *GN service*. The PAN profile prescribes how two or more Bluetooth-enabled devices can form an ad hoc network with an optional provision to reach an extended network, such as the Internet. Its future is somewhat tenuous since mobile devices inherently have the independent ability to access the wider area network when they adopt similar characteristics to a Wi-Fi access point.

14.3.1.11 Phone Book Access Profile

The *Phone Book Access Profile* (PBAP) defines two roles, *Phone book Server Equipment* (PSE) and *Phone book Client Equipment* (PCE). The PBAP is used in conjunction with the hands-free and SIM access scenarios (see Sections 14.3.1.8, "Hands-free Profile," and 14.3.1.12, "SIM Access Profile," respectively), where the profile defines the protocols and procedures required to retrieve phone book *objects*. For example, in a hands-free context, a smartphone device would behave as PSE and the integrated hands-free car kit would act as the PCE.

14.3.1.12 SIM Access Profile

The *SIM Access Profile* (SAP) defines two roles, a *server* and a *client*. The SAP details how messages and procedures should be used to access a subscription module, such as the *Subscriber Identity Module* (SIM) card typically used in a cellular or smartphone handset.

14.3.1.13 Video Distribution Profile

The *Video Distribution Profile* (VDP) defines two roles, namely a *source* and a *sink*. The VDP defines how a Bluetooth-enabled device distributes video content to other Bluetooth-enabled devices by defining a number of protocols and procedures. It prescribes how a Bluetooth-enabled camera (the source device) streams video content to a monitor (the sink device), for example. In Figure 14.11, we illustrate a camera (**1**) as the video source device, which is transmitting video content to the monitor (**2**), the sink device. The illustration also depicts VDP's dependency on GAP, GAVDP, and A2DP.

14.4 Bluetooth wireless technology and its Competitors

Originally, Bluetooth wireless technology and Wi-Fi were touted as competing technologies; a decade or so later, both technologies have individually shaped their respective applications and market sectors. However, many industry pundits perceive the recent introduction of Wi-Fi Direct as a direct threat to Bluetooth. We discussed Wi-Fi Direct earlier in Chapter 13, "The 802.11 Generation and Wi-Fi." Likewise, some have envisaged *Near Field Communications* (NFC) competing with Bluetooth technology, but NFC has a very niche market, something we discussed earlier in Chapter 12, "Just Touch with NFC." Nevertheless, Bluetooth has a very unique and diversified application space

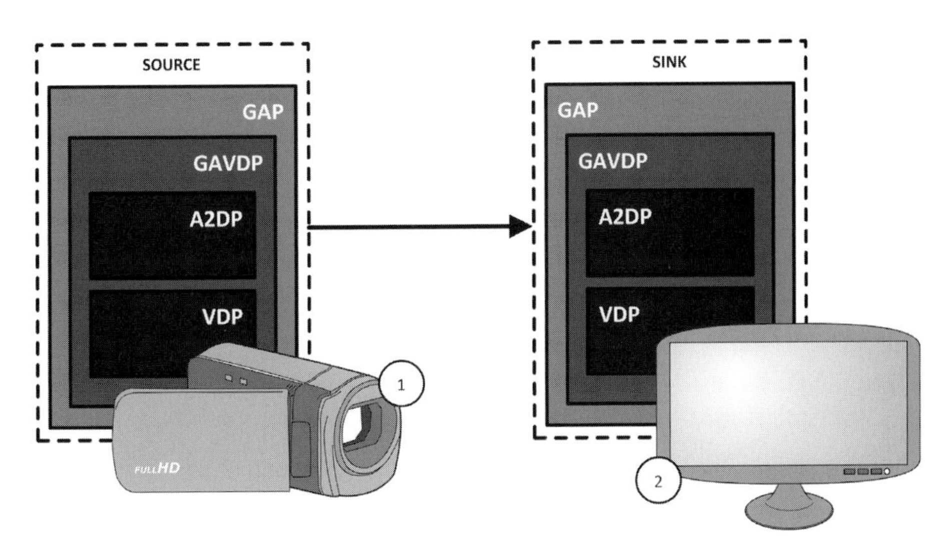

Figure 14.11. The source device (a digital camera) transmits video content over Bluetooth to the monitor (the sink device).

and remains somewhat unfazed by other technologies encroaching on its applications' domains. Many have envisaged Bluetooth low energy serving similar applications to NFC but, at the time of writing, both technologies remain firmly seated apart, posturing their respective applications.

14.5 Networking Topology

Bluetooth wireless technology supports several topologies, namely *point-to-point* (or *peer-to-peer*), *point-to-multipoint*, and *star* topologies; these are illustrated in Figure 14.12. Just as for BLE, there are two types of Bluetooth classic and high speed roles, namely a *master* and a *slave*, which we discuss later in Section 14.6.2, "The Link Controller."

14.5.1 Piconets and Scatternets

The fundamental topology used within a Bluetooth ecosystem is called a *piconet*, whereby a *master* device can communicate and interact with a maximum of seven *slaves* – typically, in a BR/EDR system, a master/slave would share a common physical channel. In Figure 14.13, we illustrate a master device (**1**) interacting with its two slaves (**2** and **3**). A *scatternet* is a topology where two or more piconets overlap and interconnect; see Figure 14.14, in which we show two overlapping piconets (**3**) with master devices (**1** and **4**) and their respective slave devices (**2**, **3**, **5**, **6**, and **7**). Unlike the LE system, a BR/EDR system supports a scatternet topology, along with the master/slave role switch. A device that initiates a connection is deemed the master, and once a piconet has been established master/slave roles can be determined (if required).

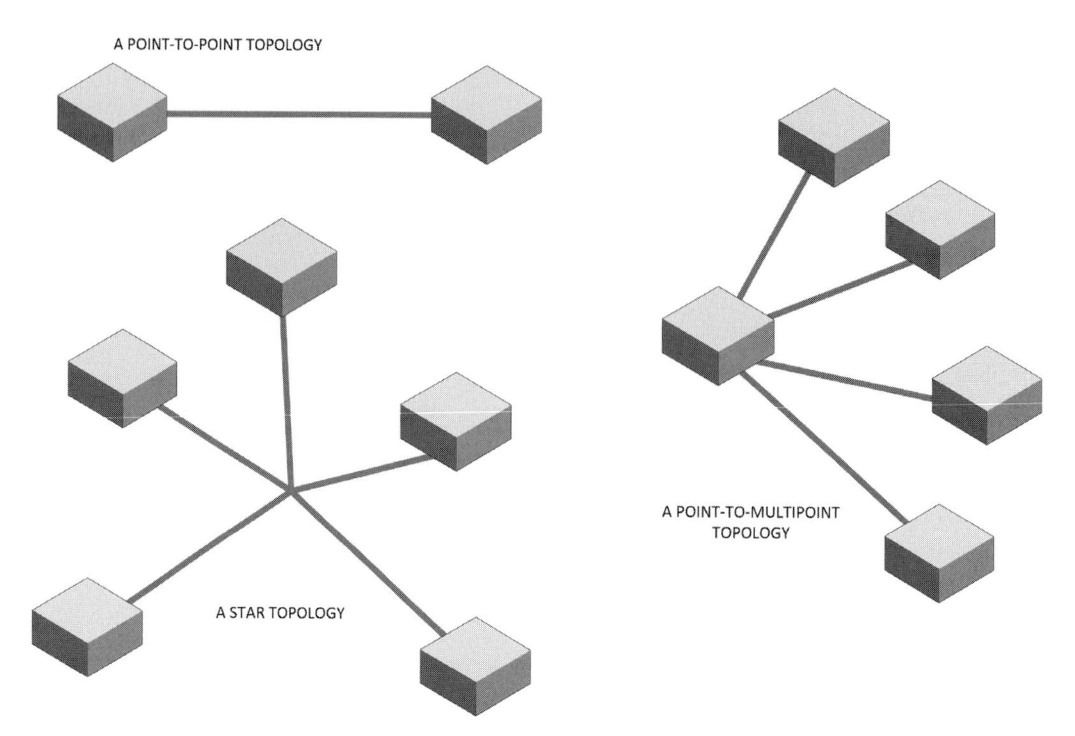

Figure 14.12. Bluetooth wireless technology supports point-to-point (or peer-to-peer), point-to-multipoint, and star topologies.

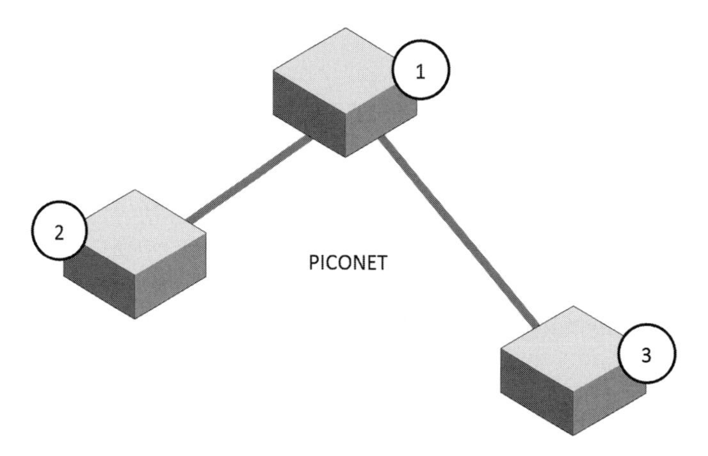

Figure 14.13. The piconet, in which a master device interconnects with a maximum of seven slaves, is the basic topology used within a Bluetooth ecosystem.

14.6 The Bluetooth Architecture

In Section 14.1, "Overview," we touched upon the fact that Bluetooth wireless technology now comprises two systems, namely LE and BR/EDR. In this chapter, we discuss

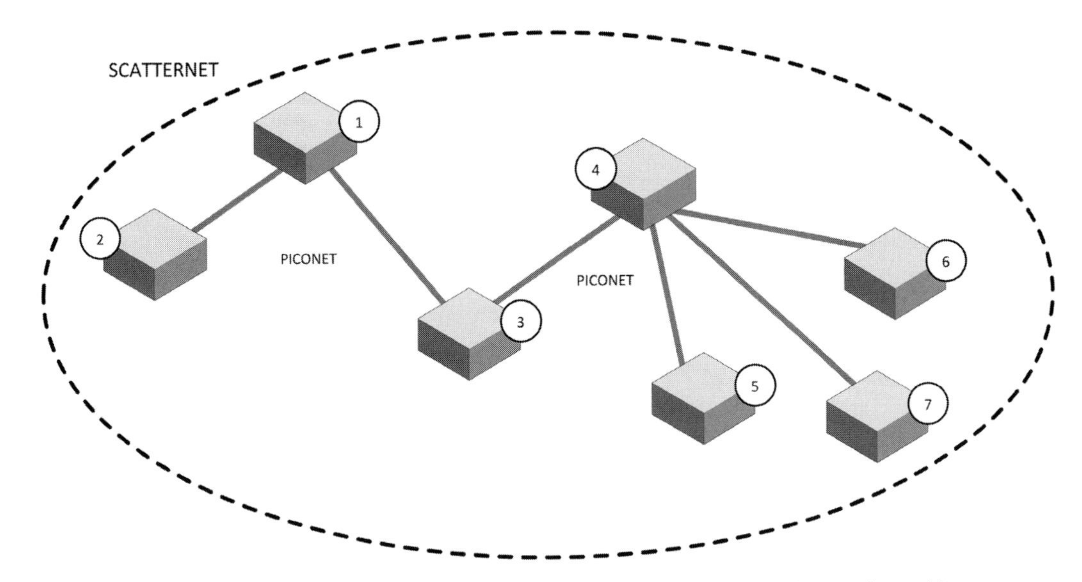

Figure 14.14. A scatternet topology comprises two or more piconets that overlap and interconnect.

the BR/EDR system, but we may often refer to the LE system just to offer an explanation on the differences between the systems – you may also want to refer back to Chapter 7. The BR/EDR system has been largely retained in terms of its architecture primarily to support the existing wealth of legacy Bluetooth products, whereas the LE system has been architected from the ground up. In Figure 14.15, we map the *Open Systems Interconnection* (OSI) model against the Bluetooth software stack. What's more, a Bluetooth system will include a *host*, an entity that sits above the *Host Controller Interface* (HCI), and zero or more *host controllers* – entities that sit below the HCI. The high speed operation is managed through an AMP secondary controller(s). In short, the BR/EDR system radio manages discover and connection establishment, along with its associated maintenance. Once a connection has been established between two devices, the AMP manager can then commence discovery of other AMPs that are available on other devices. We will return to the AMP architecture later, in Section 14.6.4, "The AMP Architecture."

In Figure 14.16, we illustrate the host, host controller, and AMP controller components of the BR/EDR system, where the HCI permits communication between the host and controller using a serial interface. We provide a review of the HCI transport layers later, in Section 14.6.5, "The Host Controller Interface." In the sections that follow, we discuss the BR/EDR software stack from the bottom up, and in Figure 14.15 our BR/EDR stack layers have been numerically labeled to aid in identification of the software building blocks that make up the Bluetooth software stack.

14.6.1 The Physical Layer

We start with the *Physical* (PHY) layer (**1**) that has the responsibility of managing the BR/EDR transceiver and operating the radio medium in the unlicensed 2.4 GHz

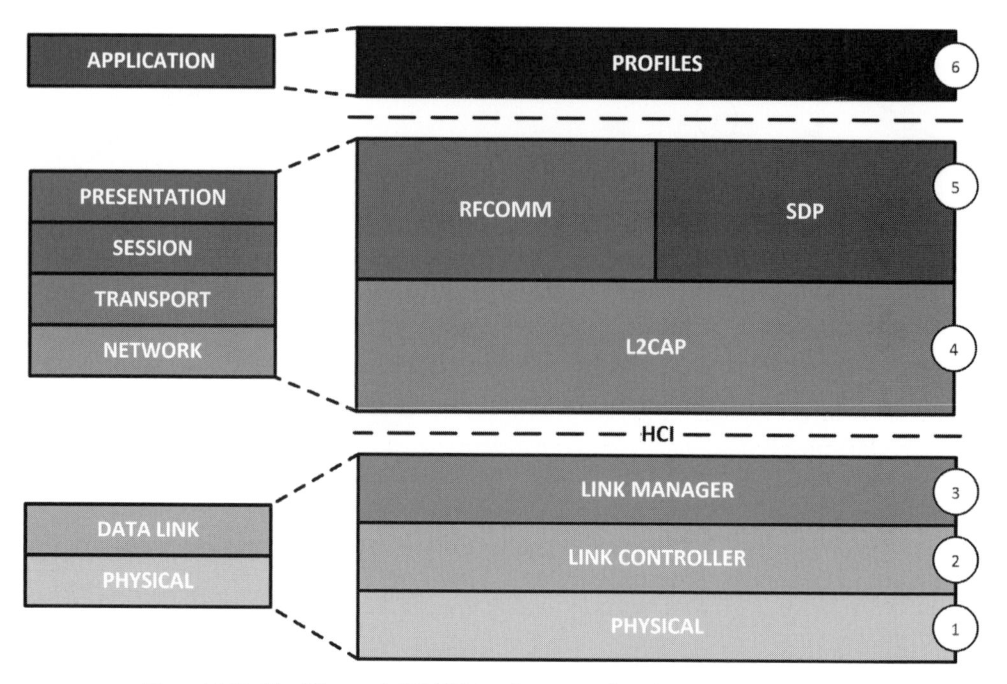

Figure 14.15. The Bluetooth BR/EDR software stack architecture mapped against the OSI model.

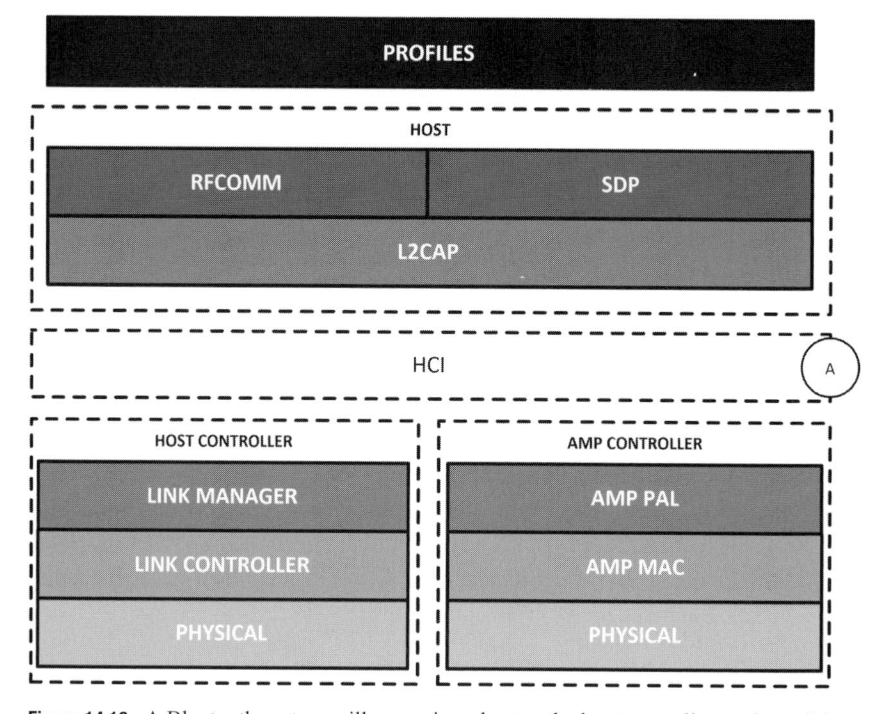

Figure 14.16. A Bluetooth system will comprise a host and a host controller, and possibly one or more AMP controllers, which uses the HCI to communicate.

(2400–2483.5 MHz) *Industrial, Scientific and Medical* (ISM) frequency band, which uses 79 RF channels spaced at 1 MHz. The BR/EDR system utilizes two modulation modes; that is, the BR operation uses a mandatory shaped, binary FM modulation scheme, and the EDR operation utilizes a *Phase-shift Keying* (PSK) scheme, which is based on two variants. The *symbol* or *modulation rate* is defined as 1 megasymbol/s, where the gross air data bit-rate provided for BR operation is 1 Mbit/s, whilst the EDR operation, based on its two PSK variants, offers a data rate of 2 Mbit/s using $\pi/4$-*Quadrature Differential* PSK (QDPSK) and operates at 3 Mbit/s for 8DPSK. Both BR/EDR modulation modes require a *Time Division Duplex* (TDD) scheme to undertake full duplex transmission.

Table 14.5. Bluetooth devices are classified into three power classes

Class	Minimum output power (mW)	Nominal output power (mW)	Maximum output power (mW)
1	1 (0 dBm)	n/a	100 (20 dBm)
2	0.25 (−6 dBm)	1 (0 dBm)	2.5 (4 dBm)
3	n/a	n/a	1 (0 dBm)

14.6.1.1 The Bluetooth Transceiver

Three *power classes* are used to classify Bluetooth devices, and this classification is based on modulation mode with the highest output power. We provide in Table 14.5 the three power classes detailing minimum, nominal, and maximum output power. A class 1 device will support power control requests, whereas power control support for class 2 and 3 devices is optional.

14.6.1.2 Frequency Hopping

Like the LE system, BR/EDR uses a *frequency hopping* transceiver to overcome *interference* and *fading*. The frequency hop is enabled during an active connection, during which the frequency hopping pattern is calculated based on certain criteria within the Bluetooth address and the clock of the master device. The *Frequency Hopping Spread Spectrum* (FHSS) technique, which spreads the RF power across the spectrum, aids in the reduction of interference and divides the spectrum into 79 1 MHz (wide) data channels. What's more, an *Adaptive Frequency Hopping* (AFH) scheme further improves co-existence with other similarly ISM-enabled products.

14.6.2 The Link Controller

The *Link Controller* (LC) in Figure 14.15 (**2**), or, as it's often referred to, the *baseband* layer, manages the physical connection and links between two or more Bluetooth devices. In this section, we discuss most of the baseband aspects, including the Bluetooth clock; physical channels and links; logical transports and links; device addressing; and packet formats and types. In Section 14.6.2.8, "LC Operation," we summarize and look at the overall operational overview of the link controller.

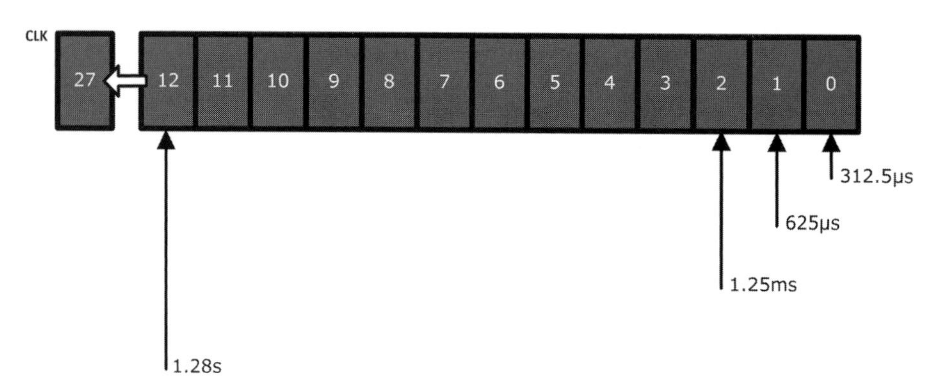

Figure 14.17. The Bluetooth clock represented as timer bits (CLK_n).

14.6.2.1 The Bluetooth Clock

A Bluetooth device has its own native clock, which is derived from a free-running system clock. The clock is used to enable synchronization with other devices and bears no correlation to the actual time of day, although the clock generally has a cycle of about a day. Put simply, a 28-bit counter is required if the clock has been implemented as a counter and wraps around $2^{28} - 1$. Each "tick" or unit is 312.5 μs, which represents half a *time slot* – we'll come back to slots later in this section. The clock is used to ascertain critical periods, which, in turn, trigger events within the device. A Bluetooth device has four critical periods, namely 312.5 μs, 625 μs, 1.25 ms, and 1.28 s; these can be represented as "timer bits," that is, CLK_0, CLK_1, CLK_2, and CLK_{12}, as illustrated in Figure 14.17.

Table 14.6. The Bluetooth clock has three different appearances

Guise	Description
CLKN	Native clock.
CLKE	Estimated clock.
CLK	Master clock.

A Bluetooth device can be in various modes and/or states during operation and, as such, the clock can appear in three guises, as listed in Table 14.6. The master device never adjusts the native clock when a connection has been established within a piconet – CLKN denotes the native clocks.

14.6.2.2 Device Addressing

Like the LE system, the BR/EDR system uses a 48-bit device address, which uniquely identifies a device within a piconet – the *Bluetooth Device Address* (BD_ADDR). The public address must conform with the "48-bit Universal LAN MAC addresses" as specified by the *Institute of Electrical and Electronic Engineers* (IEEE) 802–2001 standard, which will use a valid *Organizationally Unique Identifier* (OUI) sourced from the IEEE

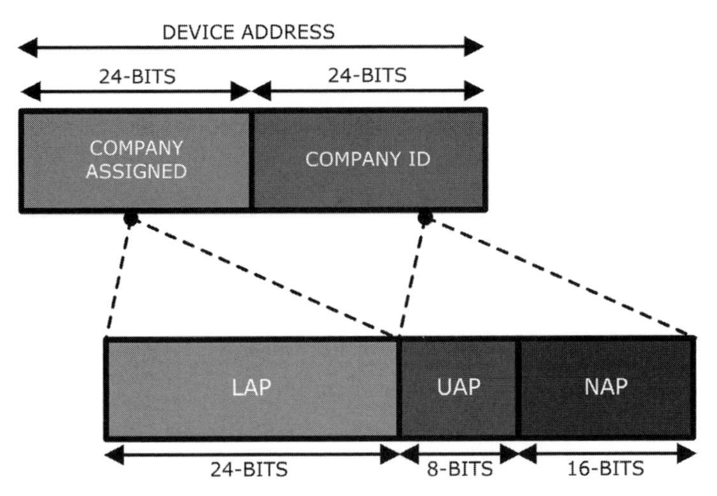

Figure 14.18. The format of the device address, also showing the LAP, UAP, and NAP formats.

Registration Authority. In Figure 14.18, we illustrate the two fields that make up the device address field.

You will also note from Figure 14.18 that the structure has been divided into three further parts, namely the *Lower Address Part* (LAP), *Upper Address Part* (UAP), and *Non-significant Address Part* (NAP). Sixty-four contiguous LAPs (`0x9E8B00` to `0x9E8B3F`) are reserved for inquiry operations, something we discuss in the following section. However, one particular LAP address (`0x9E8B33`) is reserved for general inquiry, leaving the other 63 LAPs for dedicated inquiries. The UAP value is derived from the *Default Check Initialization* (DCI) when a reserved LAP address is used – the DCI value in this instance is `0x00`.

Table 14.7. There are three different access codes in use

Code	Description
DAC	Device Access Code.
CAC	Channel Access Code.
IAC	Inquiry Access Code.

14.6.2.3 Access Codes

The physical channel, which we discuss in Section 14.6.2.4, "Physical Channels and Links," begins with an *access code*, and we list and describe these codes in Table 14.7. As already discussed in Section 14.6.2.2, "Device Addressing," an access code is derived from the LAP or an inquiry address. A paged device's BD_ADDR is used to determine the DAC during *page*, *page scan*, and *page response*, while the CAC, which is derived from the LAP of the master's BD_ADDR, is used in the *connection* state. Lastly, the IAC is used in the *inquiry* state.

Furthermore, one *General* IAC (GIAC) is used for general inquiry, whereas there are 63 *Dedicated* IACs (DIACs) available for dedicated inquiries. In short, an access code is used by the receiving device to notify it of an incoming packet, and it also aids in timing synchronization.

Table 14.8. There are four physical channels

Physical channel type	Description
Basic piconet	The basic piconet physical channel is divided into 625 μs time slots and is defined by the master of the piconet, where the device manages overall traffic using a polling scheme.
Adapted piconet	This channel type may be used for a connected device that has AFH enabled.
Page scan	A paging device is assumed to be the master; it may become the master in the connection state whilst a device is a slave during page scanning. The page scan physical channel uses a slower hopping pattern when compared with the basic piconet channel.
Inquiry scan	An inquiry device is assumed to be the master; it may become the master in the connection state whilst a device is a slave during inquiry scanning. The inquiry scan physical channel uses a slower hopping pattern when compared with the basic piconet channel.

14.6.2.4 Physical Channels and Links

In Table 14.8, we list and describe the four physical channels that are available. Each physical channel is subdivided into time slots and is constructed with an access code, as we have already discussed, a packet header, and a packet (slot) timing for the transmission. A physical channel is further characterized by a pseudo-random RF frequency channel hopping sequence, which is determined by the UAP and LAP of the BD_ADDR, where the exact phase in the sequence is derived from the Bluetooth clock. We will come back to the hop selection sequence shortly, and in Section 14.6.2.6, "Packet Format," we discuss the format of the various physical channels. A physical link is representative of a *baseband* connection, which is associated with a physical channel. As such, these physical links contain properties that apply to logical transports – we discuss logical transports and links in Section 14.6.2.5, "Logical Transports and Links." These properties include power control, link supervision, encryption, channel quality, and multi-slot packet control.

There are six types of hopping sequence: one is used for an adapted series of hop locations specifically used for AFH, and the remaining five are used within the basic hop system. We discuss the available sequences in Table 14.9.

14.6.2.5 Logical Transports and Links

A connection established between a master and slave device may use different types of logical transport. There are five logical transports available, namely *Synchronous Connection-oriented* (SCO), *Extended* SCO (eSCO), *Asynchronous Connectionless* (ACL), *Active Slave Broadcast* (ASB), and *Parked Slave Broadcast* (PSB). Both SCO

Table 14.9. There are six hopping sequences

Hopping sequence	Description
Page	With a period length of 32, this sequence provides 32 wake-up frequencies, which are divided evenly over the 79 MHz.
Page response	The page response sequence offers 32 response frequencies; this bears a direct correlation to the current page hop sequence.
Inquiry	With a period length of 32, this sequence provides 32 wake-up frequencies, which are divided evenly over the 79 MHz.
Inquiry response	The inquiry response sequence offers 32 response frequencies; this bears a direct correlation to the current inquiry hop sequence.
Basic channel	The basic channel provides a long period length, which, in turn, distributes the hop sequence evenly over the 79 MHz.
Adapted channel	The basic channel hopping sequence can be used instead of the basic channel sequence.

and eSCO are time-sensitive (primarily used for audio or video) and represent a point-to-point connection between a master and a slave. The master device in this instance will reserve slots at regular intervals, whereas the eSCO may also use a retransmission window to ensure quality. The ACL logical transport represents a point-to-point connection, but conversely it does not reserve time slots – this type of transport is used for generic data. Finally, the ASB is used by a master device to communicate with active slaves, and the PSB logical transport is used by the master to communicate with all parked slaves.

Furthermore, five logical links are defined, *Link Control* (LC), ACL-*Control* (ACL-C), *User Asynchronous/Isochronous* (ACL-U), *User Synchronous* (SCO-S), and *User Extended Synchronous* (eSCO-S). LC is used at the link control (baseband) level, and transports data, such as flow control and payload characterization, which is mapped on to the packet header. ACL-C is used at the *Link Manager* (LM) level (see Section 14.6.3, "The Link Manager"), and transports control information with peer-to-peer LM entities. ACL-C will only use *Data – Medium-rate*[2] (DM1) or *Data Voice* (DV) type packets. We cover packet formats and types in Section 14.6.2.6, "Packet Format," and Section 14.6.2.7, "Packet Types." ACL-U carries L2CAP user data, which we discuss later, in Section 14.6.6, "L2CAP." Both the SCO-S and eSCO-S logical links transport synchronous data.

14.6.2.6 Packet Format

The BR/EDR system provides two generic packet formats, which we share in Figure 14.19 and Figure 14.20. The BR packet comprises three fields, as shown: the access code, a header, and the payload. In Figure 14.20, we show the general EDR packet format; it consists of the access code and header, which is identical to the BR packet, along with guard, sync, payload, and trailer fields.

As we have touched upon previously, every packet starts with an access code; its size, that is, 68- or 72-bits, is dependent on whether a packet header is included. If a packet

[2] The DM1 packet type is used to transport control messages, but may also carry regular data.

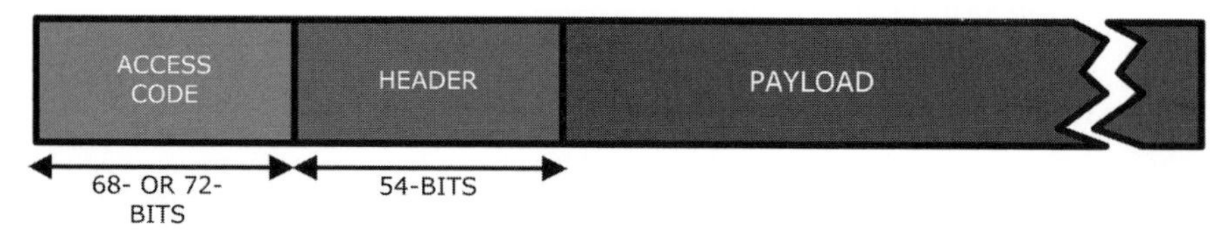

Figure 14.19. The general basic rate packet format.

Figure 14.20. The general EDR packet format.

header is not included, the access code is 68-bits in length, is referred to as a shortened access code, and is used for paging, inquiry, and park. The access code is broken down further into a *preamble*, *sync word*, and an optional *trailer*, as we illustrate in Figure 14.21.

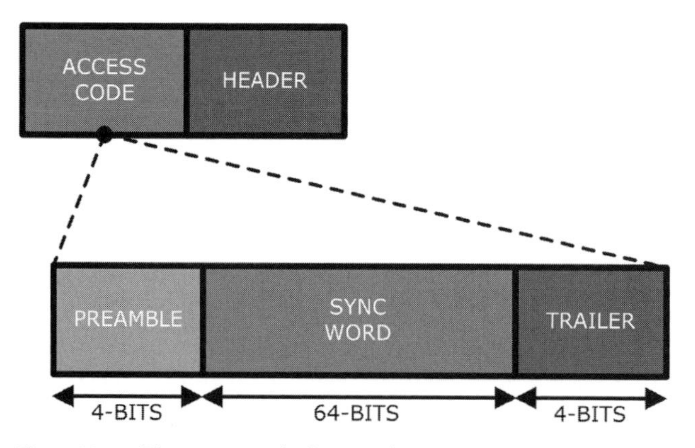

Figure 14.21. The access code format; the access code and header are common to both general packet formats.

We discussed earlier, in Section 14.6.2.3, "Access Codes," the various access codes that are available. The CAC is 72-bits in length and includes the trailer field. The trailer is only used with the DAC, GIAC, and DIAC when these are transported using *Frequency Hopping Synchronization* (FHS) packet types during paging and inquiry responses.

The header field contains LC-specific data and comprises six fields, *Logical Transport Address* (LT_ADDR), *type*, *flow*, a basic *Acknowledgement/Repeat Request* (ARQN), *Sequential Numbering Scheme* (SEQN), and *Header Error Check* (HEC), as shown in Figure 14.22. In Table 14.10 we look at the specific fields that make up the header, along with the size of each field in bits.

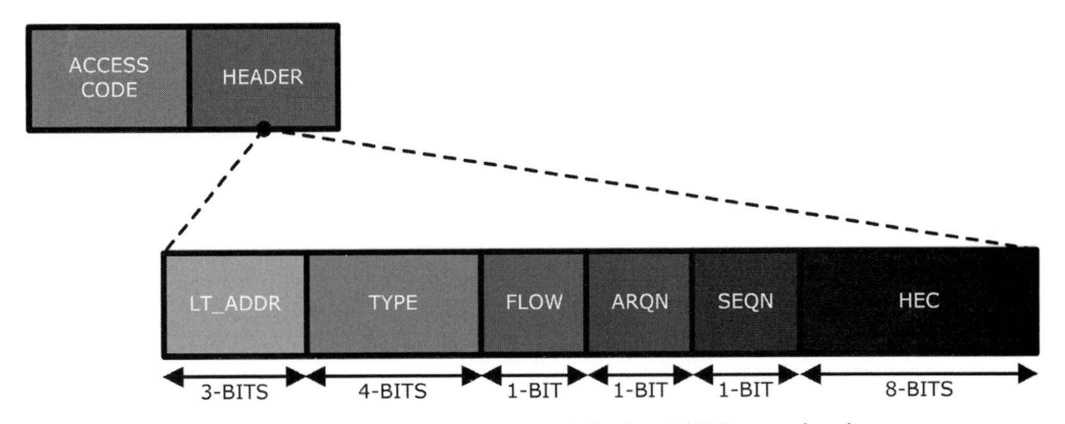

Figure 14.22. The format of the header field within the BR/EDR general packet.

Table 14.10. The fields that make up the BR/EDR general packet header

Field	Description	Size
LT_ADDR	The LT_ADDR is 3-bits in length and contains the address for the destination slave packet.	3
Type	The type field is 4-bits in length and specifies the packet type currently in use (see Section 14.6.2.7, "Packet Types").	4
Flow	The flow field is 1-bit in length and is used for flow control.	1
ARQN	The ARQN field is 1-bit in length and is used to inform the source device of a successful transfer of data.	1
SEQN	The SEQN field is 1-bit in length and offers a sequential numbering scheme for the data packet stream.	1
HEC	This 8-bit field provides an error check mechanism to ensure the integrity of the header.	8

14.6.2.7 Packet Types

The type field within the header specifies the type of packet in use over the logical transport and is segmented into four categories: the first is control (occupying a single time slot); the second segment is classified for general packets occupying a single time slot; the third segment is for packets occupying three time slots; and lastly, the fourth segment is used for packets using five time slots. Moreover, there are five common packet types, which we summarize in Table 14.11, and we describe other packet types in Table 14.12.

14.6.2.8 LC Operation

In Figure 14.23, we illustrate a state diagram portraying the different states used within the link controller. As you can see, there are three significant states, namely *standby*, *connection*, and *park* – we describe these major states in Table 14.13. What's more, there are a further seven sub-states, which include *page*, *page scan*, *inquiry*, *inquiry scan*, *master*, *slave*, and *inquiry* responses. The number of sub-states prescribes the ability for

Table 14.11. The five common packet types

Packet type	Description
ID	The *Identifier* (ID) packet type includes either the DAC or IAC.
NULL	The NULL packet type does not include a payload and is not acknowledged.
POLL	The POLL packet type does not include a payload and does not require acknowledgement.
FHS	The FHS packet is a control packet and contains real-time clock information.
DM1	DM1 occupies the first segmentation category, as it supports control messages over the logical transport.

Table 14.12. Other packet types

Packet types	Description
SCO	
HV1 HV2 HV3	*High-quality Voice* (HV) packet types, HV1, HV2, and HV3, are used on the SCO logical transport and may be used for 64 kbit/s speech.
DV	The DV packet type is combined data and voice packet.
eSCO	
EV3 EV4 EV5	EV packet types, EV3, EV4, and EV5, are used on the eSCO logical transport and may be used for 64 kbit/s speech transmission.
2-EV3	Similar to EV3, but uses $\pi/4$-DQPSK modulation instead.
2-EV5	Similar to EV5, but uses $\pi/4$-DQPSK modulation instead.
3-EV3	Similar to EV3, but uses 8DPSK modulation instead.
3-EV5	Similar to EV5, but uses 8DPSK modulation instead.
ACL	There are seven packets included in the BR operation, DM1, DH1, DM3, DH3, DM5, DH5, and AUX1; of course, six packets are additionally defined for EDR operation, 2-DH1, 3DH-1, 2-DH3, 3-DH3, and 3-DH5, using $\pi/4$-DQPSK or 8DPSK modulation schemes.

Table 14.13. LC has three major states

State	Description
Standby	This default state may place the device into low power mode, in which the native clock is running. A controller may leave this state to scan for page or inquiry messages.
Connection	A device in this state represents an established connection, such that a device can freely exchange packets.
Park	A slave device will enter this state if it no longer needs to participate within a piconet.

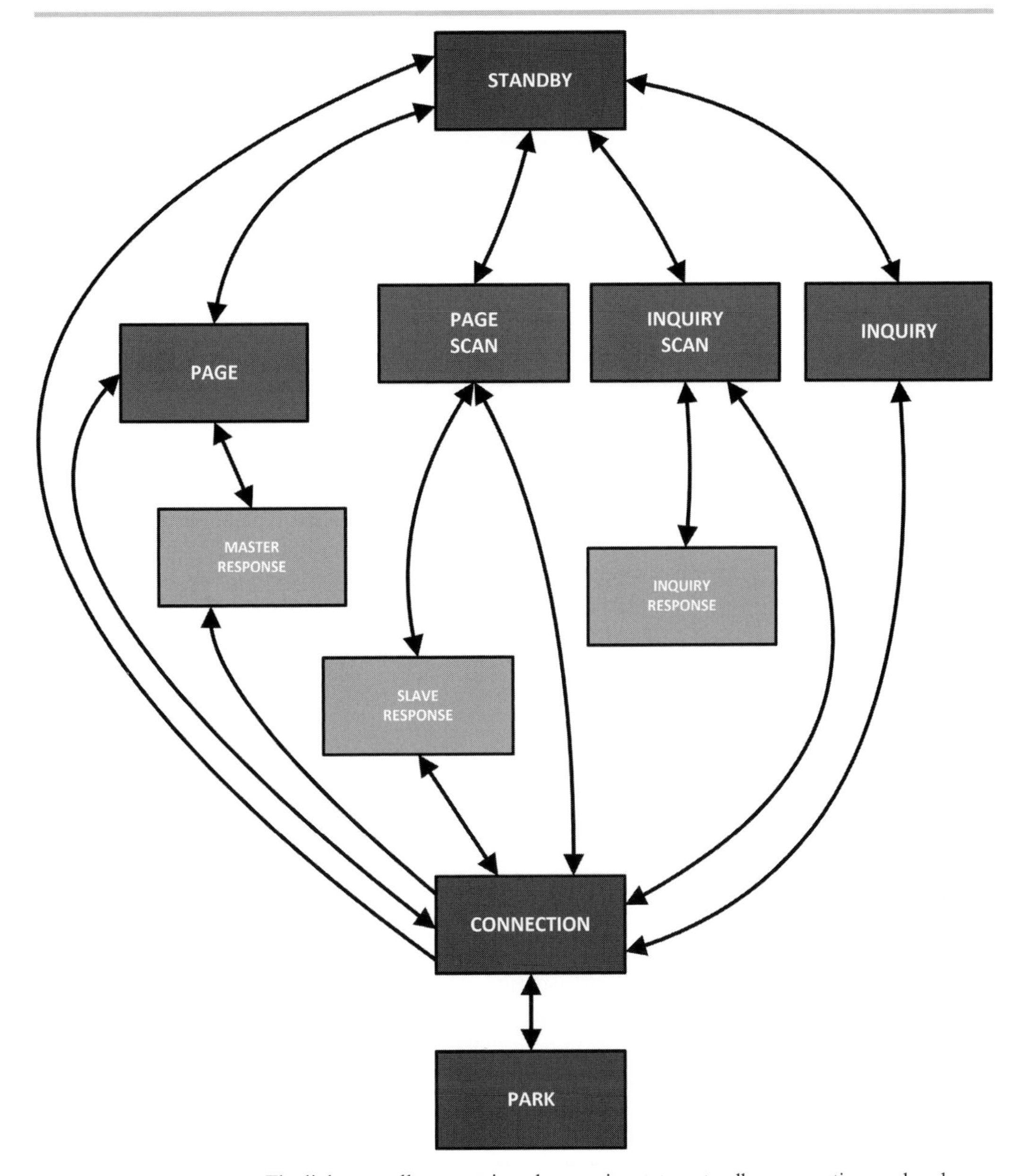

Figure 14.23. The link controller comprises three major states: standby, connection, and park.

a device to establish a connection and to perform device discovery. The LM is used to move from one state to the next using commands within the link controller.

14.6.3 The Link Manager

The *Link Manager* layer in Figure 14.15 (**3**) has the overall responsibility for the management and negotiation of the connection between two Bluetooth-enabled devices. The LM, using the underlying services provided by the LC, will set up and manage physical and logical links, and will control the logical transports. The *Link Manager*

Figure 14.24. The LMP PDU using a 7-bit OpCode, along with the TID that identifies the originating party, that is, master or slave.

Protocol (LMP) is a transactional scheme used to permit communication between peer LM entities. These devices are connected through an ACL logical transport, whereby the LM protocol uses a series of messages that are transferred over the ACL-C logical link (over the default ACL logical transport). The ACL-C has a higher priority than ACL-U traffic, which transports L2CAP and user data. We say more about how L2CAP ACL-U data is transported and formatted in Section 14.6.6, "L2CAP."

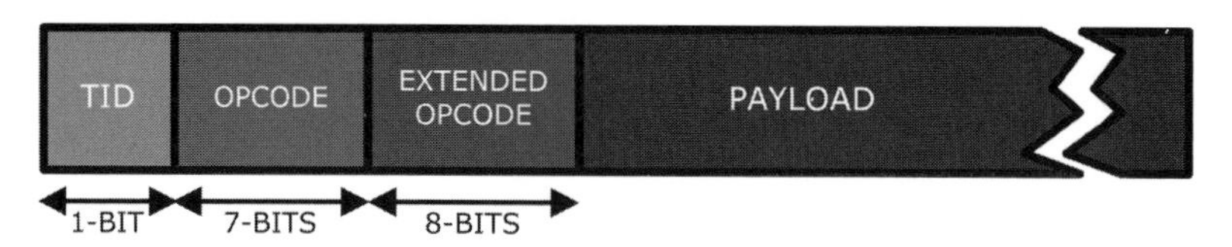

Figure 14.25. The LMP PDU using a 15-bit OpCode (Extended), along with the TID that identifies the originating party, that is, master or slave.

14.6.3.1 LMP Packet Format

LMP uses a series of messages, which are packaged in a *Protocol Data Unit* (PDU), the formats of which are illustrated in Figure 14.24 and Figure 14.25. A LMP PDU will either be assigned a 7-bit or 15-bit *Operation Code* (OpCode), which is used to identify uniquely the message being transported. The *extended OpCode*, as shown in Figure 14.25, is used if the initial 7-bit OpCode uses a special escape value, in turn ensuring that the message is identified. In both PDUs, a *Transaction ID* (TID) is set to 0 if the master originally initiated the transaction; conversely, if the bit is set to 1, then the slave device initiated the transaction sequence.

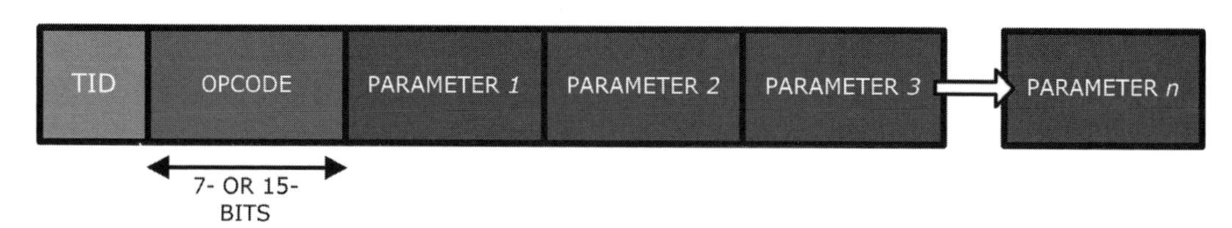

Figure 14.26. The LMP PDU carries a number of parameters that are associated with the message and occupy the first byte following the OpCode.

The PDU will contain one or more *parameters* in the payload field that are carried in either the 7- or 15-bit OpCode, which commences in the first byte that follows the OpCode, as we illustrate in Figure 14.26. The length of the payload will be subject to the type of message being transported and the associated number of parameters used. LMP messages are transported using the DM1 packet, although if an HV1 SCO connection is currently in use and the overall length of the payload does not exceed nine bytes, then DV packets may be used instead. We discussed the various packet types earlier, in Section 14.6.2.7, "Packet Types."

The LMP *response timeout* imposes a certain window during which transactions of PDUs carried over the LC should timeout. The LMP response timeout is initialized for each PDU that requires a reply. However, if the response timeout is exceeded or if a link loss is detected, the device awaiting a response will concede, which leads to the procedure terminating unsuccessfully.

What's more, if the LM receives an unrecognized OpCode, it will respond with either `LMP_not_accepted` or `LMP_not_accepted_ext` accompanied by the error code *unknown LMP PDU*. Likewise, if the LM receives a PDU with invalid parameter(s), it will respond with either `LMP_not_accepted` or `LMP_not_accepted_ext` accompanied by the error code *invalid LMP parameters*. The LM may also receive a PDU that's not allowed and, as such, will respond with either `LMP_not_accepted` or `LMP_not_accepted_ext` accompanied by the error code *PDU not allowed*. In instances where the PDU doesn't expect a response, the PDU will be ignored. However, in general, errors may occur over a link and, as such, it may be prudent to cease the current connection. The implementer may wish to monitor erroneous activity within the LM, so as to determine an acceptable threshold.

14.6.4 The AMP Architecture

The AMP provides zero or more secondary controller functions to the Bluetooth BR/EDR system and it primarily serves the *high speed* capabilities of Bluetooth wireless technology, something we've already touched upon. Initially, it was expected that *Ultra-wideband* (UWB) would deliver[3] the high speed variant of Bluetooth wireless. However, the collaboration between the WiMedia Alliance (wimedia.org) and the Bluetooth SIG was hindered by a lack of agreement surrounding issues of assignment of intellectual property. As such, Wi-Fi, more specifically 802.11, was chosen to deliver high speed Bluetooth whilst the issues surrounding intellectual property were resolved. Nonetheless, even at the time of writing, no such resolution has been finalized. The BR/EDR system first establishes a connection between devices using its primary radio, wherein it can then discover other AMP controllers that might be available on the other device. The architecture provides a *Protocol Adaptation Layer* (PAL) for each AMP controller, which adequately maps the underlying MAC and PHY to the upper layers of the Bluetooth protocol stack.

14.6.4.1 The Protocol Adaptation Layer

In Figure 14.27, we provide an overview of the architecture of the 802.11 PAL, although the AMP PAL has been architected such that any wireless technology, like UWB for

[3] Gratton, D. A., "The Only Alternative: Bluetooth and Ultra-wideband," 2009.

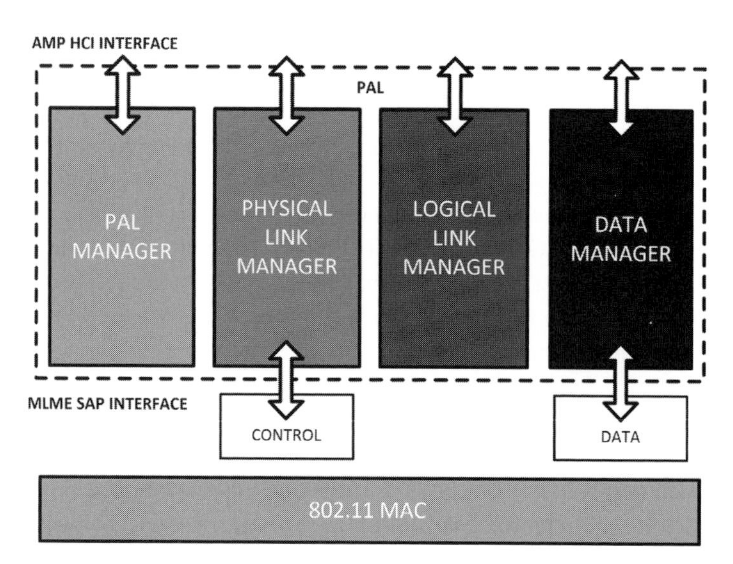

Figure 14.27. An architectural overview of the management components of the PAL.

example, can be used to provide high speed connectivity. The architecture shown in Figure 14.27 portrays an upper- and lower-edge interface, where the upper edge uses several HCI operations to engage with the layers of the Bluetooth stack, whilst the lower edge utilizes the services provided by the 802.11 MAC (as shown). In Table 14.14, we summarize the PAL management components, along with their responsibilities.

Table 14.14. The PAL management components

Manager	Description
PAL	The PAL manager provides the global operation.
Physical link	The physical link manager offers the operation of the physical connections, such as creation, acceptance, and deletion.
Logical link	The logical link manager offers the operation of the link connections for creation and deletion.
Data	The data manager takes care of the operations associated with data packets. Such operations include send and receive, and buffer management.

14.6.5 The Host Controller Interface

The presence of the HCI is solely dependent on the specific Bluetooth implementation, as we have already discussed (see Section 14.6.3, "The Link Manager"). In Figure 14.28, we illustrate two possible types of implementation: **A** represents a device that utilizes one processor to execute both the host and controller components, whereas **B** represents a device that uses two processors, one to execute the host and the other the controller. The HCI is present in **A** as the implementer has chosen to retain HCI as a conceptual interface.

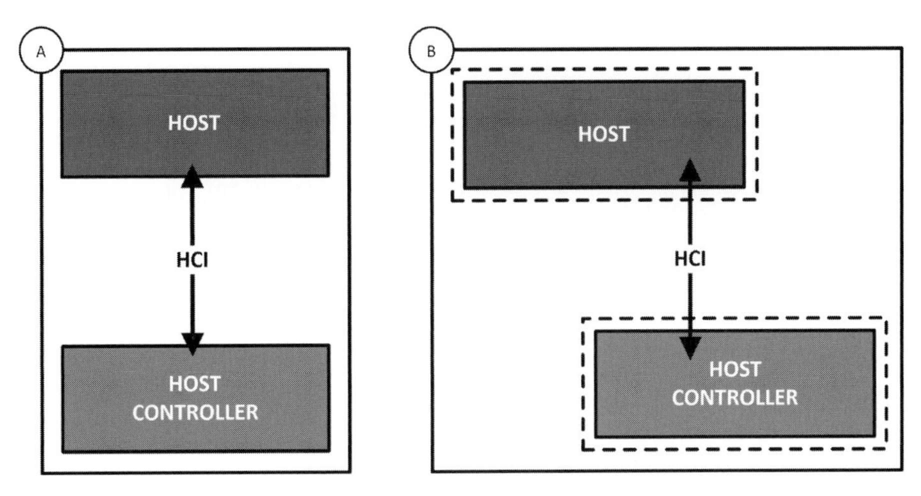

Figure 14.28. The HCI permits communication between a host and controller, but some implementations using a single processor may retain the HCI as a concept interface.

The HCI transport layer provides four interfaces to enable communication between a host and controller; we summarize these interfaces in Table 14.15. The *Universal Asynchronous Receiver/Transmitter* (UART) interface provides a basic RS232 null-modem-like connection between the host and controller entities. The 3-Wire UART interface offers an opportunity to use the HCI between two UARTs. The USB interface would permit a USB dongle, something we touched upon earlier, when discussing retro-fitting a notebook or when, for example, an independent USB module is integrated onto a *Printed Circuit Board* (PCB). Finally, the *Secure Digital* (SD) interface creates a pathway between a Bluetooth-enabled SD device (controller) and a Bluetooth host.

Table 14.15. The HCI has four types of interface

Interface	Description
UART	An RS232 null-modem-like interface.
USB	The HCI-USB interface allows a USB dongle or an independent USB module, which may be integrated onto a PCB, to communicate.
SD	The SD creates an interface between a Bluetooth-enabled SD device (controller) and a host.
3-Wire UART	The 3-Wire interface provides an opportunity for a system to use the HCI between two UARTs.

The HCI uses a *command/event* paradigm and exchanges certain messages over the interface. What's more, the HCI provides a consistent method for accessing a controller's capabilities. In Figure 14.29, we illustrate the lower layer components of the HCI for both the host and controller; these layers are unaware of the content of the data being exchanged.

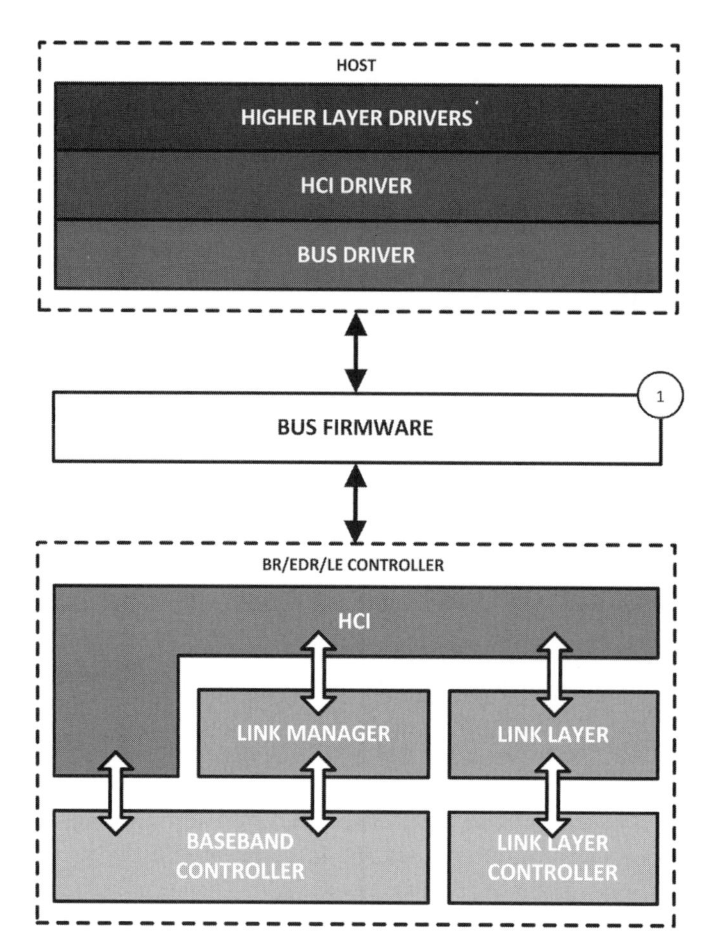

Figure 14.29. The lower layers of the HCI at both the host and host controller.

14.6.6 L2CAP

The L2CAP layer, as shown in Figure 14.15 (**4**), has the overall responsibility of providing both *connection-oriented* and *connectionless* data services to the layers above; it does so using protocol multiplexing and *Segmentation And Reassembly* (SAR) capabilities. Additionally, L2CAP also has the responsibility of providing per-channel flow control as well as error control. The L2CAP is common to both BR/EDR and LE systems, although the latter system operates in *basic mode* and does not support connection-oriented data services.

In Figure 14.30, we provide the conceptual architecture for L2CAP. Within the BR/EDR system, L2CAP enables the higher layers of the Bluetooth stack, for both protocols and applications, to send and receive upper layer packets via L2CAP *Service Data Units* (SDUs) (**A**) with a maximum size (length) of 64 Kbyte. The *resource manager* comprises three further components, namely *segmentation & reassembly* (**1**), *retransmission & flow control* (**2**), and *encapsulation & scheduling* (**3**). The resource

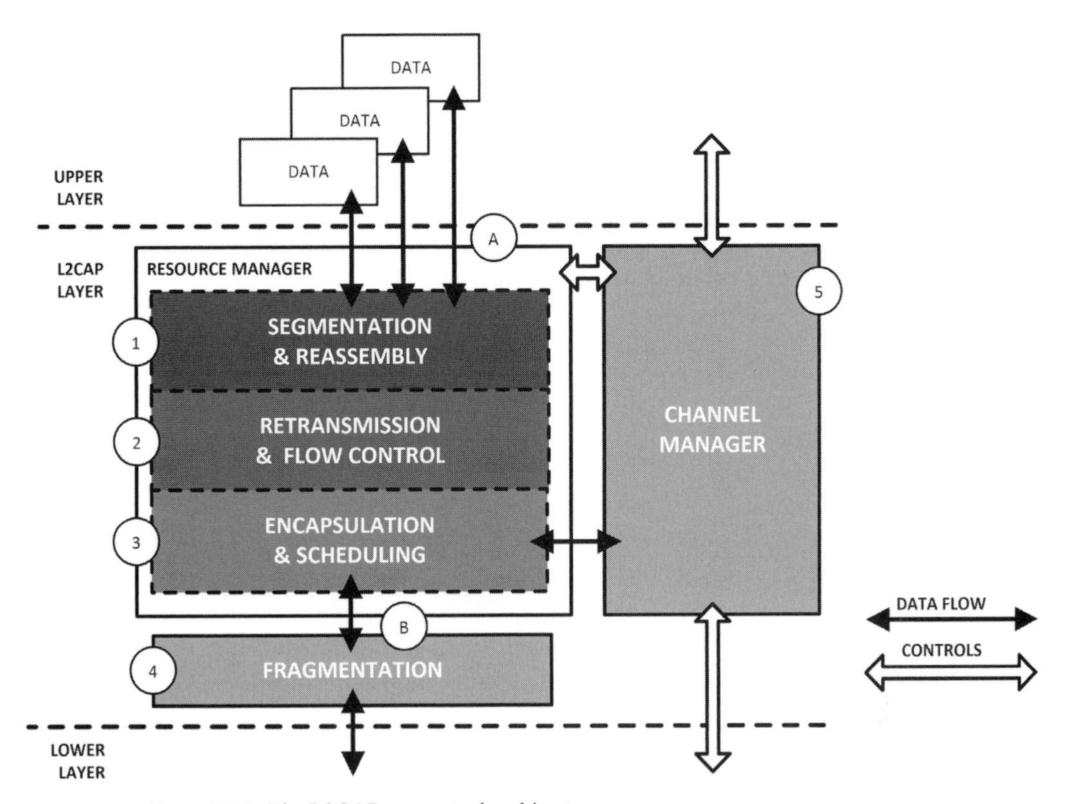

Figure 14.30. The L2CAP conceptual architecture.

manager provides a control mechanism, more specifically a frame relay service, to the *channel manager* (**5**), as indicated. In short, the resource manager provides the overall responsibility for managing the reception and transmission of data packets for multiple L2CAP channels to the facilities offered at the lower layer interface (as shown). With the frame relay service in mind, the application has the primary responsibility for defining the size of transport frames over L2CAP – nevertheless, it's preferable if L2CAP, in turn, optimizes the PDU length for the lower layer (**B**), as shown in Figure 14.30. As such, L2CAP provides several benefits, such as optimization of latency, buffer and memory management, and error correction.

In Figure 14.30, *fragmentation and recombination* (**4**) supports the ability for some controllers that don't have the bandwidth capability. This architecture component provides fragmentation of L2CAP PDUs, in turn accommodating any limitations imposed by that controller. The L2CAP has the additional responsibility of establishing *Quality of Service* (QoS) requirements between two Bluetooth devices; whilst the connection is maintained, L2CAP monitors QoS contracts and ensures they are sustained.

14.6.6.1 Channel Identifiers

The philosophy of the L2CAP layer is based on the notion of *channels*, as we have already intimated, where each *endpoint* of an L2CAP channel is referred to as a *Channel*

Table 14.16. The fixed channel type values for both the BR/EDR and LE systems

CID	Channel type
`0x0000`	Null (not used).
`0x0001`	Signaling (BR/EDR-only).
`0x0002`	Connectionless channel.
`0x0003`	AMP manager protocol.
`0x0004`	Attribute protocol.
`0x0005`	Signaling (LE-only).
`0x0006`	Security manager protocol.
`0x0007–0x003E`	Reserved.
`0x003F`	AMP test manager.
`0x0040–0xFFFF`	Dynamically allocated.

Identifier (CID). In Table 14.16, we list the CIDs for a number of *fixed* channels for the BR/EDR operation; CID `0x0000` is never used as a destination endpoint, whereas CIDs `0x000A` to `0x003F` are reserved for various L2CAP functions and are fixed channels. Each fixed channel possesses certain attributes, such as the ability to change parameters; configuration parameters for reliability; *Maximum Transmission Unit* (MTU) size; QoS; and security. Other channel assignments are left to the discretion of the implementer.

A number of CIDs have been reserved, for example the L2CAP signaling channel (`0x0001`), which is used to create, establish, and negotiate changes in connection-oriented data channels. It is also responsible for determining the characteristics of connectionless channels operating over the ACL-U logical link. The L2CAP connectionless channel (`0x0002`) is used for all incoming and outbound data for both broadcast and unicast traffic. The L2CAP signaling channel and all supported fixed channels are available once the ACL-U logical link has been established. In Table 14.17, we identify the available channel types along with their corresponding CIDs.

Table 14.17. Available channel types with corresponding CIDs

Type	Local CID	Remote CID
Connection-oriented	Dynamically allocated and fixed.	Dynamically allocated and fixed.
Connectionless data	`0x0002` (fixed).	`0x0002` (fixed).
L2CAP signaling	`0x0001` and `0x0005` (fixed).	`0x0001` and `0x0005` (fixed).

Further to our L2CAP conceptual architecture provided earlier in Figure 14.30, we provide in Figure 14.31 the transaction model that should be adopted in most implementations. In general, implementers should provide signaling or event mechanisms between layers, such as providing event management where the notification upper layers are alerted to new events or messages. Likewise, lower and upper layers should also accept or receive notifications.

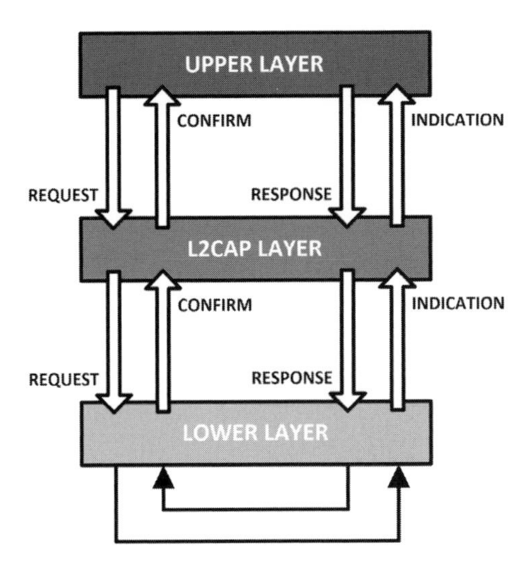

Figure 14.31. The L2CAP transaction model.

14.6.6.2 L2CAP Modes of Operation

L2CAP provides five different modes of operation as selected for each L2CAP channel; we list and describe these modes in Table 14.18.

Table 14.18. L2CAP operates in one of five modes

Mode	Description
Basic	This default mode is used when no other mode has been agreed.
Flow control	This mode is used when L2CAP entities do not support either enhanced retransmission or streaming modes.
Retransmission	This mode is used when L2CAP entities do not support either enhanced retransmission or streaming modes.
Enhanced retransmission	This L2CAP mode is used for all reliable connections created over AMP-U and ACL-U logical links.
Streaming	This L2CAP mode is used for all streaming applications created over both AMP-U and ACL-U logical links.

14.6.6.3 L2CAP Data Packet Formats

As we have already mentioned, channels within the L2CAP entity may either be connection-oriented or connectionless. So, normally all fixed and dynamically assigned channels are considered connection-oriented, with the obvious exception of the L2CAP connectionless channel (`0x0002`) and the two signaling channels, that is, CIDs `0x0001` and `0x0005`. In Figure 14.32, we illustrate the L2CAP PDU format for a connection-oriented packet in basic L2CAP mode, which is also referred to as a *B-frame*.

As you can see, the structure comprises the basic L2CAP header and payload. The L2CAP payload is of variable length and will represent data to or from the upper layer

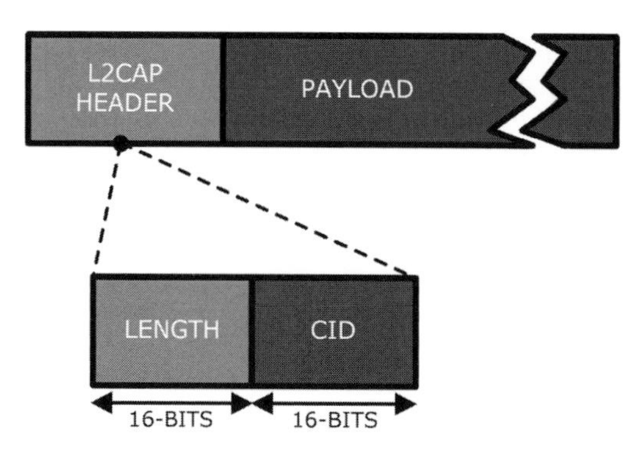

Figure 14.32. The L2CAP PDU format for a connection-oriented channel used in basic L2CAP mode.

protocol. In Table 14.19, we look at the specific fields that make up the L2CAP header, along with size of each field in bits.

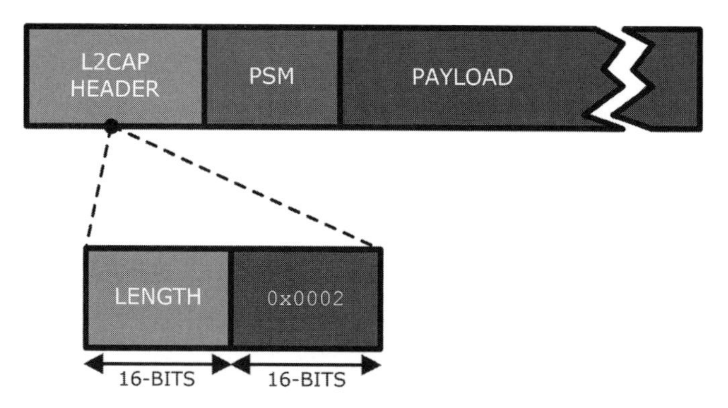

Figure 14.33. The L2CAP PDU format for a connectionless channel used in basic L2CAP mode.

In Figure 14.33, we illustrate the L2CAP PDU format for a connectionless packet in basic L2CAP mode, which is also referred to as a *G-frame*. Like the connection-oriented L2CAP frame format, the structure comprises the basic L2CAP header and payload. The L2CAP payload is again of variable length and contains information to be distributed to all slaves within a piconet (broadcast connectionless traffic) or to a specific Bluetooth device. In Table 14.19, we looked at the specific fields comprising the L2CAP header. The length field indicates the size of the payload, which will include the

Table 14.19. The fields that make up the L2CAP header

Field	Description	Size
Length	The length field is 16-bits in length and indicates the size of the payload excluding the header.	16
CID	The CID is 16-bits in length and identifies the destination channel endpoint of the payload.	16

16-bit *Protocol/Service Multiplexer* (PSM) field. Naturally, the CID is set to `0x0002`, representing the fixed connectionless channel.

In Figure 14.34 and Figure 14.35, we illustrate the L2CAP PDU formats, which support flow control, retransmission, and streaming using the basic L2CAP header, along with other protocol elements. In the first illustration, the *Supervisory-frame* (S-frame) is used to acknowledge and request retransmission of *I-frames*, the format of which is depicted in our second illustration (see Figure 14.35).

Both I- and S-frame PDUs include the L2CAP header, as used in the basic L2CAP mode. The length field determines the variable length of the payload within the I-frame where the FCS field is optional. The additional protocol elements are listed and discussed further in Table 14.20.

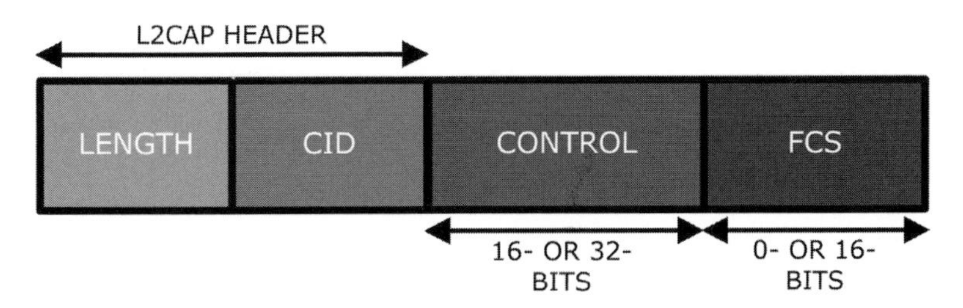

Figure 14.34. The L2CAP PDU format for S-frame PDU.

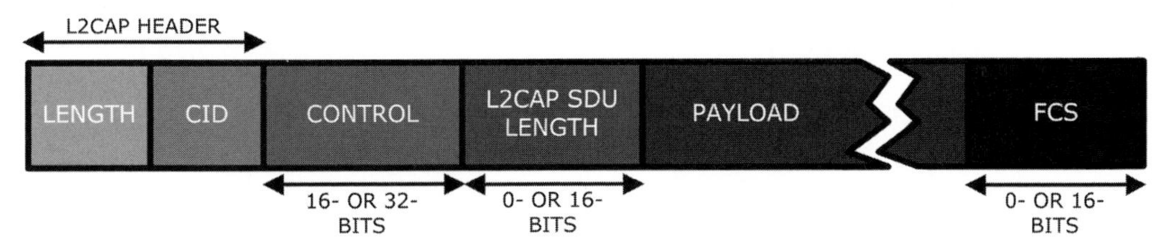

Figure 14.35. The L2CAP PDU format for I-frame PDU.

14.6.7 Service Discovery Protocol

The *Service Discovery Protocol* (SDP), as we illustrated earlier in Figure 14.15 (**5**), is used by an SDP *client* device to learn of other services and/or applications that might be available on the SDP *server* device. The server will maintain a set of *service records*, which uniquely offer characteristics describing the accessible services where at least one active SDP server is available to each Bluetooth device. However, a Bluetooth device can undertake the role of both client and server. In Figure 14.36, we illustrate the typical client/server interaction. As such, a client device can discover certain attributes of the "discovered" services, using service records, and the associated mechanisms needed to utilize such services. Furthermore, each service is an instance of a service class, which provides all associated attributes contained within service records. We discuss the service class later, in Section 14.6.7.3, "Service Class."

Table 14.20. The additional fields that make up the S- and I-frames

Field	Description	Size
Control	The control field is 16- or 32-bits in length and identifies the type of frame, namely *standard*, *enhanced*, and *extended* control fields.	16 or 32
L2CAP SDU length	The first I-frame sequence is denoted as *start of L2CAP SDU* (SAR = 01), if the SDU is segmented, where this 16-bit field specifies the total number of octets that comprise the entire SDU. If the SDU is not segmented (SAR = 00), the L2CAP SDU length field is not needed.	0 or 16
FCS	The *Frame Check Sequence* (FCS) field is 16-bits in length and is optional.	0 or 16

14.6.7.1 Service Record

A service record provides information relevant to a single service that, in turn, offers information relevant to how to execute an action or manage a resource. A service record is uniquely assigned a 32-bit handle, which identifies the specific service available on the SDP server. What's more, a service record handle with a value of `0x00000000` represents the actual SDP server itself, where the record will contain attributes relating to the SDP server and its associated protocols it supports. In Figure 14.37, we illustrate a typical service record, along with its potential attributes.

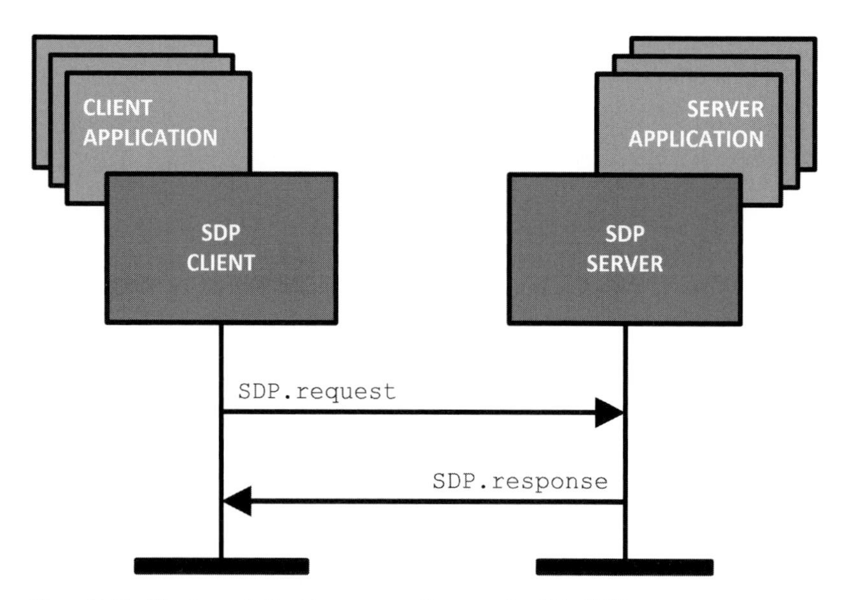

Figure 14.36. The typical client/server paradigm used within SDP.

14.6.7.2 Service Attribute

In Figure 14.38, we illustrate the service attribute field, which comprises two further fields, namely an attribute *ID* and attribute *value*. In Table 14.21, we describe the possible service attributes, each of which individually describes the characteristics of that service.

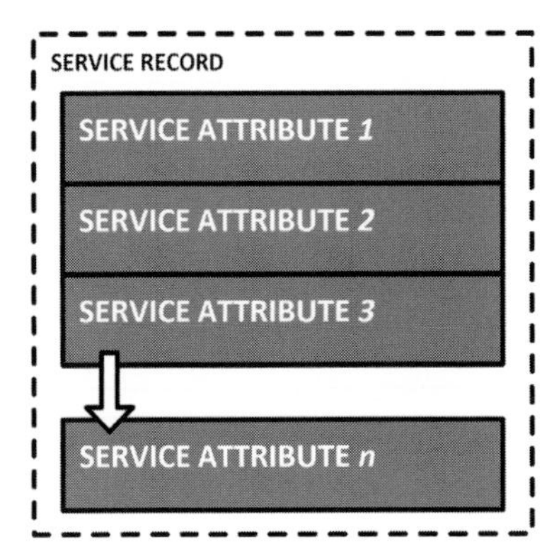

Figure 14.37. A typical service record with its potential service attributes.

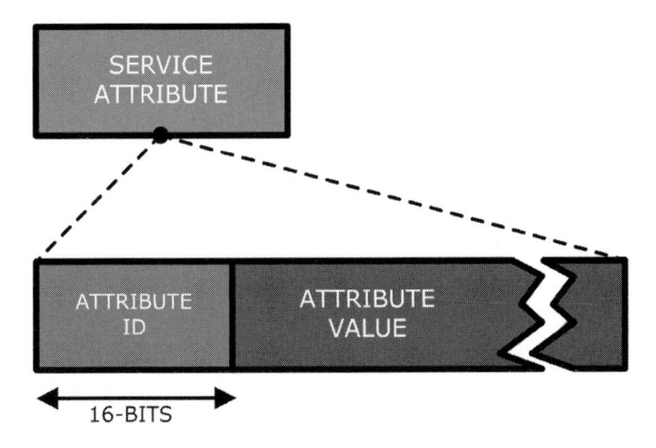

Figure 14.38. The service attribute field, which further comprises the attribute ID and value fields.

Table 14.21. The available service record attributes

Attribute	Description
`ServiceClassIDList`	This service record attribute identifies the type of service that's represented by a service record.
`ServiceID`	A unique instance of a service.
`ProtocolDescriptorList`	This service record provides a list of protocols that may be employed to use a service.
`ProviderName`	A text name of the company or individual that provides the service.
`IconURL`	An image that is used to represent the service visually.
`ServiceName`	A text name of the service provided.
`ServiceDescription`	A text name describing the available service.

Table 14.22. The fields that make up the service attribute

Name	Description	Size
ID	This 16-bit field uniquely identifies each service type within a service record.	16
Value	This is a variable length field and is dependent on the meaning of the attribute ID, along with the service class, something we discuss in the following section.	

Furthermore, manufacturers may create their own service attributes. In Table 14.22, we describe the attribute ID and attribute value further.

14.6.7.3 Service Class

As we already mentioned, each service is an instance of a service class, which provides definitions of all attributes contained within service records; each attribute, for that matter, defines the numerical value of the attribute ID and how the attribute value is specifically formatted within the field. The service class itself is also assigned a unique identifier, `ServiceClassIDList`, as we described earlier, in Table 14.21.

14.6.8 Generic Access Profile

As we touched upon earlier in Section 14.3.1, "Bluetooth Profiles," Bluetooth profiles achieve interoperability for a host of manufacturers with the use of *capabilities*, as prescribed in numerous SIG-profile specifications (see Figure 14.15 (**6**)) that, in turn, define such capabilities from the physical layer or air-interface through to L2CAP. A profile will also prescribe specific interactions at the application layer – in other words, a profile will describe naming conventions and offer visual cues by defining commonality in behavior and user experience. The *Generic Access Profile* (GAP) is a core profile used for all device types, as shown in Table 14.23. GAP specifically describes the essential requirements for all Bluetooth-enabled devices, and provides a single defining role. In short, a Bluetooth device may incorporate either initiating or accepting procedures, wherein the peer device must support the corresponding functionality, although a BR/EDR *controller* device is expected to support all functionality. Within an LE system, however, as you may recall from Chapter 7, "Bluetooth low energy: The *Smart* Choice," we define *four*

Table 14.23. GAP is supported by all device types

Type	Description
BR/EDR	Bluetooth-enabled devices that support the basic rate operation.
LE-only	Bluetooth-enabled devices that support the low energy operation.
BR/EDR/LE	Bluetooth-enabled devices that support a combination of basic rate and low energy operations.

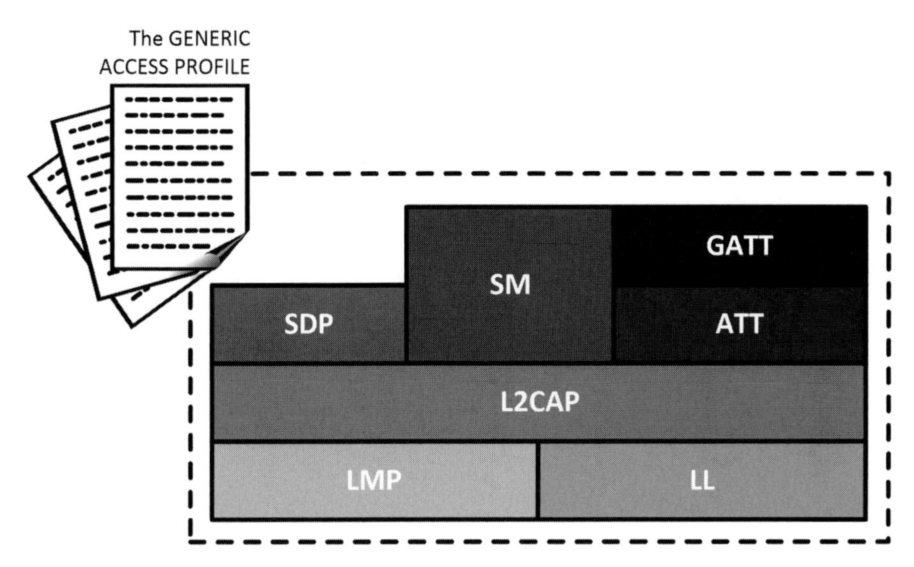

Figure 14.39. The all-encompassing GAP provides common functionality, capabilities, and user experience for the BR/EDR system.

roles, namely *broadcaster*, *observer*, *peripheral*, and *central*, where only one role can be supported at a given time.

What's more, GAP describes specific roles that are undertaken and, further, defines discoverability, connection and security modes, and procedures. In the sections to follow, we discuss roles, user expectations, modes of operation, security, and authentication. In Figure 14.39, we depict the overall relationship that governs GAP capabilities across the Bluetooth stack architecture. You will note that the stack architecture as shown covers both the BR/EDR and LE systems – we listed the supported device types in Table 14.23. So, looking at the architecture from the bottom up, LMP was discussed earlier, in Section 14.6.3, "The Link Manager," whilst the *Link Layer* (LL), which is specific to the LE system, was discussed earlier in Chapter 7. We discussed L2CAP earlier, in Section 14.6.6, "L2CAP," and noted that it is common to both Bluetooth systems. SDP was covered in Section 14.6.7, "Service Discovery Protocol," and the *Attribute Protocol* (ATT) and the *Generic Attribute Profile* (GATT) were discussed in Chapter 7, "Bluetooth low energy: The *Smart* Choice," along with the *Security Manager Protocol* (SM). Incidentally, device types, that is BR/EDR or BR/EDR/LE, will use SDP for service discovery; moreover, BR/EDR device types that utilize ATT will also include a GATT implementation. However, BR/EDR/LE and LE-only device types will implement GATT for service discovery, which is transported over the LE channel. Naturally, LE-only device types will implement ATT.

14.6.8.1 Initiator, Acceptor, and Paging Roles for the BR/EDR System

In Figure 14.40, we illustrate a device (**A**) responsible for initiating the establishment of a physical channel (*initiator* device) or one that may instigate a transaction over an

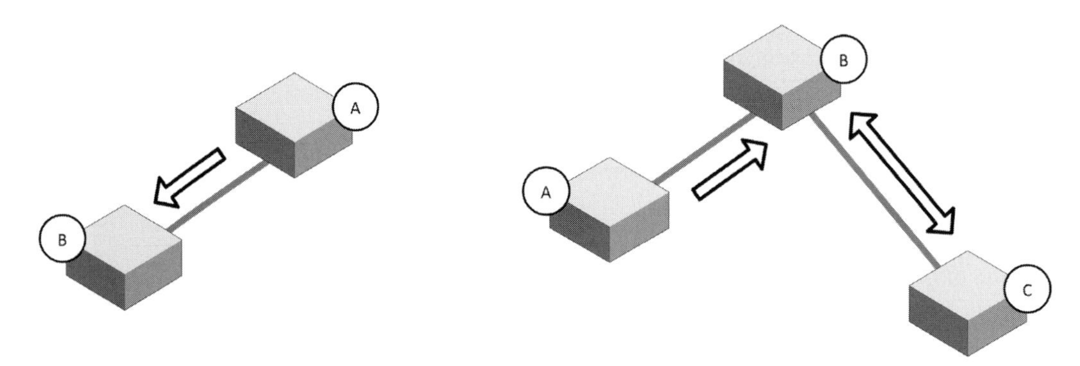

Figure 14.40. GAP provides procedures for a single device (**A**) toward another (**B**) that may or may not have an existing link.

existing channel (*paging* device). However, the party device (**B**) has the responsibility of being the *paged* device, or *acceptor*. The GAP manages the procedures for discovery, link, and connection establishment, even when the devices have an existing link or where **B** may be connected to another party (**C**). The GAP also provides procedures to manage instances where multiple profiles are supported in a single Bluetooth device.

14.6.8.2 The User Interface Expectations

The GAP establishes commonality at the user interface level, something we have already intimated, such that consumers have every opportunity to connect to other Bluetooth devices without confusion. In other words, it's important for a consumer to witness common or identical terminology or naming conventions at the user interface level when utilizing the same application across any Bluetooth-enabled device, irrespective of the manufacturer. The definition "user interface" is not limited to a device's display – it also encompasses associated dialogs used within a display context, as well as any related user manuals, product packaging, and advertising. In essence, when a consumer encounters a Bluetooth's product names, values, and any other manifestation of Bluetooth terminology, the GAP wholly embodies a common ethos. As such, it's imperative that the user's experience is not hindered by conventions that deter from the guidance provided by the GAP. Moreover, GAP offers a number of requirements that aid the developer in providing names of parameters and how the consumer should experience the user interface.

There are three significant Bluetooth parameters with which we commence our discussion here, namely the *Bluetooth Device Address* (BD_ADDR), Bluetooth *device name*, which represents a user-friendly name of the Bluetooth device, and the Bluetooth *passkey* or Bluetooth *Personal Identification Number* (PIN). You may recall from Section 14.6.3, "The Link Manager," that we discussed the format of the BD_ADDR and how it is created and used to identify uniquely a Bluetooth device. At the user interface level, the term "Bluetooth device address" is used, as shown in Figure 14.41, although in this illustration the term has been shortened to "address." Similarly, the Bluetooth device

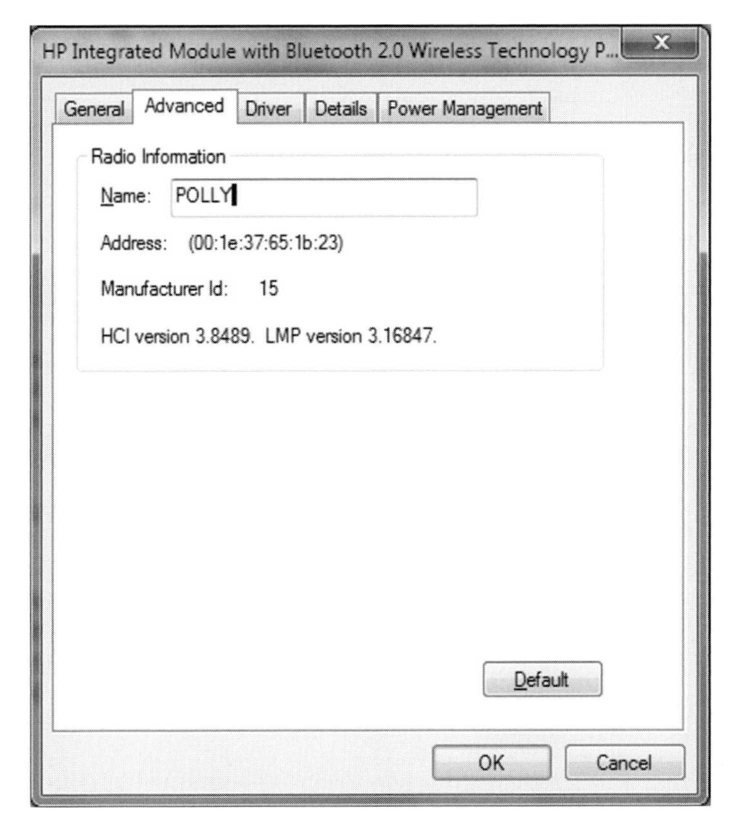

Figure 14.41. A typical user experience at the display level, where we can see both the user-friendly name along with the Bluetooth device address.

name shows "name," where "Polly" is displayed as the user-friendly name of the device. When a user wishes to connect with this device, the service discovery procedure will reveal the device's name as "Polly." A passkey or PIN will be required to connect with "Polly" successfully, which completes our third significant parameter.

14.6.8.3 Modes

In Table 14.24, we list the modes of operation permitted within GAP. In Table 14.25, however, we list the inquiry and discovery procedures, which are uniquely applicable to devices that initiate them.

14.6.8.4 Security and Authentication

At the user interface level, the term "Bluetooth authentication" is used to describe the procedures undertaken when one device initiates authentication with another. Authentication occurs either during or following link establishment. In Table 14.26,

Table 14.24. GAP supports a number of modes

Mode	Description
Discoverability	
Discoverable	During an inquiry, a device may be in either a *discoverable* or in a *non-discoverable* mode. At the user interface level, the device will display "discoverable [mode]."
Non-discoverable	During an inquiry, a device may be in either a discoverable or in a non-discoverable mode. In non-discoverable mode, a Bluetooth device will not respond to an inquiry.
Limited discoverable	The *limited discoverable* mode should be used by a device that wishes to be discovered for a certain period. At the user interface level, the device will display "discoverable [mode]" whilst within its window of discoverability.
General discoverable	The *general discoverability* mode permits a device to be discovered continuously. At the user interface level, the device will display "discoverable [mode]."
Connectability	
Connectable	During *paging*, a Bluetooth device may be in either a *connectable* or in a *non-connectable* mode, where it will respond to paging in the former instance. At the user interface level, the device will display "connectable [mode]."
Non-connectable	A non-connectable device will not respond to paging, and at the user interface level the device will display "non-connectable [mode]."
Bondability	
Bondable	During a *pairing request*, a Bluetooth device may be in either a *bondable* or in a *non-bondable* mode, where in the former instance the device will respond to the request. At the user interface level, the device will display "bondable [mode]."
Non-bondable	A Bluetooth device in the non-bondable mode will not respond to a pairing request. At the user interface level, the device will display "non-bondable [mode]."

Table 14.25. The inquiry and discovery procedures unique to GAP

Procedure	Description
General inquiry	The *general inquiry* procedure provides the initiating device with further information, such as the Bluetooth device address, clock, class of device, and so on. At the user interface level, the device will display "Bluetooth device inquiry."
Limited inquiry	The *limited inquiry* procedure provides the initiating device with the same information as during a general inquiry, but the device is only available during certain periods. At the user interface level, the device will display "Bluetooth device inquiry."
Name discovery	The *name discovery* procedure provides the initiating device the Bluetooth device name or user-friendly name. At the user interface level, the device will display "Bluetooth device name discovery."
Device discovery	The *device discovery* procedure provides the initiating device specifically with information such as the Bluetooth device address, clock, class of device, and so on. At the user interface level, the device will display "Bluetooth device discovery."
Bonding	The bonding procedure is responsible for creating a relationship between two devices that share a common link key.

Table 14.26. The security modes supported by all device types

Mode	Description
1	A Bluetooth-enabled device using *security mode 1* (non-secure) will never initiate any security procedures.
2	A Bluetooth-enabled device in *security mode 2* (service level enforced security) classifies its security requirements using authorization, authentication, and encryption attributes.
3	A Bluetooth device utilizing *security mode 3* (link level enforced security) will ensure that security procedures are in force prior to link establishment being complete.
4	A Bluetooth-enabled device using *security mode 4* (service level enforced security) mandates its security requirements using authentication or unauthentication link key requirement.

we list four possible security modes that may be invoked during link establishment; however, a device may support two modes, namely *security mode 2* and *mode 4*, as shown in Table 14.26. The latter security mode is used to support legacy devices (backwards compatibility) that do not support *Secure Simple Pairing* (SSP).

15 One Standard, All Devices

WHDI

The *Wireless Home Digital Interface* (WHDI) is a standard which has been developed by a consortium of leading consumer electronics companies, which include Amimon, Hitachi, Motorola, Samsung, Sharp, Sony, and LG Electronics. The WHDI standard offers wireless connectivity for a number of interoperable multimedia-centric devices that, in turn, deliver high-quality video and audio content within your *multimedia* network. The standard scopes the *Audio/Video Control Layer* (AVCL) in addition to both the *Medium Access Control* (MAC) and *Physical* (PHY) layers, which we'll discuss later on in this chapter. In Figure 15.1, we show WHDI's trademark and logo.

In this chapter, we discuss WHDI's inception and evolution, along with the WHDI Alliance and its membership benefits and structure. The chapter also discusses the market scope, wherein initial WHDI-enabled products have already started to emerge. We will, of course, look at how WHDI compares with other competing technologies within the same market sector. Finally, the chapter lifts the lid on WHDI and takes a closer look at its software architecture and protocol.

Figure 15.1. The WHDI trademark and logo. (Courtesy of WHDI LLC.)

15.1 Overview

WHDI (whdi.org) is a standards-based wireless technology supported and developed by a consortium of companies, as we have already touched upon. WHDI LLC is responsible for licensing and promoting the standard, and is a subsidiary of the leading founder company Amimon Incorporated (amimon.com). WHDI aims to revolutionize home entertainment systems by wirelessly delivering *Audio/Video* (A/V) content and, in turn, offering flexibility when interconnecting video sources such as notebooks, smart or cellular phones, tablets, *Set-top-boxes* (STBs), and Blu-ray/DVD players. WHDI supports the most recent *Electronics Industries Alliance* (EIA)/*Consumer Electronics Association* (CEA), video format, that is EIA/CEA-861-E, and any other formats within the same bandwidth – *High-definition Multimedia Interface* (HDMI) 1.4 also uses EIA/CEA-861-E. The standard additionally supports several audio standards, including *Linear Pulse-code Modulation* (LPCM), *Direct Stream Transfer* (DST), 1-bit audio *Direct Stream Digital* (DSD), and both compressed and uncompressed audio, which is formatted to the *International Electrotechnical Commission* (IEC) 60958 or 61937 standard. The WHDI standard delivers uncompressed HDTV (up to 1080p/60 Hz) throughout the home using the unlicensed 5 GHz frequency band, which is also shared with the *Institute of Electrical and Electronics Engineers* (IEEE) 802.11a, 802.11n (Greenfield-only), and 802.11ac standards. The WHDI standard purports superior and consistent picture quality analogous to the wired HDMI equivalent, and is a natural evolution to the cable. Later in this chapter, we will lift the lid on WHDI and explore in greater detail many of the technology's features and attributes that lend themselves aptly to delivering audio/video content wirelessly throughout the home.

15.1.1 The WHDI Consortium

The WHDI Consortium, founded in 2008 by Amimon, Hitachi, Motorola, Samsung, Sharp, Sony, and LG Electronics, collectively drives the future evolution and development of the standards-based wireless technology, which is operated by WHDI LLC. The Consortium is responsible for public education and serving certification and compliance programs, in addition to pursuing market development. The Consortium remains open to most participants and currently retains over 40 members across the Americas, Europe, and Asia.

Table 15.1. The WHDI Consortium offers potential participants a three-tier membership structure, to include Promoter, Contributor, and Adopter

Membership level	Scope
Promoter	Founding members of WHDI manage and control the future evolution of the WHDI standard. A promoter participant will be able to scope specifications: define timings for specifications and new features, as well as determine testing, compliance, trademark, and branding requirements.
Contributor	The contributor level membership affords its members the ability to lead the WHDI ecosystem and ensure that the technology satisfies market requirements. Contributors help shape the WHDI specification and roadmap future development.
Adopter	If you plan to manufacture or distribute WHDI-enabled products, then you must, as a minimum, acquire Adopter membership. An Adopter will be able to receive both WHDI and test specifications and access a number of WHDI test centers. They will also be able to use the WHDI logo on their products.

15.1.1.1 Membership and the WHDI Consortium

The Consortium offers potential participants three types of membership, as shown in Table 15.1. These three types of membership enable companies to decide on how they wish to participate within the group. Like most standards-based committees, the offer of membership is structured, so that a participant can decide on how much involvement they wish to have – from *Adopter*, an entry level into the Consortium, where members can access full specifications and have the use of the WHDI logo on their products, through to *Promoter*, where founder level membership entitles members to scope the future of the technology.

15.1.1.2 WHDI Certified

In the WHDI Compliance Program, manufacturers have the opportunity to utilize an *Authorized Test Centre* (ATC) for WHDI certification – this is in accordance with the WHDI *Compliance Test Certification* (CTS). Tests conducted at the ATC are done so at three levels: (1) compliance with the WHDI specification; (2) minimum product performance; and (3) successful interoperability between products, as well as *High-bandwidth Digital Content Protection* (HDCP) v2.0. Upon ATC approval, the WHDI logo can be used, either for product certification or module certification, as described further in Table 15.2.

Table 15.2. Two types of WHDI certification are available

Type	Description
Product	For products that are sold to the consumer.
Module	A component within the WHDI architecture, such as a chipset, board, or other system that can be tested against the CTS.

15.1.2 WHDI's Timeline

The infographic shown in Figure 15.2 provides a snapshot of WHDI's technology timeline, from its inception to where the technology is today. WHDI has had less than a decade to reach its market, which is an incredible achievement, as you can witness by comparing it with other technologies in this book. In fact, the initial *Proof of Concept* (PoC) only emerged circa 2005 by Amimon Incorporated, an Israeli start-up. The WHDI Consortium was established in 2008, and early first-generation products also appeared during that year. Later, in December 2009, the Consortium's initial v1.0 specification appeared; associated enhancements followed, including 3D support in 2011. In the same year, the market saw the emergence of WHDI-compliant technology for a range of consumer electronics such as PCs, smartphones, and tablet products. With an imminent, expected v2.0 release, the WHDI Consortium decided to steer away from the typical incremental association with version numbering to depict errata, features, and/or enhancements. Moreover, the confusion surrounding HDMI numbering led to the Consortium's decision not to label enhancements and new features with a version number – for HDMI technology this had ultimately caused some confusion for many consumers and retailers alike. So, building on HDMI's experience, the Consortium also

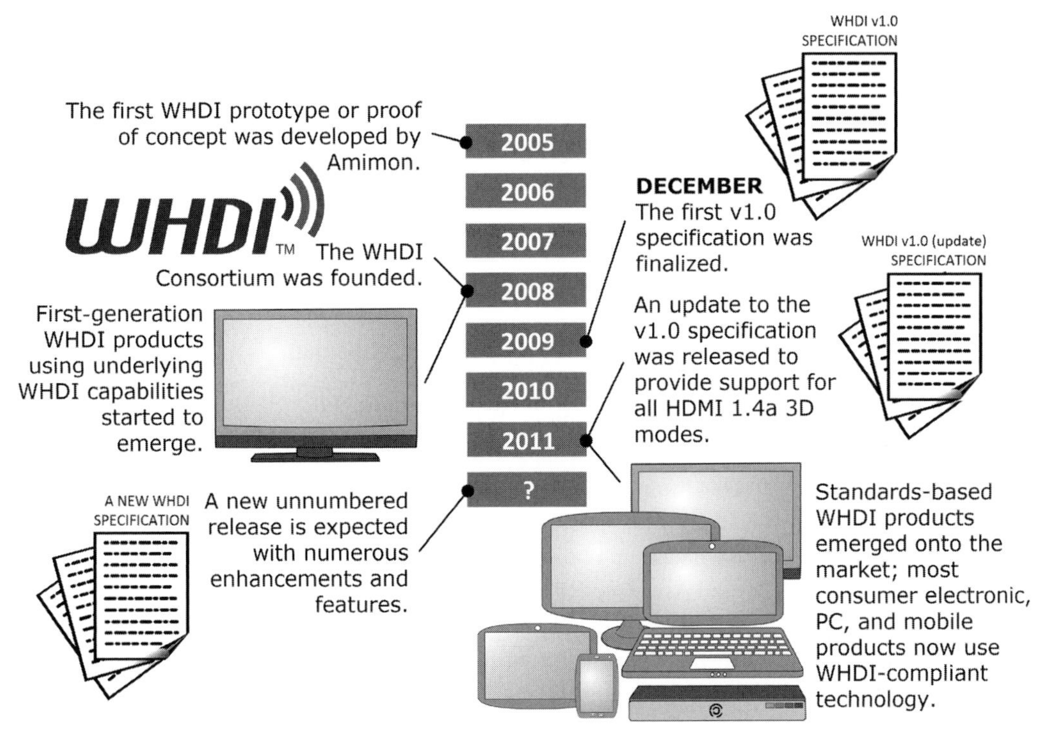

Figure 15.2. WHDI's technology timeline, from its inception to where the technology is today with its new, "unnumbered" specification.

decided to avoid version numbering to designate functionality. Nonetheless, an update is still expected at some point, irrespective of how it might be labeled.

15.2 WHDI's Market

The wireless technologies that form Part III, "The Classic Personal Area Network," are typically high-end and utilize high data rates, which are best suited for audio-, voice- and data-heavy-centric applications. WHDI is no exception, as it exclusively provides audio/video content delivery for a new generation of consumer electronics products. You may recall from Chapter 4, "Introducing the Lawnmower Man Effect," that we discussed *media convergence* and how the delivery of content has diversified over the last few years. No longer do we have to rely on traditional forms of delivery content, as the Internet has expanded our opportunity to source content from almost anywhere. What's more, with consumer electronics products packed with a wealth of connectivity technologies, we have no excuse but to remain connected to the wider area network. WHDI is yet another technology that ultimately ensures we have content delivered when and wherever we want it – further solidifying our conceptual *Lawnmower Man Effect* (LME), which we discussed in Chapter 4.

Figure 15.3. The Amimon AMN2220 chipset, a second-generation baseband wireless video modem supporting full HD and 3D A/V formats. (Courtesy of Amimon Inc.)

WHDI has an unusually short development lifespan, and has moved forward rather quickly, in just under five years in fact; we are already witnessing products using readily available WHDI-compliant technology. WHDI has essentially hit the ground running and delivered a range of chipsets that support full HD (1080p/60) with 3D A/V formats. In Figure 15.3, we show Amimon's second-generation AMN2220 transceiver chipset, which implements WHDI v1.0 and is used by multiple consumer electronics manufacturers, including LG Electronics and Philips, in addition to multiple PC manufacturers, such as Hewlett Packard and Asus (see HP's Wireless TV Connect Adapter utilizing Amimon's chipset in Figure 15.4). Amimon's range of chipsets has also seen itself introduced into markets for medical equipment, hospitals, and residential environments. Amimon is currently the largest supplier of wireless video connectivity within the consumer electronics market, and, in turn, this provides a mature and robust solution for wireless A/V connectivity. As we have already mentioned, Hewlett Packard, Asus, and others provide *Peripheral Component Interconnect* (PCI), PCI *Express* (PCIe), and *stick* devices to retro-fit existing products, such as that shown in Figure 15.4. The HP Wireless TV Connect Adapter delivers high-definition 1080p/60 media content to your TV, along with support for 7.1 surround sound.

Figure 15.4. HP's Wireless TV Connect Adapter utilizing WHDI technology. (Courtesy of Hewlett Packard.)

The market potential for wireless video streaming is enormous, and tablets and smartphones also represent a large opportunity to deliver content wirelessly. With such devices, the associated challenge surrounds the overall chipset footprint (100 mm^2) and, more importantly, its power consumption (less than 500 mW). The WHDI standard leads companies in a direction where these requirements can be adhered to.

Moreover, WHDI is starting to witness significant momentum in the professional market, where a robust, high-quality wireless system is particularly needed. For example, the medical market is developing openings, as the many video applications such as orthoscopic cameras, fiber optics video filming, and vision-aided devices, which include MRI and CT scanners, to name a few, offer opportunities for freedom within the medical sector. WHDI technology can also help reduce overall installation costs with digital signage for content broadcasting and interactive advertising. The ability to broadcast content to multiple screens for public or consumer infotainment and advertising within shopping malls and airport terminals also presents a breadth of opportunities for WHDI technology. Similarly, WHDI is used in the educational market to enable shared displays wirelessly in the classroom, whereby multiple students can connect to one display to present their work, or in business settings, where it can replace the once very familiar projector cable by a laptop to laptop scenario, something which some of us might already be familiar with!

WHDI also shows significant promise for use in professional video production. Several professional video cameras are on the market that make use of WHDI technology. These are used not only to free camera operators from wires on location, but also in multi-display conference monitoring systems for in-studio filming.

In many of the applications in the professional market, for example medical and professional video production, the need for low latency, high quality, and reliability is perhaps even greater than in the typical home-AV implementation and, as such, other wireless technologies may find it difficult to meet these specific requirements.

15.3 WHDI's Application Portfolio

As we have already touched upon, WHDI technology is still relatively young. Nonetheless, in such a short period it has become somewhat precocious and has swiftly matured to capture a diverse market space. In this section, we further explore WHDI technology's market sector and scope, along with its potential application-base, all supported by a WHDI ecosystem.

The technology's initial remit was to conquer the home video challenge,[1] since the delivery of content can be sourced from almost anywhere. The WHDI Consortium has delivered a technology that ultimately revolutionizes home entertainment systems by wirelessly delivering HD audio/video content, in turn offering a high level of flexibility when interconnecting multiple video sources, such as (see Figure 15.5) STBs (**1**), notebooks (**2**), digital cameras (**3**), tablet and smart or cellular phones (**4**), and Blu-ray/DVD players (**5**), to a single HDTV-capable display (**6**) that perhaps uses various audio and video formats. What's more, the Consortium's objective is to populate all consumer electronics devices with WHDI. The technology itself uses the unlicensed 5 GHz frequency band and supports both 20 MHz and 40 MHz channel bandwidths, along with *Multiple Input and Multiple Output* (MIMO) and *Orthogonal Frequency-division Multiplexing* (OFDM) radio transmission techniques that are comparable with Wi-Fi technology. As such, WHDI provides an effective scheme for coexistence with other devices that utilize

[1] WHDI, "WHDI Press/Analyst Briefing," 2012.

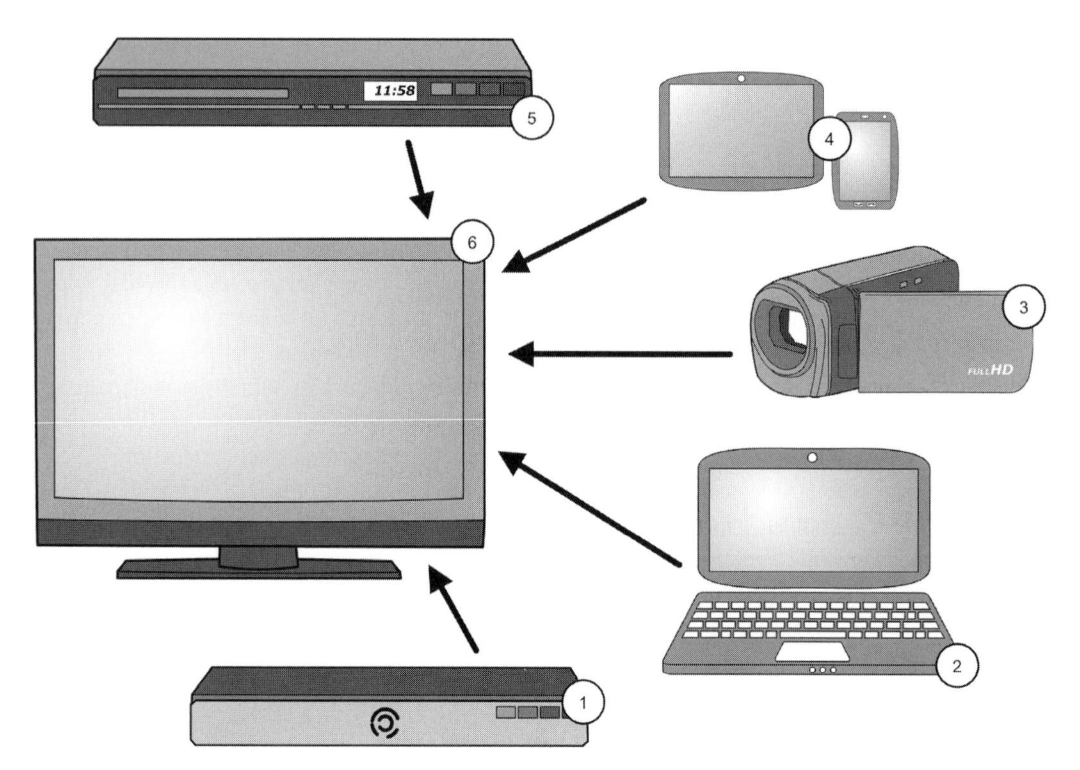

Figure 15.5. The home video challenge is to enable most consumer electronic products with WHDI to stream HD content to an HD-capable display.

the same frequency, such as Wi-Fi and cordless phones. Similarly, WHDI's transmission techniques further provide the technology with robust capabilities, which include a range of over 30 meters as well as an ability to penetrate walls and other obstacles. It can effortlessly deliver HDMI-like wired quality content at 1080p (scalable to higher resolutions) along with 3D. WHDI also supports HDCP v2.0 content protection protocol, which prevents the act of copying audio and video content when it passes between either a wired or wireless connection.

Additionally, with multiple A/V sources potentially connected within the home ecosystem, WHDI also offers the ability to control and select all video and audio sources within the home through the destination display, that is, via the display's own remote control device. This allows the user to select the streaming video to switch between multiple sources, and offers play, record, standby control, volume, and so on. The video latency afforded by the WHDI standard is less than 5 ms, whilst an audio stream has a latency of less than 10 ms.

The WHDI ecosystem typically forms an *extended Personal Area Network* (PAN). You may recall from Chapter 2, "What is a Personal Area Network?," that we discussed the divergence of the personal area networking space, which has become somewhat diluted with numerous topologies that uniquely characterize scope and function. You

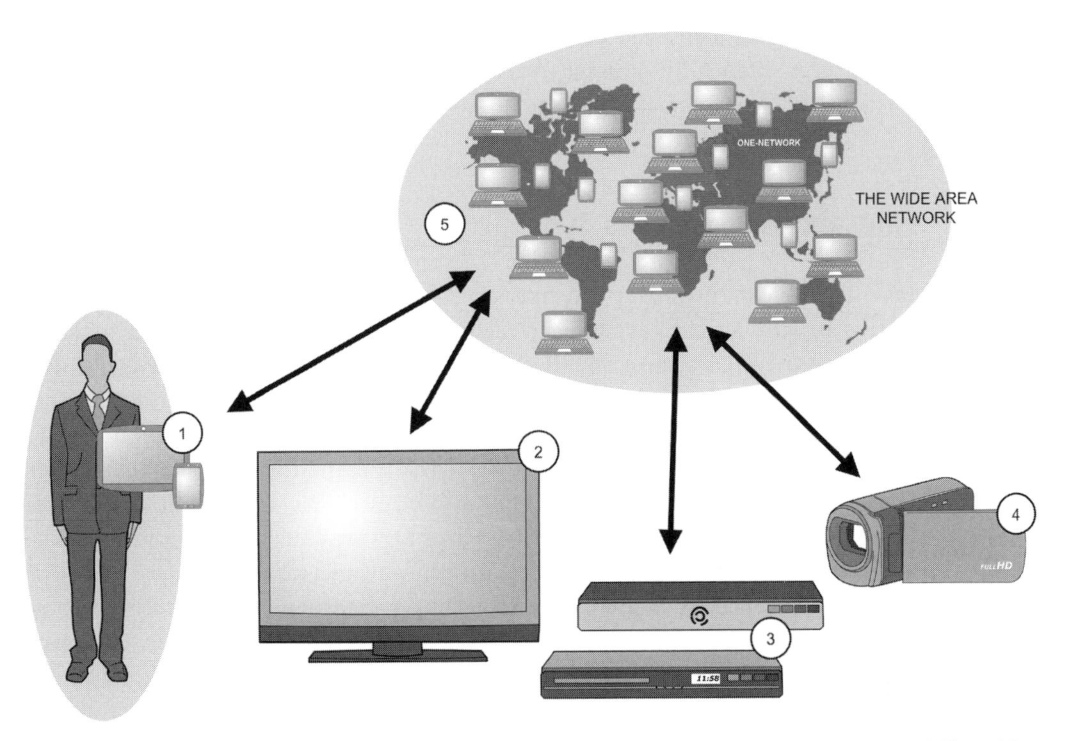

Figure 15.6. Nowadays, most consumer electronic products have an independent capability with which to access the Internet.

may also recall that we discussed and characterized the diverse number of area networking topologies that have become popular in mainstream computing. In Chapter 2, we provided several "colloquial" topologies, which are used to portray a specialized ad hoc or unique form of area network. We introduced such terms as *Vehicle Area Network* (VAN) and *Wireless Video Area Network* (WVAN) as a couple of examples. The WVAN, *Home Area Network* (HAN), and VAN are increasingly becoming part of mainstream conventionalism. Naturally, these conventions are used to characterize the function, scope, and purpose of the devices that form such a topology. What's more, in Chapter 4, "Introducing the Lawnmower Man Effect," we argued that there still very much exists a PC-centric perspective, and that, despite manufacturers attempting to steer away from PC-dependent technologies and applications, the PC still remains integral to most topologies. In Chapter 2, "What is a Personal Area Network?," we suggested that the WVAN, HAN, and VAN are merely extended PAN topologies. Nevertheless, the products portrayed in Figure 15.5 may have the ability to connect directly to the *Wide Area Network* (WAN) or Internet, in turn potentially eliminating the need to rely on a PC to source content. In Figure 15.6, we portray the collection of consumer electronics products having direct access to the WAN or Internet (**5**). The tablet and smartphone (**1**) have the ability to source content, along with televisions (**2**), STBs and DVD players (**3**), and cameras (**4**), for example. Nowadays, television manufacturers are integrating "net"-capable features, such as LoveFilm,[2] Netflix,[3] YouTube,[4] and other dedicated TV services offering TV content for pay-per-view channels, all offered through the Internet-capable TV, something we discussed earlier, in Chapter 4.

15.4 WHDI and its Competitors

Inevitably, with the popularity and prevalence of wireless technology and the freedom it affords, it seems a natural evolution to take away the HDMI cable and make it wireless; we compare WHDI against HDMI later, in Section 15.6, "Comparing WHDI with HDMI Systems." This analogy of taking away the cable and making it wireless is very much comparable to what has already happened with Ethernet, *Wireless Local Area Network* (WLAN) or Wi-Fi (see Chapter 13, "The 802.11 Generation and Wi-Fi"), and USB. It seems relevant to mention Wi-Fi and WirelessUSB in this section, as a number of wireless technologies have emerged, offering varying capabilities, that may present alternative methods for enabling wireless delivery of content. Furthermore, *WirelessHD* and *WiGig* both present the ability to deliver HD content wirelessly – we introduce these two technologies later, in Chapter 16, "Future and Emerging Technologies." However, both WirelessHD and WiGig are bound by line-of-sight operation and, alas, are limited to single-room use. Wi-Fi is synonymous with connecting to the Internet or wirelessly to your local area network (WLAN). What's more, Wi-Fi also unintentionally offers its users the ability to stream content either directly from the Internet or from a PC; however, latency and quality[5] issues may disrupt your viewing experience and, as mentioned previously, may be especially problematic in the professional video sector. Moreover, Wi-Fi was never intended to deliver wireless streaming of audio/video content, but many consumers use the wireless technology as a streaming alternative. We compare WLAN with WHDI later, in Section 15.7, "Comparing WHDI with WLAN Systems." With the higher bandwidth 802.11n and 802.11ac, along with improved latency and quality, the inherent use of Wi-Fi technology as a streaming mechanism may increase. In Figure 15.7, we illustrate D-Link's Boxee Box, which enables its users, via Wi-Fi, to access content directly from the Internet or to use *Digital Living Network Alliance* (DLNA) to stream content from their PC. Other third-party services and applications, such as Connect, Video Stream,[6] provide a service that enables tablet or smartphone devices to source HD content from the Internet typically served over a Wi-Fi connection.

[2] A British provider and an Amazon subsidiary (lovefilm.com, circa 2002), lovefilm offers streaming video on demand in Northern Europe.

[3] An American provider (netflix.com, circa 2007) of streaming video on demand services across the United States, Canada, and Latin America. Netflix launched their services in the United Kingdom and Ireland in 2012, competing against British provider LoveFilm.

[4] YouTube also provides rental service for the United Kingdom, Canada, and the United States.

[5] *Latency* and *quality* refer to the over the air-interface. If a user is streaming content using the Internet or via DLNA, then the quality of the image shown on the display is limited to the inherent quality of the originating content.

[6] Source: collect3.com.au/videostream.

Figure 15.7. The Boxee Box by D-Link (DSM-380) provides streaming content from the Internet or with DLNA. (Courtesy of D-Link.)

15.5 Networking Topology

The WHDI ecosystem supports two topologies, namely a *point-to-point* and a *point-to-multipoint* combination; these are illustrated in Figure 15.8. Likewise, there are also two types of WHDI device or entity, namely a *source* and a *sink*, which are summarized in Table 15.3. Furthermore, two types of bridges, *wired-WHDI* and *WHDI-wired*, are used to maintain compatibility and interoperability in a mixed wired/wireless environment. We explain in Table 15.4 the context in which these bridges are used, but we shall discuss

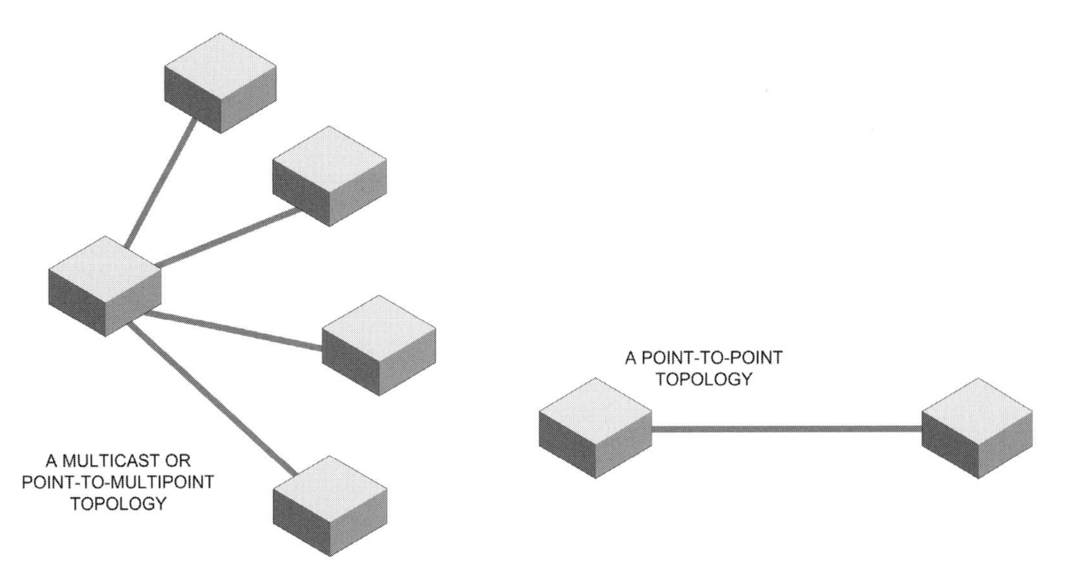

Figure 15.8. The WHDI ecosystem supports a multicast, or point-to-multipoint, and point-to-point topology configurations.

Table 15.3. WHDI defined entities

Entity	Description
Source	A WHDI network device is responsible for the transmission of A/V content, data, and control information to the sink entity, as well as for determining transmission timings.
Sink	A WHDI network device that receives A/V content, data, and control information from a source entity.

Table 15.4. WHDI defined bridges

Bridge	Description
Wired-WHDI	A WHDI source may have a wired input, which forwards the wired-received A/V content to WHDI, but must also convert and forward the control information.
WHDI-wired	A WHDI sink may have a wired input, which forwards the wired-received A/V content to the wired interface. It must also convert and forward the control information.

WHDI entities and bridges in more detail in the following sections. Additionally, within a WHDI ecosystem, each WHDI-enabled device is allocated a unique *device ID* that is assigned solely by the WHDI LLC such that no two devices will ever receive the same ID. This is akin to device manufacturers assigning *Media Access Control* (MAC) addresses to a *Network Interface Card* (NIC), for example – these addresses are typically stored in hardware such as non-volatile memory. The device ID, like the MAC address size, is a 48-bit identifier used to recognize and address WHDI devices uniquely within a network.

15.5.1 The WHDI Network

WHDI-enabled devices within an ecosystem form a *registration* network, for example all WHDI devices within an end-user's home. Using different RF channels, concurrent transmissions are permitted within the same WHDI ecosystem, where an *active* network is characterized by the subset of devices engaged in a single transmission. The active network will include one *active source* device and one or more *passive sources* and *sinks*. Referring back to Table 15.3, we know that a source device is responsible for transmitting A/V content, data, and control information, as well as regulating transmission timings, and it remains the only device capable within the network that can transmit A/V streams. However, a passive source will only exchange data and control information with devices in the active network, but is not permitted to stream A/V content. What's more, a WHDI-enabled device may be connected to a certain active network, which may operate in the same environment using different RF channels. But a WHDI-enabled device is allowed to engage, connect, and disconnect with/from other active networks.

In Figure 15.9, we illustrate an example of a registration network which comprises three active networks. You should note, as we discussed and illustrated earlier (see Figure 15.8), that active networks **A** and **B** utilize a multicast, or point-to-multipoint, topology

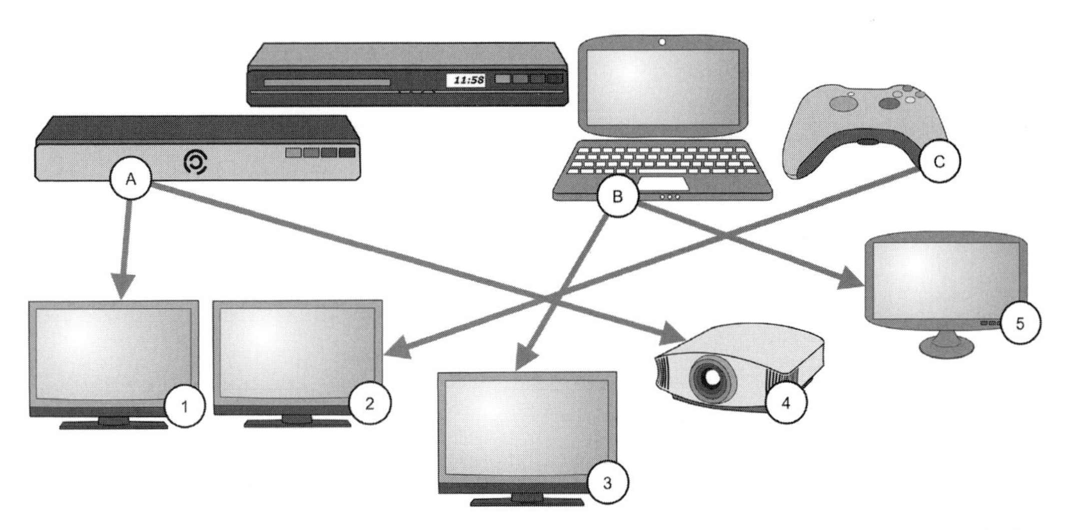

Figure 15.9. A typical registration network demonstrating point-to-point and multicast topologies.

and active network **C** is utilizing a point-to-point topology. In active network **A** we show an STB (an active source) streaming content to both a TV (**1**) in the living room and to a projector (**4**) – the TV and projector are passive sinks in the topology; in active network **B**'s multicast topology, we show a PC (an active source) streaming content to both a TV (**3**) that's located in the children's bedroom whilst streaming content to a monitor (**5**); again, the TV and monitor are both passive sinks. Lastly, in active network **C** the gaming console (an active sink) is streaming content solely (point-to-point) to the bedroom TV (**2**), a passive sink. A passive sink can only receive a single A/V stream from an active source; however, the end-user may select another source using the remote control device, for example. Finally, in Figure 15.10 we demonstrate a WHDI network with an active source (**A**), along with two passive sources, **B** and **C**. As we have already mentioned earlier, data and control information may be carried either along with the A/V stream or simply between active and passive sources and sinks. Normally, if one sink and/or passive source are connected to the active network, then the bandwidth available for data and control information is shared between the devices. The data/control network provides the ability for the connected sources and sinks to be managed and/or controlled.

15.5.1.1 Assigning Active Network Addresses

As you are already aware, each WHDI device carries a unique 48-bit device ID; however, once a device is connected to an active network, it is assigned a single-byte *Active Network Address* (ANA) which is unique within the active network. The ANA is used to address devices within the active network. The ANA is temporary, as once a device disconnects from the active network the address is discarded, although if the same device reconnects to the network it *may* be assigned the same ANA (you should not assume that the device will receive the same ANA once it reconnects). Naturally, if the device were to connect to a different active network, then the same method applies. Since there's always one active source within a network, the active source device has the ANA value

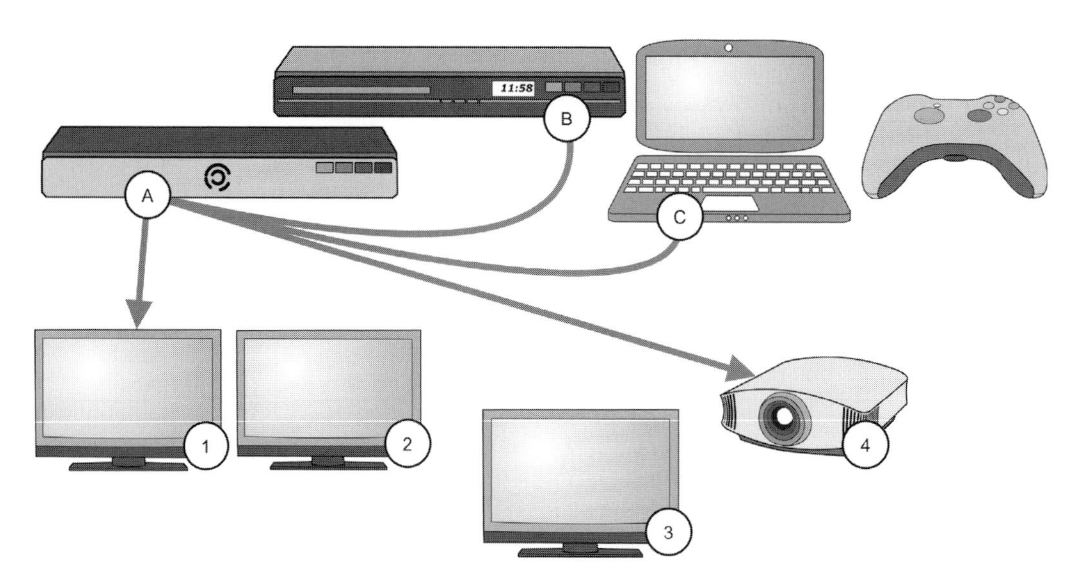

Figure 15.10. A WHDI network illustrating one active source with two passive sources, which, in essence, forms a data/control network.

`0x00`. As each sink or passive source connects to the active network, the active source will assign a unique ANA value arbitrarily or contiguously.

15.5.2 Network Management

A WHDI-enabled device can be in one of four network states within a registration network, and we describe these states in Table 15.5. In Figure 15.11, we illustrate the transition between states, applicable to both source and sink devices.

Table 15.5. A WHDI-enabled device may be in one of four states

State	Description
Unavailable	In this state, a device may characterize the device as switched off or in power-down mode. A device in this state is unable to initiate or accept a connection. The offline transition may typify the device being switched off or placed into a wireless-disabled mode.
Listening	In this state, the device is not in communication with any other devices but rather is available to connect. The power consumption in this state is low, and the device will periodically scan its spectrum for connection requests.
Link set-up	In a link set-up or "setting up a connection" state, the device is in the process of executing the link set-up protocol. If connection establishment fails, the device reverts to the listening state.
Connected	The device is connected and is actively participating within the network with one or more devices.

The WHDI MAC layer, which we discuss later in Section 15.9.2, "The WHDI MAC Layer," provides seven network related *services*, and we summarize these in Table 15.6. In short, these services provide access to the network and offer confidentiality of

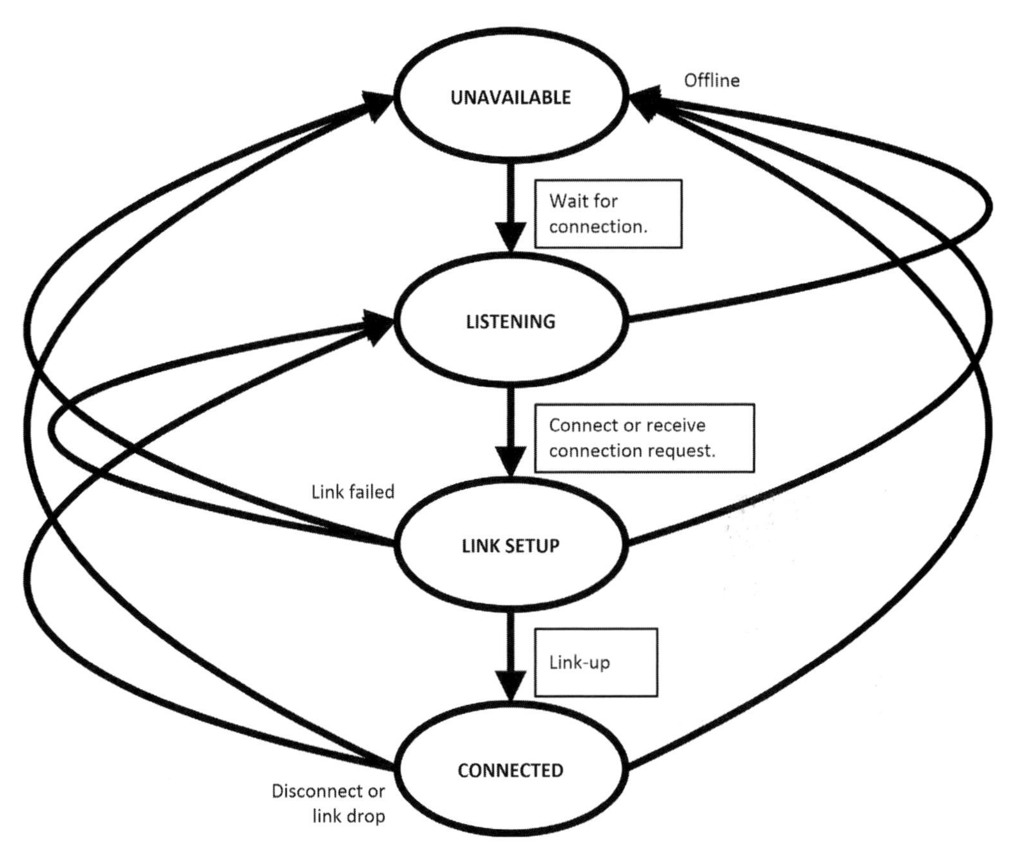

Figure 15.11. A source or sink device can be in one of four network states.

Table 15.6. Services that aid in the creation of a registration network and the formation of the active network

Service	Description
Registration	The process of registration is the act of introducing one device to another such that it is known within the network. This process helps with reducing unauthorized access to the network.
Unregistration	The act of unregistration is to remove or exclude a device from the registered network.
Discovery	The act of discovery allows a device to learn about neighboring devices prior to association.
Association	The association service provides the ability for a source and sink device to create a *session*.
Disassociation	A device may use this service to cease its connection with the active network.
Authentication	Encryption is used when a source and sink device become associated. Both devices authenticate each other to ensure the correct identity.
Privacy and encryption	This service ensures the privacy of communication between devices active within the network, and also protects content.

Table 15.7. A WHDI-enabled device may be in one of three states

States	Description
Unregistered	A source and sink device have not completed a registration operation.
Registered (unassociated)	A source and sink device have completed a registration operation and are registered with each other and may then set up a connection (associated).
Associated (and registered)	The source and sink device are part of the same active network exchanging A/V content, data, and control information.

information shared across the network. Executing a network service ultimately changes the relationship between two devices, whereby the relationship determines the level of communication permitted. A network service is invoked to change the relationship between two devices, where the relationship determines the level of permitted communication or demonstrates a high level of connectivity between two devices. These services enable the formation of a registration network and the creation of an active network. A source and a sink device will have one of the relationships shown in Table 15.7, where Figure 15.12 demonstrates the sequence of transition states. Furthermore, Table 15.8 depicts in a little more detail the communication that is permitted for a device in a given relationship state and shows the transition between states.

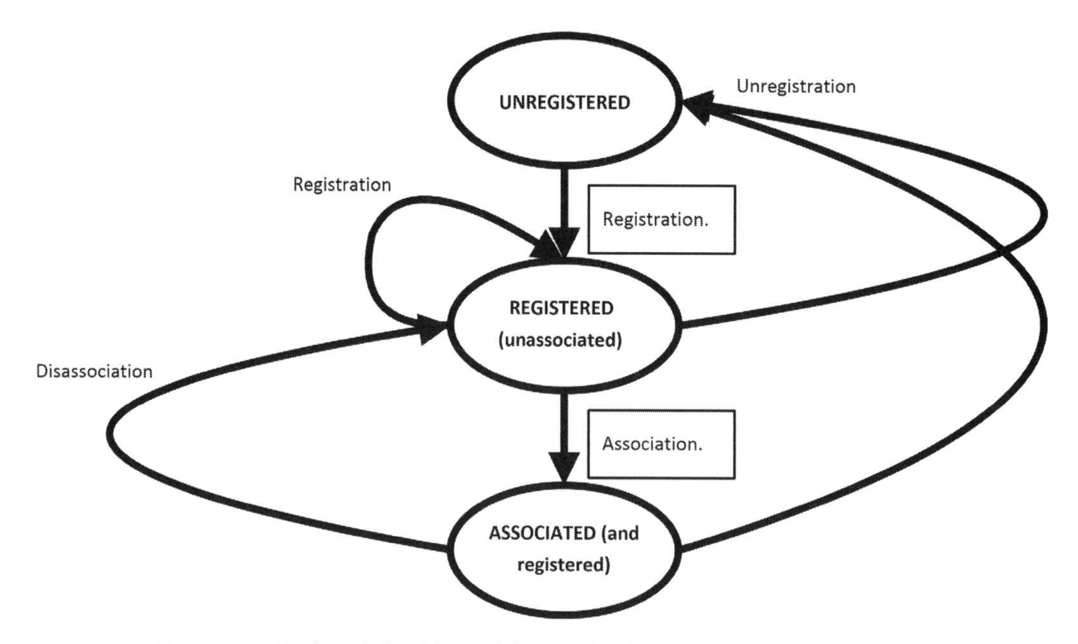

Figure 15.12. Device relationships and the associated transitions.

Table 15.8. Permitted communication, along with possible transitions

States	Permitted communication	Possible transition	
		To state	By service
Unregistered	Exchange registration information.	Registered (unassociated).	Registration.
Registered (unassociated)	Associated transactions; exchange registration information.	Unregistered Registered (unassociated). Associated (and registered).	Unregistered. Registration. Association.
Associated (and registered)	Any A/V content, data, control transmission.	Unregistered. Registered (unassociated).	Unregistration. Dissociation.

15.6 Comparing WHDI with HDMI Systems

It's worthtaking a moment just to compare WHDI and HDMI systems. The WHDI wireless system has to modulate A/V content that's acceptable over the wireless channel without (or with minimal) degradation, as well as extrapolate audio and video timings that would ordinarily be present over a HDMI cable. Nonetheless, the advantage afforded by the WHDI wireless system enables content to be multicast transmitted, along with over the air remote control switching between numerous sources, all using the same physical medium and interface.

15.7 Comparing WHDI with WLAN Systems

Likewise, it is also certainly relevant to take a moment to compare WHDI and WLAN systems. In a WHDI topology, a dedicated channel, always originating from one source, is used and occupied most of the time, whereas in a WLAN network many devices can share the same channel. The transmission of A/V content within a WHDI topology is also divided into consecutive frames, which are aligned with the video frame itself. WLAN systems tend to separate their transmission into independent data packets; however, with 802.11ac we shall witness a wealth of improvements. The WHDI method of transmission favors the streaming content buffering, which, in turn, eliminates latency.

15.8 WHDI's Video-Modem

As you may already know, A/V content delivered between a video source and its display over a hard-wired link such as HDMI is uncompressed. Likewise, a wireless interface will also need to deliver uncompressed content. Normally, delivery of compressed video between a source, such as Blu-ray/DVD players, STBs, etc., and a display is rarely

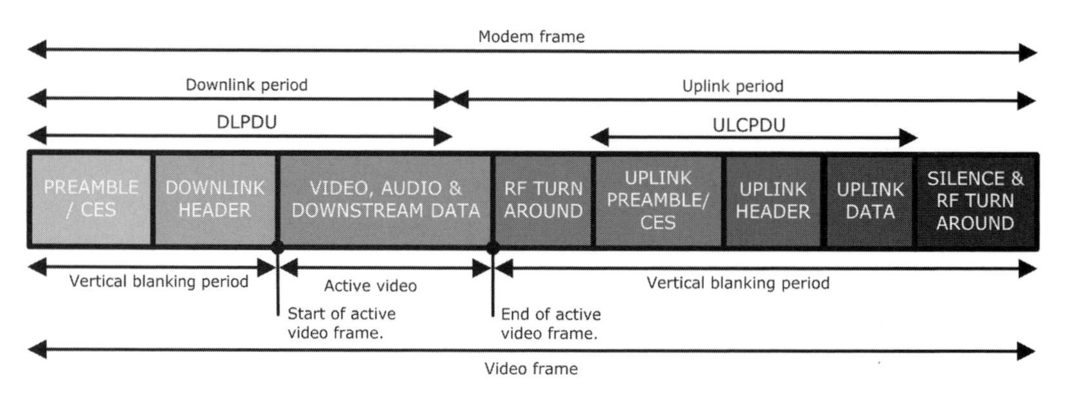

Figure 15.13. The PHY frame structure.

used since compressing this content can be subject to copy-protection infringement. Moreover, complexity surrounds the interoperability of the numerous codecs available to compress content, and it remains somewhat inconceivable for consumer electronic products to support them all. WHDI uses Amimon's video-modem technology and, as such, "the video coding and modulation are jointly optimized to enable capabilities far beyond those of traditional wireless modems that have been optimized for data."[7]

Downlink and uplink transmission share the same RF channel during a session, which is analogous to *Time Division Multiplexing* (TDM). In fact, typically 98% of the bandwidth is used for downlink traffic, whilst the remaining 2% is used for uplink data, although the actual percentage split is always subject to transmitted video timing. Uplink transmissions occur during a *vertical blanking period*, that is, a period where there's no video to transmit – the whole transmission format is divided into modem frames. During a downlink transmission, *Downlink PHY Data Units* (DLPDUs) are transmitted containing A/V and control information; conversely, during an uplink transmission, only *Uplink Initialization PHY Data Units* (UPIPDUs) are sent. When video frames are transmitted, the modem frames are aligned with the video frame, as we illustrate in Figure 15.13, so both video and modem frames have identical rates. Each frame offers a downlink and an uplink period contained within the DLPDU during synchronized bi-directional access, where the active source determines and publishes the use of each uplink period. The active source may decide to offer these periods as an uplink transmission window for a specific device, where *Uplink Control PHY Data Units* (ULCPDUs) can be transmitted by the same connected device. As we have already mentioned, each modem frame directly corresponds with a video frame, that is, the video information of just one video frame.

The *active video period* is defined as the time between the first pixel in a video frame and the last pixel, which includes any horizontal periods. So, the downlink data transmission is typically transmitted during the active video period, as shown in Figure 15.14. Using this active video period helps with buffering requirements and reduces latency

[7] Source: amimon.com.

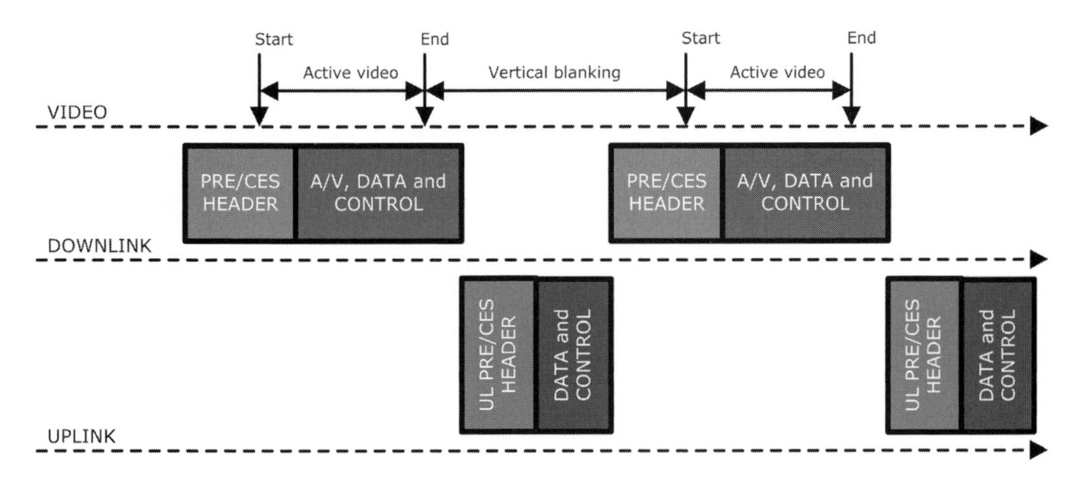

Figure 15.14. The video-modem frame structure.

in the system. What's more, the interval for the vertical blanking period is specifically used for downlink preamble and header transmission, as well as the entire uplink transmission, again as shown in Figure 15.14. If we take a look back at Figure 15.13, we can see that a *silence period* is used to space the uplink data transmission and the downlink transmission, to await the subsequent video frame. The silence period varies in length to accommodate the variances between the modem frame length and the video frame length.

15.9 The WHDI Architecture

In the sections that follow, we take a top-down approach to the WHDI protocol stack architecture – essentially, we take the technology from the top to the bottom and explore in greater detail the technology architecture and its software and application building blocks. WHDI specifies just three layers, and in Figure 15.15 we map the *Open Systems Interconnection* (OSI) model against the WHDI stack, along with its associated high-level modules, to gain a better understanding of the significant components that make up the overall stack architecture. You should also note that these building blocks have been labeled numerically to ease identification and classify responsibility. So, looking at Figure 15.15 again, block (**1**) represents the WHDI PHY layer, which is Consortium responsible as we discuss later, in Section 15.9.3, "The WHDI PHY Layer"; block (**2**) represents the WHDI MAC layer (see Section 15.9.2, "The WHDI MAC Layer") and the *Audio/Video Control Layer* (AVCL) (see Section 15.9.1, "The Audio/Video Control Layer"), which are both Consortium responsible; and finally (**3**) represents building blocks that are manufacturer-specific, namely the end-application.

The application layer is the responsibility of the end-user or manufacturer, as they will utilize the WHDI specification and, in turn, engineer a WHDI-compliant product. The application layer provides communication with the AVCL in order to control and

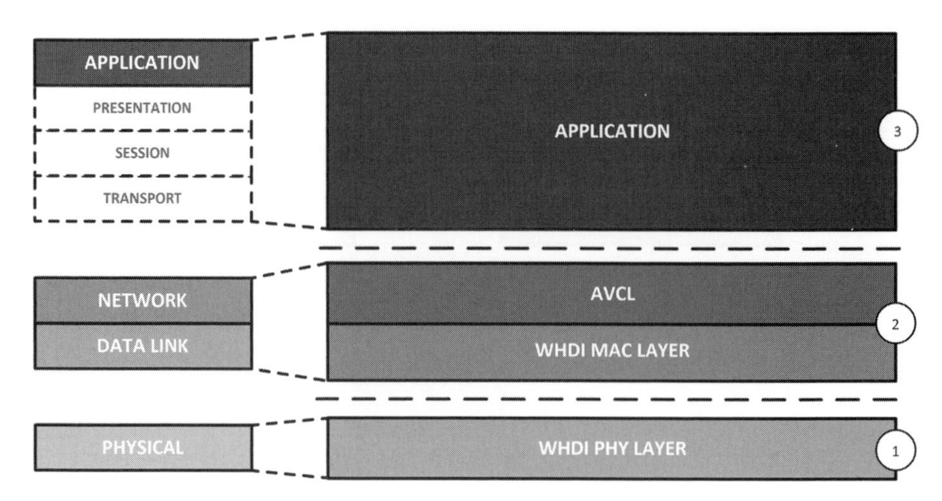

Figure 15.15. The building blocks that form the WHDI architecture.

realize specific WHDI functionality. Moreover, the application layer will also deliver A/V signals directly to the WHDI PHY layer, as we depict in Figure 15.16. In the sections that follow, we briefly explore the relative responsibilities managed by the AVCL, MAC, and PHY layers.

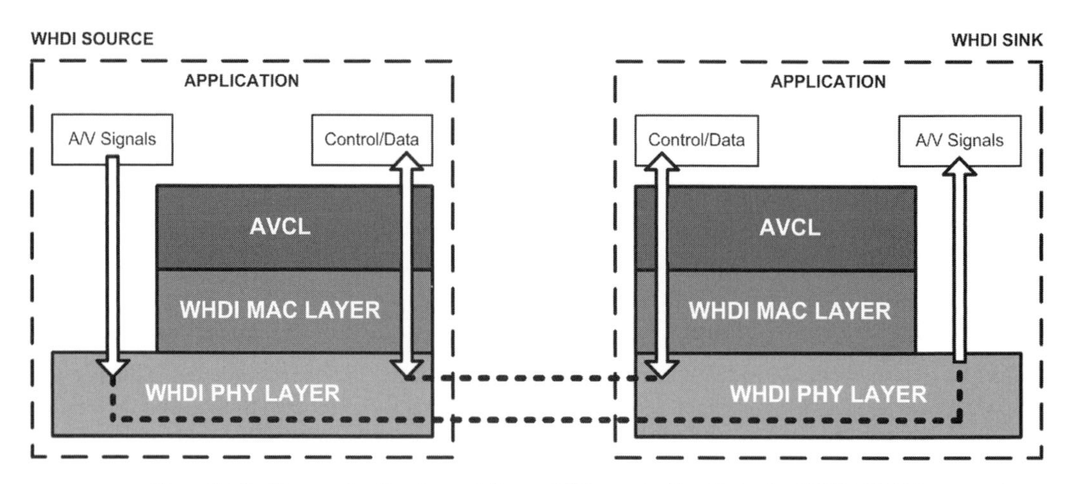

Figure 15.16. The application layer delivers A/V signals directly to the WHDI PHY layer, and control/data information is passed up and down the WHDI stack.

15.9.1 The Audio/Video Control Layer

The WHDI AVC layer presents to the application level a "package" of protocols to enable varying A/V-specific processes for newer and existing protocols. Primarily, connection management provides a unique mechanism to distribute both the A/V and wireless connectivity information to all the devices that form the registration network – this also includes the capability for HDMI *hot-plug* detection. In the WHDI ecosystem, it is important for the source device to learn of the video and audio capabilities of the active video and audio systems. As such, *Extended Display Identification Data* (EDID) support is also offered so that a digital display can share its capability know-how with a video source. *InfoFrame* support is also provided at this layer, along with *Remote Device Control* (RDC), something which we discussed earlier. The AVCL uses the conceptualization *subdevice* rather than device, since each subdevice represents an independent entity in terms of A/V processing and its associated management. A WHDI device comprises one subdevice, but a number of subdevices may be present on one WHDI device if the WHDI device possesses multiple functionalities such as a tune and a digital video recorder.

A sink device will inform the active source via the AVC layer of its A/V-specific capabilities (EDID) that it wishes to receive an A/V stream, is currently receiving and rendering a stream, or that it no longer wishes to receive a stream. Furthermore, the AVCL is used by all devices to execute RDC-related functions, such as playback and changing channels.

15.9.2 The WHDI MAC Layer

The WHDI MAC layer manages the interface to the PHY layer as well as the AVCL. It provides the automatic set-up of a communication channel as well as interference avoidance, and ensures coexistence with other technologies in proximity. Like the PHY layer, the MAC layer is also responsible for encryption and the registration, association, and authentication services as we discussed earlier. Finally, the MAC layer also has the responsibility for frequency reuse, frequency selection, and *Dynamic Frequency Selection* (DFS), as well as for transmit power control.

15.9.3 The WHDI PHY Layer

The WHDI PHY layer has the responsibility for communication of A/V and data over the air-interface. More specifically, it manages the uni-directional transmission of A/V as well as the internal bi-directional data channel for uplink and downlink communication. It also provides encryption over the wireless link. Additionally, the PHY layer receives and transmits (source and link) raw video samples in 4:4:4 Y'CbCr, a subsampling technique used in high-definition equipment and audio, in various formats. We discuss video and audio formats in more detail in Section 15.10, "Audio and Video."

15.10 Audio and Video

In this section, we take a snapshot view of the requirements provided by the WHDI v1.0 specification for both audio and video support.

Table 15.9. The WHDI audio encapsulation formats

	Format
1	Audio stream formatted to either IEC 60958 or 61937 standards.
2	Up to eight channels of L-PCM audio stream with up to 24-bits per sample.
3	Up to eight channels of L-PCM audio stream with 16-bits per sample.
4	IEC 61937 data block stream only.
5	DSD (1-bit audio) streams.
6	DST (compressed DSD) streams.
7	Vendor-specific audio bytestream.

15.10.1 Audio Requirements Snapshot

WHDI itself specifies a number of formats suitable for transporting audio streams over a connection. More specifically, as we already touched upon earlier in this chapter, and as we summarize in Table 15.9, these are supported within WHDI. Audio sample words and synchronization data are formatted into a single-bytestream for all formats listed in Table 15.9 – the stream is encoded prior to transmission with the highest level of protection.

WHDI *basic audio* is defined as a two-channel L-PCM audio stream at a sample rate of 32 kHz, 44 kHz, or 48 kHz, along with a sample size of 16-, 20-, or 24-bits in either IEC 60958/61937 or raw L-PCM encapsulation – this ensures compatibility with EIA/CEA-861-E. As such, if a source device offers support for audio transmission across any output, then it will provide support for WHDI basic audio. A sink device will receive any audio format, even if a sink device doesn't support that format, and, as such, the sink device will not affect any video or control handling information. A sink's subdevice will support all audio capabilities, as indicated by the *Vendor-specific Data Block* (VSDB) EDID structure. You may recall that WHDI supports a multicast topology (see Section 15.5, "Networking Topology"), in which an active source transmits the same A/V stream to multiple sinks; therefore, it is necessary for some sinks within the ecosystem to disregard audio formats that it neither supports nor offers adverse effects, such as loss of video or control information.

15.10.1.1 Concurrent Audio and Video Support

In the same transmission stream, a WHDI device will concurrently support minimum audio and video requirements. However, manufacturers of WHDI-compliant products should consider a trade-off when implementing combined audio and video formats. The trade-off specifically refers to the audio bit-rate and the video quality, with reference to resolution, frame rate, incurred visual noise/distortion, and so on. For example, a 3 Mbit/s audio stream with any video format up to 1080p, 60 *frames per second* (fps), will normally offer good quality video, although a 24 Mbit/s audio stream with 1080p, 60 fps, over a 20 or 40 MHz channel may cause poor to moderate video quality.

15.10.1.2 Audio Latency and Lip-sync Correction

In general, audio latency in a WHDI source (when delivered on any output port) and a WHDI source that is a bridge should be less than 10 ms. Likewise, latency within a sink device should also remain less than 10 ms. The WHDI specification offers two methods for lip-sync correction. The first method relies on the rendering device's internal capability to provide a sufficient audio delay, whereas the second method relies on the source device's capabilities of delaying the audio stream. However, these two methods should be chosen for use according to context. For example, the initial method is recommended for multicasting environments, whereas the latter should be used in an environment where a single audio receiver and video display is used. Lip-sync correction is recommended when the video latency of the expected rendering path is greater than 20 ms.

Table 15.10. A WHDI source supports at least one of these video format timings

	Format
1	640×480p at 59.94/60 Hz
2	720×480p at 59.94/60 Hz
3	720×576p at 50 Hz

15.10.2 Video Requirements Snapshot

The WHDI specification offers support for any EIA/CEA-861-E, *Video Electronics Standard Association* (VESA) defined, or any other video format timing to be transmitted, for which interoperability is maximized between products with common *Digital Television* (DTV) formats that are defined and conditionally required. The video format timings actually specify the pixel and line counts and frame rate, as well as whether the format is *interlaced* or *progressive*. What's more, WHDI permits vendor-specific timings to be used. As we mentioned earlier, in Section 15.9.3, "The WHDI PHY Layer," video data across the PHY layer is a representation of 4:4:4 Y'CbCr, but data relating to video synchronization for *Horizontal/Vertical* (H/V) sync polarities, positions, and durations is also transmitted. It is left to the WHDI source to select the video format of the transmitted signal using information specific to the source video content, along with format conversions available at the source, the format capabilities, features of the sink, and the inherent quality of the wireless connection.

Minimum requirements are specified such that maximum compatibility between the source and sink devices is provided. This means that both a source and sink device will always support 20 MHz and 40 MHz transmission bandwidths, whilst all WHDI sources will support four space time streams or two space time streams, where some categories of device only support two streams; for example, low power, battery operated type devices, such as mobile and tablet products. However, a sink device will support the reception of all four space time streams. In Table 15.10, Table 15.11, and Table 15.12, we list other video timing support combinations supported by the WHDI specification.

Table 15.11. A WHDI sink at 59.94/60 Hz video format supports all of the following video format timings

	Format
1	640×480p at 59.94/60 Hz
2	720×480p at 59.94/60 Hz

Table 15.12. A WHDI sink at 50 Hz video format supports all of the following video format timings

	Format
1	640×480p at 59.94/60 Hz
2	720×576p at 50 Hz

It is recommended that WHDI sink devices supporting 50 Hz video formats also support 720×480p at 59.94/60 Hz to encourage successful interoperability for both source and sink devices. Likewise, a device that supports HDTV capability will also support 1280×720p and 1920×1080i video timing formats either at 50 Hz or 59.94/60 Hz. A WHDI device supporting any of the video formats listed in Table 15.13 using a component analog or uncompressed digital video port must also support that video format timing across the WHDI port. As such, confusion is avoided for the end-consumer, since a product's WHDI video format support is recommended to be equivalent to any analog or uncompressed digital video port.

Table 15.13. A WHDI device supporting any of the video formats using a component analog or uncompressed digital video port must support these video format timings

	Format
1	1280×720p at 59.94/60 Hz
2	1920×1080i at 59.94/60 Hz
3	1280×7209 at 50 Hz
4	1920×1080i at 50 Hz
5	720×480p at 59.94/60 Hz
6	720×576p at 50 Hz

A source device may change the parameters associated with the transmitted video format at any time – the sink device must also support the new format. As such, these new parameters are sent to the sink via the DLPDU extended header, where the sink device adheres to the new information.

15.10.2.1 Video Latency

Overall, video latency in a WHDI source when transmitted on any output port (with no de-interlacing, scaling, or frame rate conversion) and in a WHDI source that is a bridge should be less than 3.5 ms. Likewise, latency within a sink device with no de-interlacing, scaling, or frame rate conversion should also remain less than 3.5 ms.

15.11 Security

In summary, the WHDI security system comprises two levels of security schemes, namely *content protection* and *wireless security*. The former scheme provides protection for distrusted content to sinks that are approved on the WHDI network, whereas the latter scheme ensures privacy and network access control.

15.11.1 The Content Protection Scheme

As previously discussed, the WHDI specification offers content protection for content transmitted using WHDI technology. The WHDI specification recommends HDCP v2.0, something we have already touched upon, but there are numerous content protection methods currently available from the media industry, so the WHDI specification offers a framework within which a manufacturer can easily adopt existing content protection technology.

15.11.2 The Wireless Security Scheme

It is unavoidable that removing the cable results in the introduction of a new level of complexity relating to the inherent security offered by a fixed connection, and therefore WHDI technology is architected with security in mind. As such, WHDI technology provides bi-directional authentication and bi-directional information encryption as a security mechanism to manage network accessibility; this offers privacy for the user and secures the distributed content when a content protection scheme is not available. So, the wireless security scheme prevents unauthorized sink device access to the WHDI ecosystem, which may obtain A/V content from a source device, likewise preventing unauthorized source device access from transmitting unwanted content to a sink device and preventing any unauthorized access to the network from a device that does not possess a valid WHDI device certificate.

Part IV

Forthcoming Technologies and Conclusions

16 Future and Emerging Technologies

In this penultimate chapter, we discuss future and emerging technologies that didn't make the relevant sections in this book – the technologies were either too immature or sufficient information could not be obtained at the time of writing. Perhaps we can include these technologies in a future edition if they have matured sufficiently and have been widely adopted. Nonetheless, this chapter introduces a number of technologies that are definitely worth watching, as they will ultimately redefine and enhance our user experiences with wireless connectivity. The wireless industry is notorious for advancing at a modest pace, but manufacturers are always eager to launch the latest wireless-enabled gadget, a difficult dichotomy for both industries. In this chapter, we review a number of high profile technologies that have received significant press coverage, and in Figure 16.1 we take a timeline snapshot of these technologies to demonstrate their maturity level.

16.1 802.11ac

We discussed in Chapter 13, "The 802.11 Generation and Wi-Fi," that a number of products have emerged touting .11ac "draft" capability – very much reminiscent of the 802.11n draft standard furor. 802.11n is only several years' old and provides backwards compatibility with .11a, .11b, and .11g legacy products. Likewise, .11ac also provides backwards compatibility with the same legacy products – this is primarily due to the incumbent base of products that still very much perform, even today, the function that they were originally designed to do. If you recall from Chapter 13, 802.11b was introduced as far back as 1999, and undoubtedly it may take another decade or so for consumers to exchange their access points for newer products.

16.1.1 Super Wi-Fi or 5G?

What's more, the industry has erroneously acclaimed this new generation of Wi-Fi technology as "5G," something which is typically attributed to mobile or cellular generation technology. For example, we have witnessed the cycle of Apple products evolving generation after generation, seemingly aligned appropriately with a cellular technology, yet Apple avoided the 4G labeling, as the cellular technology was far from ready. A similar misrepresentation has been made of *white space radio* (see Section 16.2, "White Space

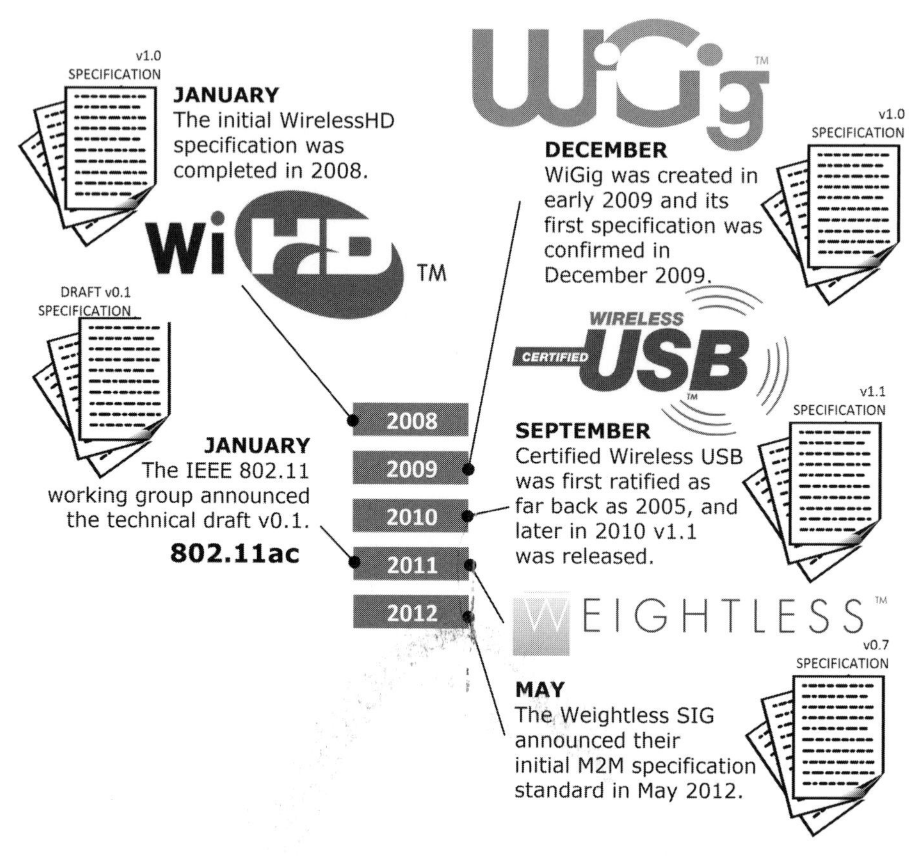

Figure 16.1. The current timeline of a select number of technologies that have received substantial press coverage and are set to become publicly available.

Radio"), where the industry has characterized the technology as "Super Wi-Fi." This inappropriate labeling of technologies often causes confusion for everyday consumers. Nevertheless, the new 802.11ac has been described as "Wi-Fi on steroids," and can be more realistically regarded as "Super Wi-Fi."

16.1.2 The Obvious Market Potential

Wi-Fi has nowadays become synonymous with the ability to connect to the Internet; in fact, some consumers incorrectly regard Wi-Fi *as* the Internet. 802.11ac is to improve on speed and reliability, something which the *Institute of Electrical and Electronic Engineers* (IEEE) has successfully maintained and evolved since introducing the initial standard. Naturally, Internet speeds vary across *Internet Service Providers* (ISP) and location, so the need to deliver a theoretical and rumored 1 Gbit/s capability does raise the question "Where are the IEEE hoping to align the technology's market potential?"

Perhaps more obviously, due to its robust and reliable delivery mechanism, 802.11ac can realistically shift into not only providing Internet access, but also enabling content

Figure 16.2. It is anticipated that the IEEE is leaning toward not only a faster Internet experience, but also the possibility of opening up a range of consumer electronics devices that share content with one another.

streaming, which, in turn, inevitably puts it into competition with the likes of WHDI (discussed earlier, in Chapter 15, "One Standard, All Devices: WHDI"), WiGig, WirelessHD, and Certified Wireless USB – something we discuss later in this chapter. The IEEE is expected to finalize the 802.11ac specification during 2013, so early "draft" .11ac products may not necessarily interoperate with the official specification. In Figure 16.2, we illustrate an example ecosystem, in which an 802.11ac access point (**1**) still provides Internet access for the number of like-enabled devices as shown (**2** to **7**); moreover, each device may have Internet capability, and the devices may also be capable of connecting and streaming audio/video content to an HDTV (**8**).

16.2 White Space Radio

We have all had the experience of watching a portable TV in our bedroom, employing a very flexible antenna, normally a metal coat-hanger, to enable us, along with some dexterous bending and positioning, to receive a picture. Well, that's it in a nutshell – that essentially sums up the excitement surrounding white space radio.[1] It can travel great distances and can penetrate walls and, of course, the deep recesses of our homes. Radio

[1] Gratton, D. A., "White Space Radio: Revisited," 2011.

frequencies are assigned to specific functions for broadcasting purposes – typically, government agencies or bodies are responsible for their assignment. When these frequencies are assigned, channels are spaced so that interference does not occur between them, or perhaps a guard band is used to ensure delimitation. As a consequence, potential radio spectra are becoming available. In one such example, the switchover from analog to digital TV will ultimately render the analog frequencies redundant, increasing the availability of unused parts of the radio spectrum. With digital transmission techniques, channels can now be compressed, enabling them to be transmitted adjacently whilst confidently ensuring that there is no interference. White space refers to this unused set of frequencies and, in fact, a lot of attention has been given to those frequencies that are normally reserved for television broadcast.

16.2.1 Market Opportunity

The excitement surrounding white space radio is primarily focused on its ability to travel great distances, along with its enviable penetration of walls and whatnot. Indeed, other wireless PAN technologies may look upon white space radio with envy, as a white space-enabled access point would cover an area ten times greater than Wi-Fi. In fact, white space radio provides connectivity opportunities that appeared in our disruptive technology discussion, something we covered in Chapter 3, "Disruptive Topologies through Technology Convergence." The lower range, 50 MHz to 700 MHz, are available, and proponents of white space radio have turned their attention to the 600 MHz frequency. The focus on this particular frequency is pertinent, as it possesses exceptional propagation characteristics. The antagonists of white space radio have voiced their concerns, claiming that use of the now redundant frequencies may cause interference with existing TV broadcasting, whilst others have suggested that it may never happen.

Figure 16.3. WeightlessSIG logo. (Courtesy of the WeightlessSIG.)

A new *Special Interest Group* (SIG), the WeightlessSIG (weightless.org) – we show its logo in Figure 16.3 – has been formed to help promote and support the future of the technology. Several influential industry leaders have already provided the right momentum for this to move forward, not just with an interest in the technology, but a realization of the potential commercial possibilities. These primary pundits are attempting to establish credible evidence to ensure that there is a sustainable future for white space radio and that wireless broadband can be delivered to homes that are affected by poor traditional broadband coverage.

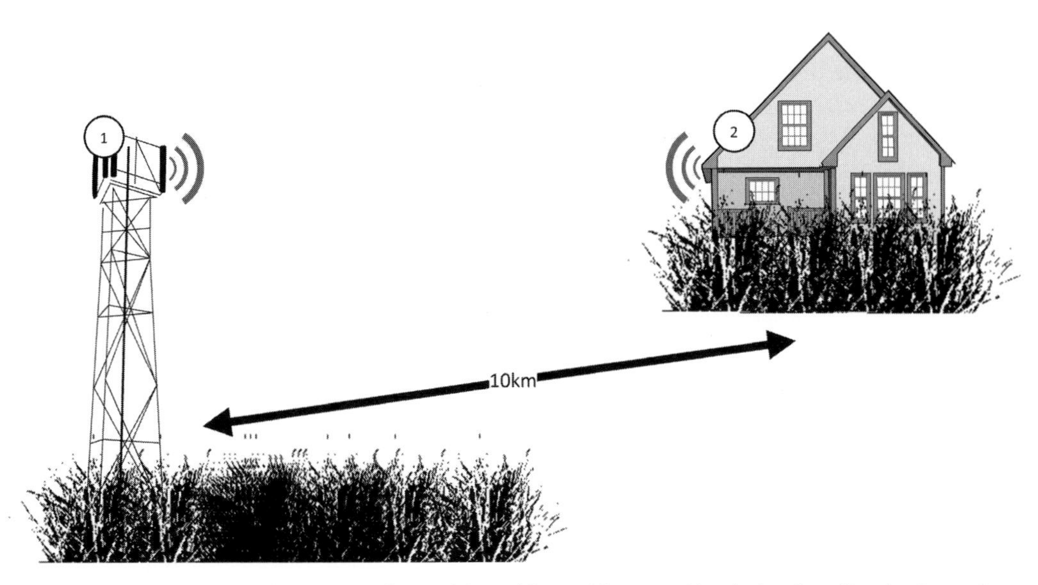

Figure 16.4. White space radio would provide rural homes with wireless broadband; a host of other use cases exist, such as M2M connectivity.

In Figure 16.4, we depict a scenario offered by white space radio whereby the provision of wireless broadband can be supported by such a white space ecosystem; here we can see the base station (**1**) providing a wireless broadband service to a rural home using a terminal product (**2**) over a distance of 10 km. Other potential user scenarios include *Machine-to-Machine* (M2M), *smart grid*, and miscellaneous "smart" applications.

Figure 16.5. The USB trident symbol has become synonymous with everyday wired connectivity. (Courtesy of the USB-IF Administration Team.)

16.3 Certified Wireless USB

The *Universal Serial Bus* (USB) wired connection is an industry standard that was developed as far back as the mid 1990s by a consortium of companies, including Intel, IBM, and Microsoft – the standard and technology are ubiquitous. The wired technology has been integrated into over two billion consumer electronic products and has become the de facto standard, widely used to interconnect a myriad of devices. The USB trident symbol, as shown in Figure 16.5, has become synonymous with simple USB cabled connectivity. So it seems a natural evolutionary step to cut the cable and to enable USB wirelessly, but agreeing on a wireless technology that would enable wireless connectivity for the USB equivalent has been an onerous and protracted mêlée.

16.3.1 Choosing Ultra-wideband for Wireless USB

Ultra-wideband (UWB) has had an arduous journey over its relatively short developmental timeframe.[2] The ultra-wideband standard is open and is available to manufacturers – such that these manufacturers are able to participate in the development and growth of the technology. Nonetheless, the process is occasionally fraught with pitfalls, as there will always be an endless debate on the use of various techniques in the design of the radio, hardware, and software architectures. Fundamentally, manufacturers must agree upon an individual solution; a way forward that is adopted by the majority – a single technique that will prevail and become the blueprint from which all other manufacturers will develop and interoperate their future UWB-enabled products.

The onerous and protracted mêlée that we mentioned in our opening paragraph has, for UWB, originated as far back as 2003. In its early days, the technology faced a dichotomy of radio modulation techniques, and at the time it was suggested that the two solutions were viable for market and could potentially coexist. The debate surrounding the *Direct Sequencing* (DS or DS-UWB) modulation technique versus the *Orthogonal Frequency Division Multiplexing* (OFDM or OFDM-UWB) approach took UWB in a reverse direction in terms of delivering a viable technology that would successfully empower Wireless USB. The WiMedia Alliance (wimedia.org) was formed in 2002, merged with the *MultiBand OFDM Alliance Special Interest Group* (MBOA-SIG) in 2005, forming the umbrella WiMedia Alliance, and took overall responsibility for defining and evolving the future of UWB. However, the debate endured for four years or so, creating a deadlock amongst contributing manufacturers – a stalemate that left the Alliance with two incompatible radio schemes.

So, the WiMedia Alliance chose to step away from the task group and sought an arbitrator who would hopefully resolve the rift between the two camps, namely DS-UWB and OFDM. What's more, the respective technical marketing engines of the various big-player manufacturers initiated a charm offensive, knowing that eventually there could only be one way forward. Inevitably, the two opposing technology camps commenced an aggressive political strategy to sweeten the various key players. The charm ploy was clearly an attempt at saving grace, as various companies had already invested heavily in their chosen radio scheme and, of course, in what they thought was the best performing radio modulation technique. An arbitration committee was formed to attempt to choose between the two incompatible technologies; the committee eventually adopted the OFDM technique as the emphatic choice for the future evolution of UWB. In essence, Wireless USB (Certified Wireless USB) uses WiMedia's *Common Radio Platform* (CRP) to fulfill the next evolution of wireless-enabled USB devices.

[2] Gratton, D. A., "The WiMedia Alliance: What Do We Have in Common?," 2009.

Figure 16.6. The Certified Wireless USB logo is synonymous with true Wireless USB. (Courtesy of the USB-IF Administration Team.)

16.3.2 Certified Wireless USB and 802.11ad

The incessant debate as to when UWB will be available has inevitably left many manufacturers impatient. So much so that some have turned to Wi-Fi (802.11n and 802.11ac), WHDI, WiGig, and WirelessHD as alternatives to provide a transport for delivering streaming content. *Certified Wireless USB* (W-USB) is mandated by the USB *Implementers Forum* (or USB-IF), and we illustrate its logo in Figure 16.6. The USB-IF has made the distinction between "Certified" Wireless USB and "Wireless USB" (as some other manufacturers have labeled their products) to provide consumers with the confidence that they are indeed using a product that has been certified and developed by the USB-IF. Numerous technologies, as we have already mentioned, currently compete with W-USB, and unfortunately the technology hasn't been widely adopted despite it being the obvious choice. Moreover, other technologies have had the opportunity to establish their own market whilst the debate ensued. As such, competing technologies have inadvertently placed W-USB on the "top shelf," so to speak. Nevertheless, with the new 60 GHz in development by the IEEE, 802.11ad, we may witness a partnership between USB-IF, W-USB, and 802.11ad, which, in turn, will become a viable union for the future of "Certified Wireless USB." Moreover, with the recent announcement of the partnership between the Wi-Fi and WiGig Alliances, some are doubting whether W-USB will ever materialize.

16.4 WiGig

As we have already touched upon, the *Wireless Gigabit Alliance* (WiGig) competes with WHDI, WirelessHD, Certified Wireless USB, and 802.11ac; we illustrate in Figure 16.7 WiGig's logo. Utilizing the 60 GHz frequency band, the Alliance was formed to create a specification for delivering multi-gigabit speeds, meeting the needs of today's digital audio/video content hungry consumers. Both the WiGig *Physical* (PHY) and *Media Access Control* (MAC) layers are capable of delivering up to 7 Gbit/s, operating in the 60 GHz spectrum, which, in turn, permits wider channels that support faster transmission speeds. Recently, however, the Wi-Fi and WiGig Alliances have announced[3] their joint effort to promote 60 GHz wireless technology.

[3] O'Brien, T., "WiFi and WiGig Alliances Become One, Work to Promote 60GHz Wireless," 2013.

Figure 16.7. The WiGig logo. (Courtesy of the Wireless Gigabit Alliance.)

16.4.1 WiGig and 802.11ad

As we have already mentioned, the WiGig specification provides the PHY and MAC layers and is based on the 802.11 standard. This motivation inherently provides IP networking over the 60 GHz spectrum, in turn providing cost effective integration which can interoperate with 60 GHz devices, as well as existing Wi-Fi tri-band radios, that is, 2.4 GHz, 5 GHz, and 60 GHz. The WiGig Alliance will also provide a *Protocol Adaption Layer* (PAL), which will offer support for data and display standards, such as *High-definition Multimedia Interface* (HDMI), DisplayPort, and other *input/output* (I/O) devices.

16.4.2 WiGig Beamforming

The use of the 60 GHz RF band presents the challenge of *propagation loss*, which is higher when compared with 2.4 GHz and 5 GHz frequency bands. As such, the WiGig specification overcomes this shortcoming by using a technique called *adaptive beamforming*, which permits a more robust communication at distances greater than 10 m. It uses *directional antennas*, which reduce interference and ultimately bind the signal between two devices into a "concentrated beam." Essentially, this provides a line-of-sight transmission, in turn providing faster data transmission; however, if the line-of-sight transmission is interrupted, for example by someone walking in between the two devices, then a new communication pathway may be established using beams that reflect off walls.

16.5 WirelessHD

The WirelessHD (wirelesshd.org) consortium of companies, which include leading technology proponents from the consumer electronics industry, have collectively developed an industry standard specification for a wealth of electronic products, boasting high speed data rates of between 10 and 28 Gbit/s and utilizing the 60 GHz RF band. The WirelessHD specification openly uses the *Wireless Video Area Network* (WVAN), something we touched upon back in Chapter 2, "What is a Personal Area Network?," supports lossless HD audio/video and data, and provides multi-gigabit support and *smart antenna* technology. In a similar manner to WiGig, WirelessHD is imposed by a line-of-sight operation, but overcomes this limitation by using a protocol that adapts to environmental changes. Using smart antennas, high speed transmissions can be directed in a similar vein to WiGig's beamforming.

17 Summary and Conclusions

The wireless industry is notorious for advancing and evolving at a modest pace; a process that doesn't satisfy eager manufacturers when they wish to develop and launch the latest "all-singing-all-dancing" new wireless gadget. Moreover, consumers inevitably find themselves somewhat ambivalent to the next-generation thing. It must both appease and frustrate them, since they've presumably had ample opportunity to become familiar with one particular set of technology traits and idiosyncrasies only to be duly informed that they will soon have to embrace and understand the next generation. This process is unavoidable, and yet it can be incredibly exciting for some.

In this final chapter, we review and summarize *The Handbook of Personal Area Networking Technologies and Protocols*. We'll paraphrase our journey so far, and use an anecdotal narrative to reinforce what we have learned and perhaps discover what may happen next.

17.1 Making Sense of Wireless Technology

We started our journey with Part I, "What's in Your Area Network?," where we offered a broad perspective of what constitutes a personal area network and what external factors influence its dynamics. We explored how wireless technology existed for over 100 years whilst avoiding a lengthy history lesson. We further learned more about the nineteenth-century wireless secret and discovered that a simple development model, *Interoperability, Coexistence, and Experience* (ICE) may be used to aid innovators of future wireless products. The wireless communication industry is still mastering how to inject ease of use and transparency into their products. A seamless experience still escapes most wireless-enabled products today and, alas, we will continue to embrace and witness only modest advances in the overall usability factor. The wireless personal area network domain is arguably crowded with multiple technologies, and there seems to be a duplication of technologies and applications, all eager to grab our attention and convince us to buy-in to their particular choice; their way of working, if you like. You may also recall we played devil's advocate when we contentiously entertained the possibility of a "one-size-fits-all" technology with wireless convergence; something we have already started to witness with several semiconductor manufacturers converging multiple technologies into a single chip. Is this merely speculation; will we start to witness a shift to a single technology, or does our assorted mix have a unique function to perform?

The diversity of wireless technology has permitted consumers to remain permanently connected to the wide area network irrespective of location, and we demonstrated how the *Lawnmower Man Effect* (LME) typified our effervescent need to sustain an IP-fix. We also drew upon some examples wherein our *Personal Area Networking* (PAN) space was being drawn into various social media platforms, and our ability to curate content seemed endless. Further fueling our LME supposition, we discussed media convergence and how our ability to curate content has changed the focus of major broadcasters, who have adapted, and to whom we are no longer the audience, as such. Nowadays, we have become integral to a show's entertainment factor, allowing us to participate through social media.

17.2 Smarter Devices

In Part II, "The Wireless Sensor Network," we deluged you with a host of technologies that are responsible for scoping the smart home and creating intelligent buildings. There has been an enormous shift in how devices communicate with each other, not just from a wireless technology standpoint, but also from a perspective where devices can become responsible for engaging with other devices and even you! For example, you may have set up your smartphone device to notify you when you have reached a particular location or simply to notify you of a new tweet or Facebook message; admittedly, you have already collaborated with your smart device and offered criteria for this dialog to occur. Nowadays, your electricity, gas, and water meters are instrumental in initiating a dialog with your energy provider, and are conscious of your energy consumption and utilization. They invariably have the ability to offer regular feedback to a backend system at the utility suppliers' computer network over the Internet, very much akin to what we describe as the *Internet of Things* (IoT). Perhaps the next shift in this communication ethos is when your refrigerator incessantly bleeps at you to inform you that your milk supplies are low and that you need to purchase some more cheese, as the packet you have has reached its best before date; one of many facets in the new generation of smart devices – a shift that perhaps absolves us of rudimentary responsibility. In fact, there are companies that still tout the premise of talking refrigerators and the need to turn your television and lights on from afar, something that's very reminiscent of smart home and building automation. Moreover, the ability to use your mobile phone as some kind of wireless-overkill-enabled remote control to turn on your living room light may seem a little excessive, although Sylvester and Tweety Pie (Looney Tunes) would surely love it, "I thwought I daw a light come on"!

Placing domotics and building automation into a more realistic "light," so to speak, means that the industry is very much geared toward embracing a greener footprint. With energy prices increasing exponentially, we are all reminded to be aware of our increasing household or commercial/industry expenditure, at the same time saving the planet and avoiding our relentless depletion of our natural resources.

17.3 Keep It Unplugged

In Part III, "The Classic Personal Area Network," we covered what was penned as the classic collection of wireless PAN technologies we have grown accustomed to over the last decade or so. Our smartphones and miscellaneous consumer electronics products are perhaps becoming a little heavier as new chipsets are being introduced – the convergence of wireless technology has only reached a level of combining technologies onto a single chipset, but for now one wireless technology suits a particular application. For example, Bluetooth offers us wireless stereo headset functionality and hands-free capability; *Near Field Communications* (NFC) permits us to make small purchases, as well as grants us access to buildings, all conjured up by our smartphone; Wi-Fi technology will always provide us with a connection to the Internet; whilst WHDI allows us to stream content through our entertainment- and multimedia-enabled devices. The issue of redundancy only emerges when several wireless technologies provide duplicate scenarios, but then we will undoubtedly witness an onslaught, a challenge to the end where one technology ultimately dominates and others will undoubtedly phase out of existence.

17.4 What's Next?

We have no crystal ball, as such!

The wireless technology industry will continue to evolve and improve those technologies that have already been introduced into the everyday language. Undoubtedly, new technologies will arise that claim to be better, and possibly these claims will prove to be true. As we have already highlighted in this book, the WPAN space is relatively new, and there is so much anticipation as to what will happen next.

Watch this space!

Glossary

1BS	1-byte Communication (EnOcean)
3GPP	3 Generation Partnership Project
4BS	4-byte Communication (EnOcean)
6loWPAN	IPv6 over Low power Wireless Area Networks
A2DP	Advanced Audio Distribution Profile (Bluetooth)
AAC	Advanced Audio Coding (Bluetooth)
ABM	Asynchronous Balanced Mode
AC	Access Category (Wi-Fi)
ACL	Asynchronous Connectionless (Bluetooth)
ADT	Addressing Destination Telegram (EnOcean)
AES	Advanced Encryption Standard
AFH	Adaptive Frequency Hopping (Bluetooth)
AG	Audio Gateway (Bluetooth)
AGC	Automatic Gain Control (Bluetooth)
AGF	Aggregated Frame (NFC)
AIB	APS Information Base
AIFSN	Arbitrary Inter-frame Space Number (Wi-Fi)
AMP	Alternative MAC/PHY (Bluetooth)
AMR	Automated Meter Reading
ANA	Active Network Address (WHDI)
ANP	Alert Notification Profile (Bluetooth)
ANS	Alert Notification Service (Bluetooth)
ANT-FS	ANT File Share (ANT)
AP	Access Point
APDU	APS Protocol Data Unit
API	Application Programming Interface
APL	Application (ZigBee)
APS	Application Support Sub-layer (ZigBee)
APSDE	Application Support Sub-layer Data Entity (ZigBee)
APSDE-SAP	Application Support Sub-layer Data Entity SAP (ZigBee)
APSME	Application Support Sub-layer ME (ZigBee)
APSME-SAP	Application Support Sub-layer ME SAP (ZigBee)
ARCNET	Attached Resource Computer NETwork
ASB	Active Slave Broadcast (Bluetooth)
ATC	Authorized Test Center (WHDI)

ATRAC	Adaptive Transform Acoustic Coding (Bluetooth)
ATT	Attribute protocol (Bluetooth)
AV	Audio/Video
AVCL	Audio/Video Control Layer (WHDI)
AVCTP	Audio/Video Control Transport Protocol (Bluetooth)
AVDTP	Audio/Video Control Distribution Protocol (Bluetooth)
AVRCP	Audio/Video Remote Control Profile (Bluetooth)
B2B	Business-to-Business
BAN	Body Area Network
BAS	Battery Service (Bluetooth)
BD_ADDR	Bluetooth Device Address (Bluetooth)
BIP	Basic Imaging Profile (Bluetooth)
BLE	Bluetooth low energy (Bluetooth)
BLP	Blood Pressure Profile (Bluetooth)
BLS	Blood Pressure Service (Bluetooth)
BNEP	Bluetooth Network Encapsulation Protocol (Bluetooth)
BNF	Backus-Naur Form
BP	Blood Pressure
BPP	Basic Printing Profile (Bluetooth)
BPSK	Binary Phase Shift Keying
BR	Basic Rate (Bluetooth)
BSA	Basic Service Area (Wi-Fi)
BSS	Basic Service Set (Wi-Fi)
BSSID	Basic Service Set Identifier (Wi-Fi)
CAN	Campus Area Network
CC	Connection Complete PDU (NFC)
CCK	Complementary Code Keying (Wi-Fi)
CEA	Consumer Electronic Association
CID	Channel Identifier (Bluetooth)
CONNECT	Connect PDU (NFC)
CPP	Cross Platform Promotion
CRC	Cyclic Redundancy Check
CRP	Common Radio Platform (WiMedia)
CSCP	Cycling Speed and Cadence Profile (Bluetooth)
CSCS	Cycling Speed and Cadence Service (Bluetooth)
CSMA/CA	Carrier Sense Multiple Access with Collision Avoidance (Wi-Fi)
CTC	Compliance Test Certification (WHDI)
CTS	Current Time Service (Bluetooth)
CW	Contention Window (Wi-Fi)
DCI	Default Check Initialization (Bluetooth)
DFS	Dynamic Frequency Selection (WHDI)
DIS	Device Information Service (Bluetooth)
DISC	Disconnect PDU (NFC)
DLNA	Digital Living Network Alliance

DLPDU	Downlink PHY Data Unit (WHDI)
DM	Disconnect Mode PDU (NFC)
DNS	Domain Name Server
DoC	Declaration of Compliance (Bluetooth)
DS	Distribution System (Wi-Fi)
DSAP	Destination Service Access Point (NFC)
DSD	Direct Stream Digital (WHDI)
DSM	Distribution System Medium (Wi-Fi)
DSSS	Direct Sequence Spread Spectrum
DST	Daylight Saving Time
DST	Direct Stream Transfer (WHDI)
DS-UWB	Direct Sequence UWB
DTV	Digital Television
EAP	Extensible Authentication Protocol (Wi-Fi)
EC	European Community
ECMA	European Computer Manufacturers Association
EDID	Extended Display Identification Data (WHDI)
EDR	Enhanced Data Rate (Bluetooth)
EEP	EnOcean Equipment Profiles (EnOcean)
EIA	Electronics Industries Alliance
EIB	European Installation Bus (EnOcean)
EPA	Environmental Protection Agency
EPL	End Product Listing (Bluetooth)
ERP	EnOcean Radio Protocol (EnOcean)
ERP	Extended Rate PHY (Wi-Fi)
ESP	EnOcean Serial Protocol (EnOcean)
ESS	Extended Service Set (Wi-Fi)
ESSL	EnOcean System Software Layer (EnOcean)
EU	European Union
FCS	Frame Check Sequence
FDMA	Frequency Division Multiple Access
FE	Fitness Equipment
FFD	Full Function Device (ZigBee)
FHS	Frequency Hopping Synchronization (Bluetooth)
FHSS	Frequency Hopping Spread Spectrum (Bluetooth)
FMP	Find Me Profile (ZigBee)
FRMR	Frame Reject PDU (NFC)
FTP	File Transfer Profile (Bluetooth)
FTP	File Transfer Protocol
GAP	Generic Access Profile (Bluetooth)
GATT	Generic Attribute Profile (Bluetooth)
GAVDP	General Audio/Video Distribution Profile (Bluetooth)
GLP	Glucose Profile (Bluetooth)
GLS	Glucose Service (Bluetooth)

GN	Group ad hoc Network (Bluetooth)
GNSS	Global Navigation Satellite System Profile (Bluetooth)
GOEP	Generic Object Exchange Profile (Bluetooth)
GPS	Global Positioning System
HA	Home Automation
HAL	Hardware Abstraction Layer
HAN	Home Area Network
HCI	Host Controller Interface (Bluetooth)
HCRP	Hardcopy Cable Replacement Profile (Bluetooth)
HDCP	High-bandwidth Digital Content Protection
HDMI	High-definition Multimedia Interface
HDP	Health Device Profile (Bluetooth)
HES	Home Electronic Systems (HES)
HF	Hands-free (Bluetooth)
HF	High Frequency
HFP	Hands-free Profile (Bluetooth)
HID	Human Interface Device
HID	Human Interface Device Profile (Bluetooth)
HIDS	HID Service (Bluetooth)
HOGP	HID over GATT Profile (Bluetooth)
HRM	Heart-rate Monitor
HRP	Heart-rate Profile (Bluetooth)
HRS	Heart-rate Service (Bluetooth)
HSP	Headset Profile (Bluetooth)
HT	High-throughput (Wi-Fi)
HTP	Health Thermometer Profile (Bluetooth)
HTS	Health Thermometer Service (Bluetooth)
HTTP	Hypertext Transfer Protocol
HVAC	Heating Ventilation and Air Conditioning
IAS	Immediate Alert Service (Bluetooth)
IBSS	Independent BSS (Wi-Fi)
ICE	Interoperability, Coexistence, and Experience
IEC	International Electro-technical Commission
IEEE	Institute of Electrical and Electronics Engineers.
IFF	Identify Friend or Foe
IoT	Internet of Things
IP	Internet Protocol
IR	Infrared
IrDA	Infrared Data Association
ISM	Industrial, Scientific, and Medical
ISO	International Organization for Standardization.
ISP	Internet Service Provider
ITU	International Telecommunications Union
KNX	Konnex (EnOcean)

L2CAP	Logical Link Control and Adaptation Protocol (Bluetooth)
LAN	Local Area Network
LAP	Lower Address Part (Bluetooth)
LBT	Listen Before Talk
LC	Link Controller (Bluetooth)
LE	Low Energy (Bluetooth)
LEV	Light Electric Vehicle (ANT)
LF	Low Frequency
LHF	Low Hanging Fruit
LL	Link Layer (Bluetooth)
LLC	Logical Link Control
LLCP	Logical Link Control Protocol
LLID	Logical Link Identifier (Bluetooth)
LLS	Link Loss Service (Bluetooth)
LME	Lawnmower Man Effect
LM(P)	Link Manager (Protocol) (Bluetooth)
LON	LonWorks (EnOcean)
LPCM	Linear Pulse-code Modulation (WHDI)
LSB	Least Significant Bit
LT_ADDR	Logical Transport Address (Bluetooth)
LTE	Long-term Evolution
M2M	Machine-to-Machine
MAC	Media Access Control (layer)
MAN	Metropolitan Area Network
MAP	Message Access Profile (Bluetooth)
MBOA-SIG	Multiband OFDM Alliance SIG (Ultra-wideband)
MBSS	Mesh BSS (Wi-Fi)\
MCAP	Multi-channel Application Protocol (Bluetooth)
MCU	Microcontroller Unit
MD	More Data (Bluetooth)
ME	Management Entity
MIC	Message Integrity Code
MIMO	Multiple Input/Multiple Output
MLME	MAC Layer Management Entity (Wi-Fi)
MNO	Mobile Network Operator
MPDU	MAC PDU (Wi-Fi)
MSB	Most Significant Bit
MSC	Manufacturer-specific Communication (EnOcean)
MSDU	MAC Service Data Unit (Wi-Fi)
MSM	Multi-sport, Sport, and Distance Monitor (ANT)
MTU	Maximum Transmission Unity
MVNO	Mobile Virtual Network Operator
MWG	Marketing Working Group
NAP	Network Access Point (Bluetooth)
NAP	Non-significant Address Part (Bluetooth)

NCIRP	NFC Forum Issue Resolution Panel (NFC)
NDCS	Next DST Change Service (Bluetooth)
NDRF	NFC Data Exchange Format (NFC)
NESN	Next Expected Sequence Number (Bluetooth)
NFC	Near Field Communications
NHLE	Next Higher Layer Entity (ZigBee)
NIB	Network Information Base (ZigBee)
NIC	Network Interface Card
NLDE	NWK Layer Data Entity (ZigBee)
NLDE-SAP	NWK Layer Data Entity Service Access Point (ZigBee)
NLME	NWK Layer Management Entity (ZigBee)
NLME-SAP	NWK Layer Management Entity SAP (ZigBee)
NSDU	Network Sub-layer Data Unit (ZigBee)
NWK	Network (ZigBee)
OBEX	Object Exchange
OFDM	Orthogonal Frequency Division Multiplexing
OOB	Out of Band
OPP	Object Push Profile (Bluetooth)
OSI	Open Systems Interconnection
OUI	Organizationally Unique Identifier
PAL	Protocol Adaptation Layer
PAN	Personal Area Network(ing)
PAN	Personal Area Networking Profile (Bluetooth)
PASP	Phone Alert Status Profile (Bluetooth)
PASS	Phone Alert Status Service (Bluetooth)
PAX	Parameter Exchange PDU (NFC)
PBAP	Phone Book Access Profile (Bluetooth)
PBC	Push Button Configuration (Wi-Fi)
PC	Personal Computer
PCB	Printed Circuit Board
PCE	Phonebook Client Equipment (Bluetooth)
PDU	Protocol Data Unit
PHY	Physical (layer)
PIN	Personal Identification Number
PLCP	PHY Layer Convergence Procedure (Wi-Fi)
PMD	PHY Medium Dependent (Wi-Fi)
PN	Personal Network(ing)
PoC	Proof of Concept
POS	Personal Operating Space (ZigBee)
PPDU	PLCP PDU (Wi-Fi)
PSB	Parked Slave Broadcast (Bluetooth)
PSDU	PLCP Service Data Unit (Wi-Fi)
PSK	Pre-shared Key (Wi-Fi)
PSM	Protocol/Service Multiplexer (Bluetooth)
PXP	Proximity Profile (Bluetooth)

QD ID	Qualified Design Identifier (Bluetooth)
QoS	Quality of Service
QPSK	Quadrature Phase Shift Keying
QR (code)	Quick response
RC	Remote Control
RC	Repeater Count (EnOcean)
RCP	Room Control Panel (EnOcean)
RDC	Remote Device Control (WHDI)
RF	Radio Frequency
RFCOMM	RS232 serial emulation (Bluetooth)
RFD	Reduced Function Device (ZigBee)
RFID	Radio Frequency Identification
RNR	Receive Not Ready PDU (NFC)
RORG	Radio (telegrams) ORGanizationally (EnOcean)
RPS	Repeated Switch Communication (EnOcean)
RR	Receiver Ready PDU (NFC)
RSCP	Running Speed and Cadence Profile (Bluetooth)
RSCS	Running Speed and Cadence Service (Bluetooth)
RSSI	Radio Signal Strength Index
RTD	Record Type Definitions (NFC)
RTUS	Reference Time Update Service (Bluetooth)
SAP	Service Access Point
SAP	SIM Access Profile (Bluetooth)
SAR	Segmentation and Reassembly (Bluetooth)
SBC	Sub-band Coding (Bluetooth)
SCO	Synchronous Connection-oriented (Bluetooth)
ScPP	Scan Parameters Profile (Bluetooth)
ScPS	Scan Parameters Service (Bluetooth)
SD	Secure Digital
SDAP	Service Discovery Application Profile (Bluetooth)
SDM	Speed and Distance Monitor
SDM	Stride-based Speed and Distance Monitor (ANT)
SDoC	Supplier Declaration of Conformity (Bluetooth)
SDP	Service Discovery Protocol
SIG	Special Interest Group
SIM	Subscriber Identity Module
SM	Security Manager (Bluetooth)
SME	STA Management Entity (Wi-Fi)
SMNP	Simple Management Network Protocol
SMP	Security Manager Protocol (Bluetooth)
SMS	Short Message Service
SN	Sequence Number (Bluetooth)
SNEP	Simple NDEF Exchange Protocol (NFC)
SNL	Service Name Lookup PDU (NFC)
SoC	System-on-Chip

SOHO	Small Office/Home Office
SPI	Serial Peripheral Interface
SPP	Serial Port Profile (Bluetooth)
SSAP	Source Service Access Point (NFC)
SSID	Service Set Identifier (Wi-Fi)
SSP	Secure Simple Pairing (Bluetooth)
SSP	Service Security Provider (ZigBee)
STA	Station (Wi-Fi)
STB	Set-top-box
STK	Short-term Key (Bluetooth)
SUN	Smart Utility Network
SYMM	Symmetry PDU (NFC)
SYNC	Synchronization Profile (Bluetooth)
TCP	Transmission Control Protocol
TDM	Time Division Multiplexing
TDMA	Time Division Multiple Access
TG	Task Group
TIP	Tile Profile (Bluetooth)
TK	Temporary Key (Bluetooth)
TKIP	Temporal Key Integrity Protocol (Wi-Fi)
TPS	Tx Power Service (Bluetooth)
TSP	Telecommunications Service Provider
TWG	Technical Working Group
TXOP	Opportunity to Transmit (Wi-Fi)
UAP	Upper Address Part (Bluetooth)
UART	Universal Asynchronous Receiver/Transmitter
UDP	User Datagram Protocol
UHF	Ultra-high Frequency
UI	Unnumbered Information PDU (NFC)
UI	User Interface
ULCPDU	Uplink Control PHY Data Unit (WHDI)
ULP	Ultra-low Power
UPIPDU	Uplink Initialization PHY Data Unit (WHDI)
UPnP	Universal Plug and Play
URI	Universal Resource Identifier
USB	Universal Serial Bus
UUID	Universal Unique Identifier
UWB	Ultra-wideband
VAN	Vehicle Area Network
VAS	Value-added Service
VDP	Video Distribution Profile (Bluetooth)
VESA	Video Electronics Standard Association (WHDI)
VLC	Visible Light Communication
VLD	Variable Length Data (EnOcean)

VoIP	Voice over IP
VSDB	Vendor-specific Data Block (WHDI)
WAN	Wide Area Networking
WEP	Wired Equivalent Privacy (Wi-Fi)
WHDI	Wireless Home Digital Interface
WiGig	Wireless Gigabit Alliance
WLAN	Wireless Local Area Network
WMM-PS	Wi-Fi Multimedia-Power Save (Wi-Fi)
WPA(2)	Wi-Fi Protected Access (Wi-Fi)
WPAN	Wireless Personal Area Network(ing)
WPS	Wi-Fi Protected Setup (Wi-Fi)
WSN	Wireless Sensor Network
WSP	Wireless Short-packet (EnOcean)
W-USB	Certified Wireless USB
WVAN	Wireless Video Area Network
ZDO	ZigBee Device Object (ZigBee)

References and Bibliography

Amimon Incorporated, "WHDI Video Modem Technology: Key Principles," February 2012. [1]

ANT Wireless, "ANT+ and Bluetooth low energy Concurrent Combo Chip Solution," September 2012. [Online] available through ANT Wireless, http://www.thisisant.com/news/stories.

Ashton, K., "That 'Internet of Things' Thing," *RFID Journal*, June 2009. [Online] available through the *RFID Journal* website, http://www.rfidjournal.com/article/view/4986.

Barnowski, S., "ANT Protocol Basics," ANT Wireless, September 2009. [1]

Bluetooth SIG, "Advanced Audio Distribution Profile Specification," v1.3, July 2012. [Online] available through the Bluetooth Special Interest Group, http://bluetooth.org. [2]

Bluetooth SIG, "Audio/Video Remote Control Profile," v1.5, July 2012. [Online] available through the Bluetooth Special Interest Group, http://bluetooth.org. [2]

Bluetooth SIG, "Basic Imaging Profile," v1.2, July 2012. [Online] available through the Bluetooth Special Interest Group, http://bluetooth.org. [2]

Bluetooth SIG, "Basic Printing Profile," v1.2, April 2012. [Online] available through the Bluetooth Special Interest Group, http://bluetooth.org. [2]

Bluetooth SIG, "BLE 101: Bluetooth low energy," September 2012. [Online] available through the Bluetooth Special Interest Group, http://bluetooth.org. [2]

Bluetooth SIG, "Bluetooth Qualification Program: Training – Qualification Process, Specification Naming," April 2012. [Online] available through the Bluetooth Special Interest Group, http://bluetooth.org. [2]

Bluetooth SIG, "File Transfer Profile," v1.3, July 2012. [Online] available through the Bluetooth Special Interest Group, http://bluetooth.org. [2]

Bluetooth SIG, "Global Navigation Satellite System Profile," v1.0, March 2012. [Online] available through the Bluetooth Special Interest Group, http://bluetooth.org. [2]

Bluetooth SIG, "Hands-free Profile," v1.6, May 2011. [Online] available through the Bluetooth Special Interest Group, http://bluetooth.org. [2]

Bluetooth SIG, "Headset Profile," v1.2, December 2008. [Online] available through the Bluetooth Special Interest Group, http://bluetooth.org. [2]

Bluetooth SIG, "Personal Area Networking Profile," v1.0, February 2003. [Online] available through the Bluetooth Special Interest Group, http://bluetooth.org. [2]

Bluetooth SIG, "Phone Book Access Profile," v1.1, August 2010. [Online] available through the Bluetooth Special Interest Group, http://bluetooth.org. [2]

Bluetooth SIG, "Specification of the Bluetooth System: Core, v4.0," June 2010. [Online] available through the Bluetooth Special Interest Group, http://bluetooth.org. [2]

Bluetooth SIG, "Video Distribution Profile," v1.1, July 2012. [Online] available through the Bluetooth Special Interest Group, http://bluetooth.org. [2]

CBR Mobility News, "Bluetooth-enabled Equipment Sales Will Soar to 4.2 Billion by 2015," June 2012. [Online] available through the CBR Mobility News, http://mobility.cbronline.com/news.

Cordeiro, C., "Next Generation Multi-Gbps Wireless LANs and PANs," Intel Corporation, IEEE Globecom, 2010. [1]

Coskun, V., Ok, K., and Ozdenizci., B., *Near Field Communication: From Theory to Practice*, Wiley-Blackwell, 2012.

Decuir, J., "Bluetooth 4.0: low energy," Cambridge Silicon Radio, 2010. [Online] available through IEEE Communications Society, http://chapters.comsoc.org/vancouver/BTLER3.pdf.

"Digital History: ARCNET, the First Local Area Network," (n.d.). [Online] available from http://www.old-computers.com/history/detail.asp?n=23&t=5.

Dynastream Corporation, "ANT and ANT+ Overview," ANT Wireless, April 2010. [Online] available through ANT Wireless, http://thisisant.com. [3]

Dynastream Corporation, "ANT Message Protocol and Usage," Revision 4.2, 2010. [Online] available through ANT Wireless, http://thisisant.com. [3]

Dynastream Corporation, "ANT+ Device Profile: Audio Controls," Revision 1.2. [Online] available through ANT Wireless, http://thisisant.com. [3]

Dynastream Corporation, "ANT+ Device Profile: Bicycle Power," Revision 2.2. [Online] available through ANT Wireless, http://thisisant.com. [3]

Dynastream Corporation, "ANT+ Device Profile: Bike Speed and Cadence," Revision 1.3. [Online] available through ANT Wireless, http://thisisant.com. [3]

Dynastream Corporation, "ANT+ Device Profile: Blood Pressure," Revision 1.0. [Online] available through ANT Wireless, http://thisisant.com. [3]

Dynastream Corporation, "ANT+ Device Profile: Fitness Equipment," Revision 3.2. [Online] available through ANT Wireless, http://thisisant.com. [3]

Dynastream Corporation, "ANT+ Device Profile: Geocache," Revision 1.0. [Online] available through ANT Wireless, http://thisisant.com. [3]

Dynastream Corporation, "ANT+ Device Profile: Heart Rate," Revision 1.13. [Online] available through ANT Wireless, http://thisisant.com. [3]

Dynastream Corporation, "ANT+ Device Profile: Light Electric Vehicle," Revision 1.1. [Online] available through ANT Wireless, http://thisisant.com. [3]

Dynastream Corporation, "ANT+ Device Profile: Multi-sport Speed and Distance," Revision 1.1. [Online] available through ANT Wireless, http://thisisant.com. [3]

Dynastream Corporation, "ANT+ Device Profile: Stride-based Speed and Distance," Revision 1.3. [Online] available through ANT Wireless, http://thisisant.com. [3]

Dynastream Corporation, "ANT+ Device Profile: Weight Scale," Revision 2.1. [Online] available through ANT Wireless, http://thisisant.com. [3]

ECMA International, "Near Field Communication – Interface and Protocol (NFCIP-1)," 3rd edn, February 2008. [Online] available through ECMA International, http://www.ecma-international.org/publications/files/ECMA-ST/Ecma-340.pdf.

EE Times, "Photos from the Frontier: The Internet of Things," April 2012. [Online] available through *EE Times*, http://www.eetimes.com/electronics-news/4370765/Photo-Gallery-Internet-of-Things.

EnOcean GmbH, "Dolphin Core Description," v0.96, January 2011. [1]

EnOcean GmbH, "Specification: EnOcean Serial Protocol 3 (ESP3)," August 2010. [1]

EnOcean GmbH, "System Specification: Smart Acknowledge," September 2010. [1]

EnOcean GmbH, "The Dolphin API," v2.2.5.0 (n.d.). [1]

Environmental Protection Agency, "The Power to Protect the Environment Through Energy Efficiency," energystar.gov, August 2003. [Online] available through the Environmental Protection Agency, http://www.energystar.gov/ia/partners/downloads/energy_star_report_aug_2003.pdf.

Gabriel, C., "Apple iWallet May Favor Bluetooth Over NFC," May 2012. [Online] available through RethinkWireless.com.

Gratton, D. A., "A Touch of Genius: Wi-Fi Protected Set-up," Incisor.TV, May 2008. [Online] available through Incisor.TV, http://www.incisor.tv/download.php?file=121may2008.pdf.

Gratton, D. A., "Dissolving the Boundaries: Introducing Femtocells," Incisor.TV, April 2008. [Online] available through Incisor.TV, http://www.incisor.tv/download.php?file=120april2008.pdf.

Gratton, D. A., "Femtocells: You're Either Right or Impatient!," Incisor.TV, June 2011. [Online] available through Incisor.TV, http://www.incisor.tv/download.php?file=158june2011.pdf.

Gratton, D. A., "Forecasting a Wireless-enabled 2012," Incisor.TV, February 2012. [Online] available through Incisor.TV, http://www.incisor.tv/download.php?file=166february2012.pdf.

Gratton, D. A., "Increasing Brand Awareness with NFC Technology and Social Media," September 2011. [Online] available through SocialMediaToday.com, http://socialmediatoday.com/grattonboy/362200/increasing-brand-awareness-nfc-technology-and-social-media.

Gratton, D. A., "Keeping Mum: Transforming the Healthcare Industry *Wirelessly*," Incisor.TV, August 2011. [Online] available through Incisor.TV, http://www.incisor.tv/download.php?file=160august2011.pdf.

Gratton, D. A., "NFC: A Gentle Evolution," Incisor.TV, April 2012. [Online] available through Incisor.TV, http://www.incisor.tv/download.php?file=168april2012.pdf.

Gratton, D. A., "The Myth of 4G," Incisor.TV, February 2011. [Online] available through Incisor.TV, http://www.incisor.tv/download.php?file=154february2011.pdf.

Gratton, D. A., "The Need for Speed: Introducing Long-term Evolution," Incisor.TV, April 2010. [Online] available through Incisor.TV, http://www.incisor.tv/download.php?file=144april2010.pdf.

Gratton, D. A., "The Only Alternative: Bluetooth and Ultra-wideband," Incisor.TV, September 2009. [Online] available through Incisor.TV, http://www.incisor.tv/download.php?file=138september2009.pdf.

Gratton, D. A., "The WiMedia Alliance: What Do We Have in Common?," Incisor.TV, February 2009. [Online] available through Incisor.TV, http://www.incisor.tv/download.php?file=131february2009.pdf.

Gratton, D. A., "White Space Radio: Revisited," Incisor.TV, September, 2011. [Online] available through Incisor.TV, http://www.incisor.tv/download.php?file=161september2011.pdf.

Gratton, D. A., "Whitespace Radio: The New Wireless Buzzword," Incisor.TV, April 2011 [Online] available through Incisor.TV, http://www.incisor.tv/download.php?file=156april2011.pdf.

Gratton, D. A., "Why NFC Should Keep Social Media in its Wallet!," Incisor.TV, October 2011. [Online] available through Incisor.TV, http://www.incisor.tv/download.php?file=162october2011.pdf.

Gratton, D. A., *Bluetooth Profiles: The Definitive Guide*, Prentice Hall, 2003.

Gratton, D. A., *Developing Practical Wireless Applications*, Elsevier Digital Press, 2007.

Gratton, D. A., *The Next Generation of Bluetooth wireless technology: Classic Bluetooth, High speed and low energy Technologies*, Wiley-Blackwell, 2013.

Gratton, S., *Follow Me! Creating a Personal Brand with Twitter*, John Wiley & Sons, 2012.

Gratton, S. and Gratton, D. A., *Marketing Wireless Products*, Butterworth-Heinemann, 2004.

Gratton, S. and Gratton, D. A., *Zero to 100,000: Social Media Tips and Tricks for Small Businesses*, QUE-Pearson, 2011.

IBM Social Media, "The Internet of Things," ASmarterPlanet.com, March 2010. [Online] available through YouTube, http://www.youtube.com/watch?v=sfEbMV295Kk.

IEC/ISO 18092.2004 Information technology, "Telecommunications and Information Exchange Between Systems; Near Field Communication, Interface and Protocol (NFCIP-1)."

IEEE 802.15 Working Group for Wireless Personal Area Networks, http://www.ieee802.org/15/about.html.

IEEE Computer Society, IEEE-Std 802.11–2012, "Part 11: Wireless LAN Medium Access Control (MAC) and Physical Layer (PHY) Specifications," March 2012. [Online] available through IEEE Standards Association http://standards.ieee.org/findstds/standard/802.11-2012.html.

IPSO Alliance, "IP for Smart Objects," Whitepaper, Version 1.1, July 2010. [Online] available through IPSO Alliance, http://www.ipso-alliance.org/white-papers. [4]

Jenkins, H., *Convergence Culture: Where Old and New Media Collide*, rev. edn, New York: NYU Press, 2008.

Kinney, P., "ZigBee Technology: Wireless Control that Simply Works," Kinney Consulting LLC, October 2003. [Online] available through the ZigBee Alliance, http://zigbee.org/LearnMore/WhitePapers.aspx.

Knight, S., "Televisions Overtake Computers for Online Video Streaming," September 2012. [Online] available through TechSpot.com, http://www.techspot.com/news/50337-televisions-overtake-computers-for-online-video-streaming.html.

NFC Forum, "How Does NFC Technology Work?," May 2012. [Online] available through the NFC Forum, http://www.nfc-forum.org/resources/faqs#howwork.

NFC Forum, "NFC Forum Device Requirements: High Level Conformance Requirements," Revision v1.0, January 2010. [Online] available through the NFC Forum, http://certification.nfc-forum.org/docs/NFC_Forum_Device_Requirements.pdf.

NFC Forum, "Technical Specification: Logical Link Control Protocol," LLCP v1.1, June 2011. [Online] available through the NFC Forum, http://www.nfc-forum.org/specs/spec_list/. [5]

NFC Forum, "Technical Specification: NFC Data Exchange Format," NDEF v1.0, July 2006. [Online] available through the NFC Forum, http://www.nfc-forum.org/specs/spec_list/. [5]

NFC Forum, "Technical Specification: NFC Record Type Definition," RTD v1.0, July 2006. [Online] available through the NFC Forum, http://www.nfc-forum.org/specs/spec_list/. [5]

NFC Forum, "Technical Specification: Simple NDEF Exchange Protocol," SNEP v1.0, August 2011. [Online] available through the NFC Forum, http://www.nfc-forum.org/specs/spec_list/. [5]

NFC World, "Orange UK Announces Quick Tap Treats for 200,000 NFC customers," February 2012. [Online] available through NFC World, http://www.nfcworld.com/2012/02/24/313725/orange-uk-announces-quick-tap-treats-for-200000-nfc-customers/.

Nokia Research Center, "Wibree Forum Merges with Bluetooth SIG," August 2007. [Online] available through Nokia Research Center, http://research.nokia.com/news/254.

O'Brien, T., "WiFi and WiGig Alliances Become One, Work to Promote 60GHz Wireless," *Engadget*, January 2013. [Online] available through *Engadget*, http://www.engadget.com/2013/01/03/wifi-and-wigig-alliances-become-one/.

RFIDJournal.com, "What is the Difference Between Low-, High- and Ultra-high Frequencies?." [Online] available through the *RFID Journal* website, http://www.rfidjournal.com/site/faqs#Anchor-What-28258.

Richmond, S., "iPhone 5: Price, 4G and Everything Else You Need to Know," *Telegraph*, September 2012. [Online] available through the *Telegraph*, http://www.telegraph.co.uk/technology/apple/9540423/iPhone-5-price-4G-and-everything-else-you-need-to-know.html.

Roberti, M., "The History of RFID Technology," *RFID Journal*, January 2005. [Online] available through the *RFID Journal*, http://www.rfidjournal.com/article/view/1338.

Schneider, A., "EnOcean Company Presentation," EnOcean Alliance, May 2010. [1]

Sourgen, L., "Specifications and Application Documents," NFC Forum, April 2012. [Online] available through the NFC Forum, http://www.nfc-forum.org/resources/presentations/NFCF_SpecsPresentation_WIMA_final_2.pdf.

Tripp, M., "Henry Jenkins, *Convergence Culture*," Review by Mary Tripp, University of Central Florida, ~2006. [Online] available through Hyperrhiz: New Media Cultures, http://www.hyperrhiz.net/reviews/61-henry-jenkins-convergence-culture.

WHDI, "Press/Analyst Briefing," January 2012. [1]

WHDI Promoters, "WHDI Specification," v1.0, Revision 20, August 2010. [1]

Wi-Fi Alliance, "Backgrounder: Wi-Fi Quality of Service Features Enable The Connected World," April 2006. [Online] available through the Wi-Fi Alliance, http://www.wi-fi.org/media/press-releases/backgrounder-wi-fi%C2%AE-quality-service-features-enable-connected-world.

Wi-Fi Alliance, "Connect with the Wi-Fi Alliance: Membership Overview," February 2012. [Online] available through the Wi-Fi Alliance, http://www.wi-fi.org/sites/default/files/uploads/files/Connect_with_the_Wi-Fi_Alliance_201202.pdf.

Wi-Fi Alliance, "Wi-Fi CERTIFIED for WMM – Support for Multimedia Applications with Quality of Service in Wi-Fi Networks," September 2004. [Online] available through the Wi-Fi Alliance, http://www.wi-fi.org/knowledge-center/white-papers. [6]

Wi-Fi Alliance, "Wi-Fi Certified Wi-Fi Direct: Personal, Portable Wi-Fi to Connect Devices Anywhere, Anytime," October 2010. [Online] available through the Wi-Fi Alliance, http://www.wi-fi.org/knowledge-center/white-papers. [6]

Wi-Fi Alliance, "Wi-Fi Certified Wi-Fi Direct: Personal, Portable Wi-Fi Technology," October 2010. [Online] available through the Wi-Fi Alliance, http://www.wi-fi.org/knowledge-center/white-papers. [6]

Wi-Fi Alliance, "Wi-Fi Certified Wi-Fi Protected Setup: Easing the User Experience for Home and Small Office Wi-Fi® Networks," December 2010. [Online] available through the Wi-Fi Alliance, http://www.wi-fi.org/knowledge-center/white-papers. [6]

Wi-Fi Alliance, "Wi-Fi Protected Access: Strong, Standards-based, Interoperable Security for Today's Wi-Fi Networks," April 2003. [Online] available through the Wi-Fi Alliance, http://www.wi-fi.org/knowledge-center/white-papers. [6]

Wi-Fi Alliance, "WMM Power Save for Mobile and Portable Wi-Fi CERTIFIED Devices," December 2005. [Online] available through the Wi-Fi Alliance, http://www.wi-fi.org/knowledge-center/white-papers. [6]

Williams, C., "The £300m Cable that will Save Traders Milliseconds," *Telegraph*, September 11, 2011. [Online] available through *Telegraph*, http://www.telegraph.co.uk/technology/news/8753784/The-300m-cable-that-will-save-traders-milliseconds.html.

Wireless Gigabit Alliance, "WiGig White Paper: Defining the Future of Multi-Gigabit Wireless Communications," July 2010. [Online] available through the Wireless Gigabit Alliance, http://wirelessgigabitalliance.org/specifications/.

ZigBee Alliance, "Understanding ZigBee RF4CE," July 2009. [Online] available through the ZigBee Alliance, http://docs.zigbee.org/zigbee-docs/dcn/09-5231.pdf.

ZigBee Alliance, "ZigBee Overview," 2009. [Online] available through the ZigBee Alliance, https://docs.zigbee.org/zigbee-docs/dcn/07-5482.pdf.

ZigBee Alliance, "ZigBee RF4CE Specification," Version 1.00, March 2009. [Online] available through the ZigBee Alliance, http://www.zigbee.org/Specifications/ZigBeeRF4CE/download.aspx. [7]

ZigBee Alliance, "ZigBee Specification," October 2010. [Online] available through the ZigBee Alliance, [Online] available through the ZigBee Alliance, http://zigbee.org/Specifications/ZigBee/download.aspx.

Zimmerman, T., "Personal Area Networks (PAN): Near-field Intra-body Communication," *IBM Systems Journal*, **35**, (3&4), 1996. [Online] available through University of Washington, http://www.cs.washington.edu/education/courses/cse590es/00au/papers/zimmerman.pdf.

Notes

Guide to notes

[1] Marketing documentation, technical specifications, and other ancillary material provided by the organization with their kind permission.

[2] Membership registration is required to access and/or retrieve marketing and technical documentation from the Bluetooth SIG.

[3] A basic membership account is required to access and/or retrieve marketing and technical documentation from ANT Wireless.

[4] For IPSO Whitepapers, you are required to complete a form and offer credentials to acquire information.

[5] You are required to accept and complete a specification license agreement prior to accessing technical specification from the NFC Forum.

[6] You are required to register with the Wi-Fi Alliance to download specific content.

[7] You are required to complete registration prior to accessing technical specification from the ZigBee Alliance.

Index